THE ROAD TO
TRINITY

THE ROAD TO TRINITY

Major General K. D. Nichols, U.S.A. (Ret.)

WILLIAM MORROW AND COMPANY, INC.

New York

Library of Congress Cataloging-in-Publication Data

Nichols, Kenneth D. (Kenneth David), 1907-
    The road to Trinity.
    Bibliography: p.
    Includes index.
    1. Nichols, Kenneth D. (Kenneth David), 1907-
2. United States.   Army. Corps of Engineers. Manhattan District—History.   3. Atomic bomb—United States—History.   4. Physicists—United States—Biography.
I. Title.
QC774.N45A3   1987   355'.0217'0973   87-1718
ISBN 0-688-06910-X

Printed in the United States of America

First Edition

1 2 3 4 5 6 7 8 9 10

BOOK DESIGN BY BERNARD SCHLEIFER

CONTENTS

INTRODUCTION

ON THE MORNING OF November 28, 1952, and at his request, I met with Frank Pace, Jr., secretary of the Army. As I entered his office he greeted me warmly. He was standing beside his desk in his shirt sleeves and wearing his usual colorful suspenders. He hailed from Arkansas and was proud of it.

Pace had advocated my being assigned as deputy director of guided missiles in the office of the secretary of defense (OSD) and my additional assignment as chief of research and development in the office of the chief of staff, U.S. Army. He was almost five years younger than I, but he had already achieved great success in state and federal government. Just prior to becoming secretary of the Army in 1950, he had been an outstanding director of the Bureau of the Budget. Sometimes he was called the "boy wonder."

Dwight D. Eisenhower had just been elected president. Pace knew that with the change in administration he would be leaving the Pentagon in January. During our brief discussion of his plans for the future, he said he was looking forward to achieving success in industry. "I need that to round out my career," he said. This led me to believe that his ultimate goal might be the presidency. At the rate he was going, I thought he might make it.

Suddenly looking at his watch, he abruptly changed the subject: "I sent for you because I'd like you to write for me your personal views on the political and military implications of the hydrogen bomb. I need it in a hurry—within three hours."

The request did not surprise me. Frank Pace had on several occasions sought my opinion on subjects that were not my assigned responsibility. In this case, his request was based on a current subject: At Eniwetok in the Pacific on November 1, 1952, the United States had successfully exploded its first thermonuclear device, "Mike," with a yield of 10.4 megatons (more than five hundred times the yield of World War II atomic bombs). The hydrogen bomb era had been born.

Since June 1942, I had been involved with atomic bomb development, international control of atomic energy, the hydrogen bomb controversy, and guided missiles for air defense and tactical and strategic delivery of atomic bombs. As deputy district engineer and district engineer of the Manhattan Engineer District, I had responsibility for the design, construction, and operation of the huge plants for producing U-235 and plutonium, the fissionable material required by Los Alamos as the source of explosive energy for the Alamogordo, Hiroshima, and Nagasaki atomic bombs. Following World War II, and with the beginning of the "Cold War," I was assigned as chief of the Armed Forces Special Weapons Project, deputy assistant Chief of Staff for atomic weapons in the Plans and Operations Division of the Army, and senior Army member of the Military Liaison Committee to the Atomic Energy Commission. Working in these three assignments, I had major staff responsibilities for developing doctrine for the military use of atomic weapons; developing better weapons; increasing production; and command responsibilities for training Army, Navy, and Air Force personnel to assemble atomic weapons and to participate in execution of current war plans. Also, I had ardently supported the presidential decision to develop and produce hydrogen bombs. Near the end of 1950, I had agreed to a shift in my responsibilities to the production of guided missiles because I was interested in the potential use of guided missiles as a better means for delivery of atomic and hydrogen weapons. Of greater importance, I believed it essential to develop a defense against atomic and hydrogen bomb attack. The Army Nike ground-to-air missile as well as the Air Force and Navy Falcon and Sparrow air-to-air missiles were the major air defense missiles that our OSD office was putting in production. After being assigned additional duty as chief of Army research and development, I initiated research to develop an antimissile missile that later became known as the antiballistic missile (ABM) program.

All during the period of 1942–53, besides accomplishing my primary assignments I devoted considerable thought to the solution of the tremendous social, political, and national security problems that resulted from development of atomic and hydrogen bombs. Accordingly, I had little difficulty in hastily dictating for Secretary Pace the following memorandum:

Political and Military Implications of the Hydrogen Bomb

1. The development of the hydrogen bomb, as compared to the atomic bomb, has equal or greater political than strictly military implications.

2. Although the political implications of both atomic bombs and hydrogen bombs are derived from the potential military implications, the political implications can be considered of greater importance because of the possible use of both of these weapons as a deterrent to war as well as a way of winning wars.

3. From the standpoint of utilization of both atomic weapons and hydrogen weapons, the strictly military effort is relatively far along as compared to our political efforts to utilize these weapons as means for deterring war. At the present time, we have a capability for delivering atomic weapons and we are energetically working on improving this capability both for strategic and tactical use. We are also energetically pursuing, but perhaps not to the degree necessary, our means for defense against these weapons. In this latter effort, it should be recognized that a foolproof defense against both atomic and hydrogen weapons is probably impossible and that in the event of a major war, the United States will be vulnerable to a serious or devastating attack against our industrial potential and civilian population as well as against our military forces. As a part of our effort to improve our potential to deter war by consideration of these weapons, we must continue to improve our military potential to use them for all types of attack as well as to improve our potential to defend against them. With proper encouragement and support, our military and civilian leaders within the Department of Defense can take care of this part of the problem.

4. Greater thought and greater effort must be given to the utilization of atomic and hydrogen weapons as deterrents of war. One way of doing this is to convince the world and our potential enemies that if war is instigated on either a minor or a major scale, we will utilize these weapons ruthlessly in the interest of preserving peace and the rights of individuals in a free world. To date, we

have not done this. To a limited extent we have tried to convince the world that in the case of a major war, we would utilize the atomic bomb. However, this effort to convince has been confused by those who have stressed the tactical use of atomic bombs primarily to avoid their use against civilian population (tactical use can stand on its own merits); by our failure to utilize the atomic bomb in June 1950, when we initially supported the South Koreans by air power, again by our failure to utilize the atomic bomb when the Chinese Communists initially attacked our forces in North Korea, and on other occasions throughout the conduct of the Korean war when targets of sufficient size to warrant the use of 20–50 B-50's were destroyed by the use of conventional explosives rather than the cheaper method of utilizing an atomic attack. Intentional or not, our failure to utilize atomic weapons in Korea must contribute toward convincing the Kremlin that we may never use them. In Korea, we have fought the war in a way most advantageous to our enemy. We have failed to use atomic weapons for reasons ranging from fear of debunking atomic weapons to the argument that few potential targets exist, and possibly for moral reasons. All of these are arguments that must be thoroughly considered. However, the argument that we must convince the world that we will utilize these powerful weapons whenever peace is disturbed is overriding. Another major argument used against the use of atomic weapons in Korea has been that it might precipitate a major conflict. We should take this risk in the effort to deter minor wars as well as a major war. Every minor war has the potential of developing into a major war, and a major war becomes a greater danger to the United States as time goes on rather than less dangerous. Time definitely is not on our side when the full implications of atomic and hydrogen weapons are considered.

5. Time favors the USSR when the military implications of atomic and hydrogen weapons are considered. As the USSR stockpile of atomic weapons increases and as it achieves a hydrogen weapon capability, the potential danger to the United States increases. In assessing this potential danger, the means of delivery must be considered as well as the size of the stockpile. Although we are making major improvements in our means of defense against the delivery of atomic and hydrogen weapons, there is no assurance that our means of defense will ever reach an adequate state. As time goes on we will be more and more subject to surprise

attack either by conventional means of delivery, ultimate development of guided missiles as a means of delivery, or by such type of sneak attacks as the exploding in harbors ships that contain hydrogen weapons. Also, it is questionable if time favors our retention of allies. As the USSR stockpile and potential for delivery of atomic and hydrogen weapons increase, the advantage to many of our allies to remain neutral will increase. Moreover, as time goes on, it will take stronger and better leadership to convince our own public that we are following a wise course in this maze of difficulties, with the resulting danger that we may not make a proper effort to avoid surprise or to maintain the military potential required to deter a war and to win it in case we get in one. Moreover, as time goes on, the full meaning of "win it" becomes more questionable. The successful delivery of a number of hydrogen bombs, or even a number of the king-size atomic bombs, could wreck such devastation on our cities as to require more than a generation to recover even if we do succeed in winning the military decision to the point of dictating the terms of the peace.

6. We must increase our efforts in the direction of utilizing atomic and hydrogen weapons to deter war. The following specific steps are suggested:

a. Continue to increase our potential for delivering atomic and hydrogen weapons both against industrial and civilian targets, as well as against tactical targets on the battlefield.

b. Make an even greater effort to achieve a better defense against the USSR capability and future capability for delivery of atomic and hydrogen weapons.

c. Increase our intelligence efforts to give a greater period of warning against possible surprise attacks utilizing both conventional and unconventional means.

d. Deliberately utilize atomic weapons in the present war in Korea the first time a reasonable opportunity to do so permits. One way this opportunity might be created is by serving notice to North Korea and to Communist China that we are withdrawing the bulk of the American forces from the front lines to a position of readiness in the rear and that if the Communist forces make any offensive move to proceed beyond the present front lines, each day such a move continues atomic weapons will be dropped on both North Korean and Chinese targets to include both military and industrial types. This ultimatum should be accompanied by other military and economic measures to include a complete blockade of China. An alternate way this opportunity might be created is to initiate an offensive to drive the Chinese Communists out of North Korea. Opening of this offensive could be accompanied by the utilization of atomic weapons on targets both within North Korea and military targets such as air bases in Manchuria.

7. The consequences of the above might be:

a. The first step in proving to the world that we really mean to use every potential weapon available to us to preserve peace and thereby deter war.

b. Precipitate a major war at a time when we have greatest potential for winning it with minimum damage to the U.S.A.

8. If we are to gain sufficient political know-how to properly manage this powerful political and military weapon, we must learn how by practice when the stakes are lowest and not wait for the ultimate showdown. The time to begin has already passed, but the birth of the hydrogen bomb should be the basis for a belated beginning.

When I wrote the memorandum the United States had atomic superiority over and was ahead of the USSR in the development of the hydrogen bomb. Hence, in my zeal to stop all wars, particularly the Korean War, I probably overemphasized the term ''use'' as compared to ''threat.'' But I did recognize that in the future we would lose our superiority and that a situation like we are now encountering in 1986 was a probability. President Eisenhower knew my views on many aspects of atomic warfare, but I doubt if he ever saw this particular memo. However, Lewis Strauss, adviser to the president on atomic policy and later chairman of the Atomic Energy Commission, showed it to Admiral A. W. Radford, chairman of the Joint Chiefs of Staff, and thereafter Admiral Radford requested on several occasions that I send it to his office. At that time it was still classified top secret. It wasn't until it was downgraded to confidential that he made a copy. As it turned out, President Eisenhower's own atomic policy was remarkably similar. He ended the Korean hostilities by discreetly threatening to use atomic weapons if the North Koreans and the Chinese did not agree to a cessation of hostilities.

I do not claim to be the originator of our deterrence policy. Many others, I am sure, have had similar thoughts. But I believe that much of the thinking expressed in my memo, over thirty years ago, still is applicable to the current and much more dangerous situation confronting us. I am particularly pleased that President Reagan has restored the idea that defense should be part of our deterrence policy. It is difficult for me to comprehend why we departed from the logic of developing some form of defense. President Reagan is so right in supporting SDI. Those opponents who clamor that SDI initiates "Star Wars," or claim it is impossible, should ponder why the USSR makes it such an issue and demands that the United States should not engage in SDI. Instead, the Russians continue with their own program for defense and are developing their offensive capability to knock down our satellites.

Throughout the years when discussing my experiences with friends, I often was asked, "Why don't you write a book?" Now that I have decided to do so, the question is, "Why are you telling your story now?" Both are good questions. Initially I was too busy struggling with the problems concerning the development of commercial nuclear power. I was content to grant interviews to a number of authors who were writing about various aspects of the Manhattan Project. Many of the early authors did a fair job of dealing with the information and the issues they had discussed with me. However, in more recent years, in order to gain dramatic effect, the TV docudramas, the media, and many authors have on too many occasions disregarded or warped facts. These include history, the effects of weapons, and the hazards of nuclear power plants. As a result, the public has been misinformed. Now, achieving public support for a sound national policy concerning arms control, adequate appropriations for national defense, and a sound nuclear program to provide an appropriate share of our energy needs is an extremely difficult if not an impossible task.

In 1981, Dr. Lawrence Suid, a historian and author, interviewed me for more than forty hours for the Corps of Engineers' oral history program. In 1982, he recommended and convinced me that we should coauthor my autobiography. The main reason for doing so is to provide a straightforward, firsthand, accurate account about how the atomic bomb was developed and produced, how the decision was reached to use it, the efforts made to achieve international control of atomic weapons, the early development of our atomic military doctrine and the increase in our atomic capabilities during the "Cold War," the potential confrontation with the USSR, the many great contributions and the security problems of J. Robert Oppenheimer, and the commercial development and present status of nuclear power. The chief reason for an autobiography instead of a history is to assure greater accuracy by limiting the scope to events in which I had personally participated or of which I had firsthand knowledge at the time. However, sufficient additional events have been included to make the story coherent and meaningful. The reader is given sufficient information on my background, education, early training, experiences, and good luck to appreciate why I was chosen to play a major role in the evolution of the atomic era.

I would like to acknowledge the able assistance of Dr. Suid for approximately three years. For a large part of the manuscript, his research, oral interviews, and writing made major contributions in rounding out the scope of the account and assuring greater accuracy.

I trust and hope that this account will contribute to a better understanding of the history and problems associated with the development of this new source of energy and destruction and thus will permit all nations to benefit further from the good aspects of nuclear energy and suffer less risk from the evil aspects of nuclear and thermonuclear weapons.

1945 MANHATTAN PROJECT ORGANIZATION

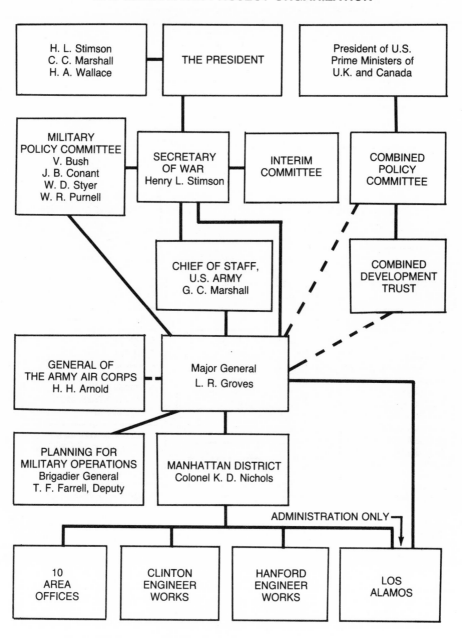

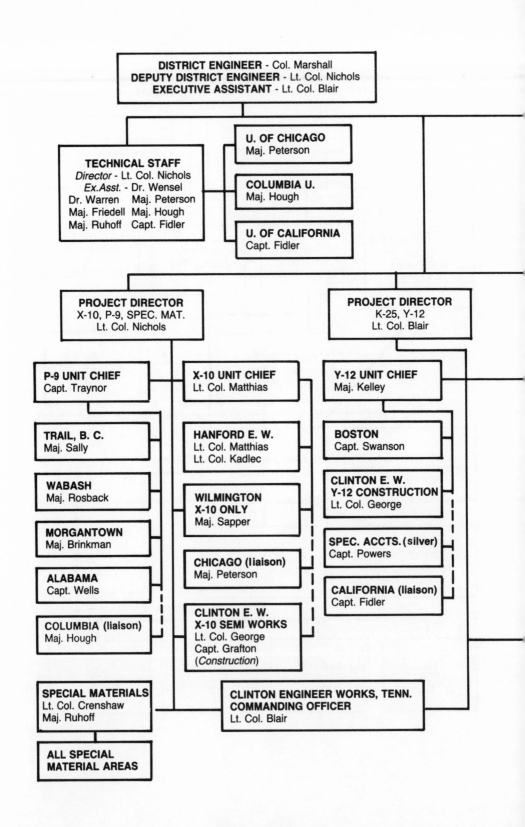

DISTRICT ENGINEER - Col. Marshall
DEPUTY DISTRICT ENGINEER - Lt. Col. Nichols
EXECUTIVE ASSISTANT - Lt. Col. Blair

TECHNICAL STAFF
Director - Lt. Col. Nichols
Ex.Asst. - Dr. Wensel
Dr. Warren Maj. Peterson
Maj. Friedell Maj. Hough
Maj. Ruhoff Capt. Fidler

U. OF CHICAGO
Maj. Peterson

COLUMBIA U.
Maj. Hough

U. OF CALIFORNIA
Capt. Fidler

PROJECT DIRECTOR
X-10, P-9, SPEC. MAT.
Lt. Col. Nichols

PROJECT DIRECTOR
K-25, Y-12
Lt. Col. Blair

P-9 UNIT CHIEF
Capt. Traynor

X-10 UNIT CHIEF
Lt. Col. Matthias

Y-12 UNIT CHIEF
Maj. Kelley

TRAIL, B. C.
Maj. Sally

HANFORD E. W.
Lt. Col. Matthias
Lt. Col. Kadlec

BOSTON
Capt. Swanson

WABASH
Maj. Rosback

**WILMINGTON
X-10 ONLY**
Maj. Sapper

**CLINTON E. W.
Y-12 CONSTRUCTION**
Lt. Col. George

MORGANTOWN
Maj. Brinkman

CHICAGO (liaison)
Maj. Peterson

SPEC. ACCTS. (silver)
Capt. Powers

ALABAMA
Capt. Wells

CALIFORNIA (liaison)
Capt. Fidler

COLUMBIA (liaison)
Maj. Hough

**CLINTON E. W.
X-10 SEMI WORKS**
Lt. Col. George
Capt. Grafton
(*Construction*)

SPECIAL MATERIALS
Lt. Col. Crenshaw
Maj. Ruhoff

**CLINTON ENGINEER WORKS, TENN.
COMMANDING OFFICER**
Lt. Col. Blair

**ALL SPECIAL
MATERIAL AREAS**

MANHATTAN ENGINEER DISTRICT
ORGANIZATION CHART
Effective 1 April 1943

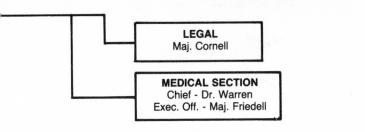

LEGAL
Maj. Cornell

MEDICAL SECTION
Chief - Dr. Warren
Exec. Off. - Maj. Friedell

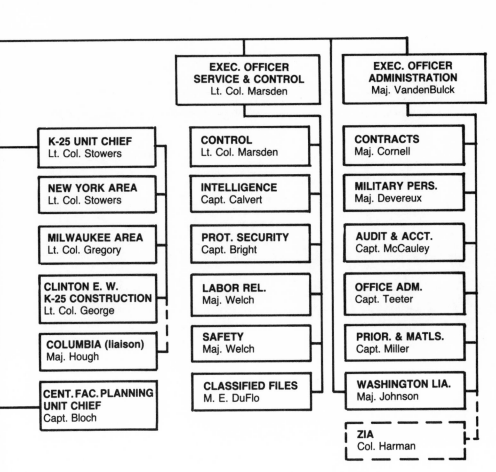

**EXEC. OFFICER
SERVICE & CONTROL**
Lt. Col. Marsden

**EXEC. OFFICER
ADMINISTRATION**
Maj. VandenBulck

K-25 UNIT CHIEF
Lt. Col. Stowers

NEW YORK AREA
Lt. Col. Stowers

MILWAUKEE AREA
Lt. Col. Gregory

**CLINTON E. W.
K-25 CONSTRUCTION**
Lt. Col. George

COLUMBIA (liaison)
Maj. Hough

**CENT. FAC. PLANNING
UNIT CHIEF**
Capt. Bloch

CONTROL
Lt. Col. Marsden

INTELLIGENCE
Capt. Calvert

PROT. SECURITY
Capt. Bright

LABOR REL.
Maj. Welch

SAFETY
Maj. Welch

CLASSIFIED FILES
M. E. DuFlo

CONTRACTS
Maj. Cornell

MILITARY PERS.
Maj. Devereux

AUDIT & ACCT.
Capt. McCauley

OFFICE ADM.
Capt. Teeter

PRIOR. & MATLS.
Capt. Miller

WASHINGTON LIA.
Maj. Johnson

ZIA
Col. Harman

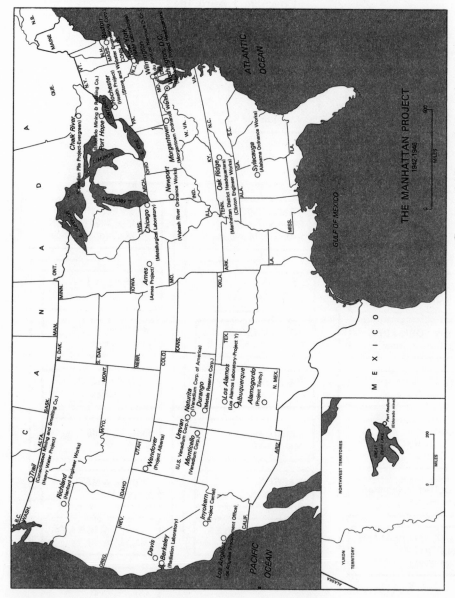

THE MANHATTAN PROJECT
1942-1946

Courtesy of Office of History, Corps of Engineers

18

PROJECTED SITE FOR
ATOMIC PRODUCTION PLANTS
Tennessee
1942

Courtesy of Office of History, Corps of Engineers

19

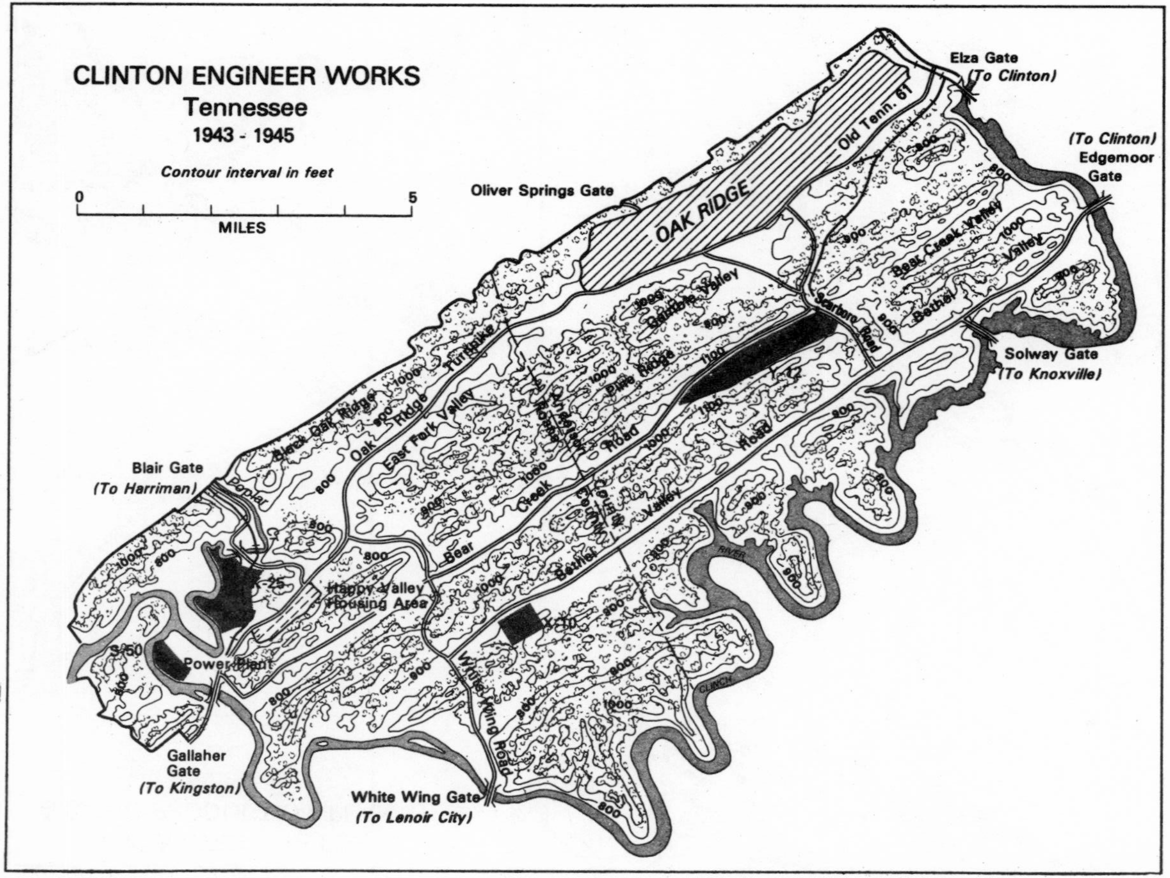

CLINTON ENGINEER WORKS
Tennessee
1943 - 1945

Contour interval in feet

MILES
0   5

OAK RIDGE

Elza Gate
(To Clinton)

(To Clinton)
Edgemoor
Gate

Solway Gate
(To Knoxville)

Old Tenn. 61

Oliver Springs Gate

Bethel Valley Road

Bear Creek Road

Administration
Road

East Fork Valley

Oak Ridge Turnpike

Louisiana Avenue

Poplar Creek

Bear Creek

East Fork Poplar Creek

Gamble Valley Road

Clinch River

County Equipment
Yard

White Wing Road

Happy Valley
Housing Area

Powerhouse

White Wing Gate
(To Lenoir City)

Blair Gate
(To Harriman)

Gallaher
Gate
(To Kingston)

Courtesy of Office of History, Corps of Engineers

20

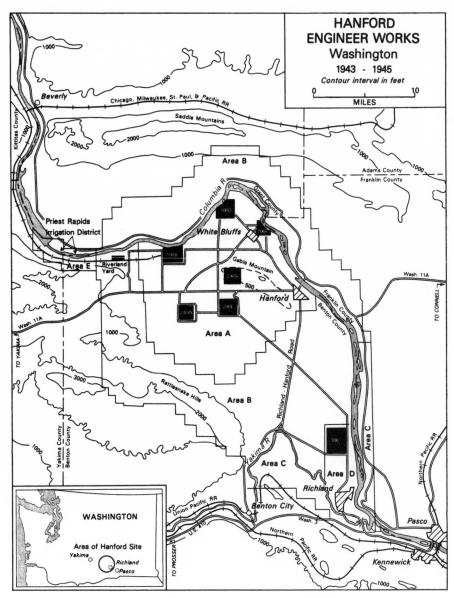

Courtesy of Office of History, Corps of Engineers

21

1

EARLY EXPERIENCES

WHILE I WAS ENJOYING some summer leave after having graduated from West Point in 1929, I received a telegram requesting that I volunteer to go to Nicaragua to help survey the route for a new interocean canal through the jungle.

I was a second lieutenant in the Corps of Engineers, I had graduated fifth in my class, and this was unusual duty. I volunteered. The assignment was a great experience. I benefited from the opportunity to have isolated commands, not many men, but far enough away from my next commander that I had to develop self-confidence, and make many decisions involving very unusual situations. I also benefited by having two very outstanding officers, Lieutenant Colonel Dan I. Sultan and Major Charles P. Gross, in charge of the project. Little did I know that First Lieutenant Leslie Groves, whom I was to meet on this jungle assignment, would become so important in my life or that a decade later we would be involved in one of the greatest secrets of World War II. At that time, the most important thing to me, having come from the small country town of West Park, Ohio, was, in addition to getting the job done, to learn how to survive in the jungle.

This I did. When our mission was accomplished, I was assigned to Cornell, where I received two degrees in engineering, and more important, married my wife of fifty-four years, Jacqueline Darrieulat. On my next assignment to the U.S. Waterways Experiment Station, Vicksburg, Miss., I took the advice of First Lieutenant Herbert D. Vogel, the director, and applied for a fellowship that would allow me to study hydraulic research methods in Berlin and Europe. With the help of Vogel and the chief of engineers, I was fortunate to receive the award. In 1934, Jackie and I sailed for Germany.

That year was an exciting time to be in Berlin. In the United States at the time Hitler was considered to be a joke. But we were able to see the many changes taking place in Germany and realize in what direction they were leading.

Besides visiting ten hydraulic laboratories throughout Europe, I also traveled with thirty German students along the entire North Sea coastline of Germany, including the island fortress of Helgoland. Gun emplacements under construction were clearly visible on many of the islands as part of a whole new system of fortification. One of my classmates, a German captain of engineers, talked freely with me about the size and range of the guns. He frankly admitted that they were rearming. And our social contacts with many other German officers in Berlin confirmed that Germany was preparing for war.

Despite my academic intentions, I did not receive a doctor of engineering degree from the Technische Hochschule, because regulations required that I go to troop duty. I returned to Vicksburg for another one-year tour of duty at the laboratory. To solve the problem of troop duty, I was assigned additional duty with the reserves.

From June 1936 to June 1937 I completed residency at the University of Iowa for a Ph.D. and also attended the engineer school at Fort Belvoir, Va. The thesis I'd written in Germany not only served for my degree, but also won an American Society of Civil Engineers Award. And I requested and received assignment as an instructor at West Point. I taught military and civil engineering, and military history. The challenge of teaching, the opportunity to serve with officers of all other branches of the Army, and the many attractive advantages of post life in an academic atmosphere added up to a meaningful and thoroughly enjoyable time from 1937 to 1941.

During the summer of 1939—just before the outbreak of World War II—Jackie and I embarked on a tour through Europe to see the political and military developments for ourselves. I felt certain that hostilities were imminent, but probably not before September. The secondary purposes of the trip were to pursue my newly acquired fascination with military history and to visit Napoleonic and World War I battlefields.

In Germany we found that Hitler's propaganda for peace had ceased, and we became more convinced by obvious troop maneuvers and demolition preparations that war was imminent. The countryside near the border was alive with soldiers, the byroads clogged with convoys.

While in France, as a result of the friendship between Colonel E. E. Farman, the librarian at West Point, and General Brousseau, the commander of the Metz sector of the Maginot Line, I was able to talk with General Brousseau and other French officers about the military situation and their strategy. General Brousseau outlined to me the major political problem France had in extending the Maginot Line to the sea. If France put the line along the French-Belgian border, then France would be relinquishing Belgium to the Germans, whereas placing the line along the Belgium-German border definitely committed Belgium and required their agreement. Consideration of Holland further complicated matters. The various governments never did resolve these political problems. In hindsight, I do not know if a line extended to the sea would have stopped the Germans, but it would have slowed their advance and provided more time to avoid the rapid collapse of the Allied forces that led to Dunkirk.

When I returned to Paris I discussed with our military attaché the formidable fortifications I had been privileged to visit in front of Metz. I described the gun emplacements, the underground installations, the well-planned layout taking advantage of terrain features to maximize the defensive fire effects, and the extensive tank obstacles. I expressed the opinion that the Germans would have to outflank the fortified line by driving through Belgium, or else incur tremendous losses in a frontal attack on the Maginot Line. I asked the attaché for terrain maps in making my report to Washington. He told me that he had been trying to get one of his officers into the Maginot Line for two years. After I told him how my visit had been arranged and assured him that I would send him my report, he cooperated fully.

The day after we returned to West Point, Hitler's war machine crashed into Poland, thereby jolting the United States out of its years of complacency. When Hitler finally struck the Western Front, the Maginot Line forced the German forces to invade Holland and Belgium to reach France. However, because of the lack of preparations, the obsolete tactics, and no real will to fight, the combined Allied forces were no match for the new German air-to-ground tactics. A point to remember is that the Germans did not take any fort in the Maginot Line until after they had entered Paris.

As the war in Europe progressed, I looked forward to commanding combat engineer troops as my next assignment. As a result of asking for it, I was very happy when I received orders to the 36th Combat Engineer Regiment at Plattsburgh, N.Y. But just before our household goods were ready for shipment, Colonel J. C. Marshall requested that I come to the Syracuse District to be the area engineer, to supervise the design and construction of the Rome, N.Y., Air Depot. He promised me almost complete independence, a choice of personnel, and said I would be free to seek another troop assignment after a year. With some reluctance I agreed.

When I started the project, acquiring the land, letting the design and construction contracts, etc., we had not yet entered the war, and the townspeople presented many objections to our operations. With the cooperation of a few of the town leaders and the help of a superior staff, we handled all local problems without too much difficulty. However, I did have difficulties with Washington for exceeding my authority. General H. H. Arnold, chief of the Army Air Corps, had ordered that no land could be acquired without his specific written authority. Colonel Marshall called me and said I was in trouble because I had bought a right-of-way for a railroad spur after the order had been issued. The chief of engineers had asked that I come to Washington to see Arnold. I responded, "I'll go, but I will prove that I have already saved more in switching charges than the spur cost us."

In Washington I was surprised at the new attitude about the war. Various officers in the office of the chief of engineers advised: "The Russians are holding the Germans. The urgency for our preparations is over. Don't get too argumentative when you see Arnold." When I told Colonel James B. Newman, Jr., Arnold's engineer, what I intended to say, he told me: "Don't do that. Just go in, take it, and forget it." I refused, and pulling out a stack of freight bills, I said, "I'm going to show him that I used good judgment and saved him plenty of money." After going into Arnold's office he returned and said, "Arnold's busy now. You will have to come back next week."

While still in bed late that Sunday I turned on the radio. Pearl Harbor had been bombed by the Japanese. Damage was still unknown, or at least unannounced. I unpacked my uniforms from my wardrobe trunk where they had been stored since I left West Point, shook out the mothballs, and arrived at the office the next day in uniform. I never heard from Arnold about the railroad spur.

Soon all construction was expedited. The Corps of Engineers mobilized for an even greater construction effort and formed additional combat units. For the first time in my career I had other officers reporting to me. First Lieutenant Harry S. Traynor was the first. He appeared at my office on March 27, 1942, saluted, and reported for duty. He was a reserve officer. I had not been informed that he was coming, but I welcomed this intelligent, personable, energetic young officer. Other reserve officers were assigned to the district. Many of our civil service employees who were reserve officers were called to active duty, and they continued in the same position. In February Marshall asked me to take on additional duty as area engineer for construction of a new TNT plant, the Pennsylvania Ordnance Works near Williamsport, Pa. He promised to relieve me as area engineer for the Rome Air Depot as soon as he could find a suitable replacement. Officers were hard to get. Troop assignments had priority.

Once again my path crossed with that of Leslie Groves, who was now a colonel and in charge of military construction. I remember in particular the advice he gave me in regard to my new project: "On this project the designs are complete. Du Pont has been responsible for designing and furnishing equipment for several of them. Stone and Webster has already built one. You have contractors and people who know what they are doing. Don't interfere with them. If there are any problems, you handle them."

I was still in charge of both projects on Friday, June 19, 1942, when Colonel Marshall unexpectedly called from Washington. "Come to see me Sunday morning in Syracuse about a very technical project," he said. "I can't tell you any more over the phone." I had planned to remind Marshall soon that my year was almost up and that I would be seeking a good combat assignment. His call perplexed me.

Sunday morning, as I came into his office he greeted me, "Nick, I am giving you fifteen seconds to volunteer for a very important technical project, or be drafted." After he had given me a very short description of the nature of the project, I didn't need the fifteen seconds. But this was going to involve another delay in getting an overseas assignment—one year again, according to Marshall.

With volunteering settled, Marshall handed me a large envelope that contained most of the top secret information he knew about the new project. He explained that Brigadier General Wilhelm D. Styer, chief of staff of Army Services of Supply, and Dr. Vannevar Bush, who headed the Office of Scientific Research and Development, had briefed him on the project and provided him with the documents in the envelope. I was to read them in his presence.

2

THE CURTAIN RISES, 1942

AFTER I FINISHED reading, Marshall continued: "We have a meeting scheduled in Washington on the twenty-fifth with Bush, [James B.] Conant, the program chiefs, and other people who will be involved in formulating definite plans for carrying out the June thirteenth report. Then we will know more, I hope."

Following our meeting, Marshall and I met with several of the staff of the Syracuse District to outline the formation of a new district and to advise Virginia Olsson, Marshall's secretary; Robert Blair, the top civilian engineer in the Syracuse District; and Charles Vanden Bulck, my chief administrative assistant, that they would be made available immediately "for practically exclusive duty" with the new organization. Marshall turned over the running of the Syracuse District to his deputy effective the next day, June 22, and we were considered "as on special duty." After wrapping up a few loose ends at Rome and Williamsport Pa., on the twenty-second and twenty-third, I left for Washington.

Marshall and I met at the Office of Scientific Research and Development (OSRD) with the S-1 executive committee, Vannevar Bush and General Styer. After listening to members of this distinguished committee (three of them were Nobel Laureates) outline the status and scope of the proposed atomic development program—and the significance of winning or losing the race with Germany—I quickly realized that working with these men on such an important project was a unique opportunity. I had volunteered properly.

Van Bush, a shrewd electrical engineer, president of the Carnegie

31

Institution, and director of OSRD, was in overall charge of the atomic bomb effort and reported directly to the president. Bush was an outspoken New England Yankee, caustic at times but with a twinkle in his eye. I grew rather fond of him.

Styer was a senior officer in the Corps of Engineers. I had met him only once before, but all my friends who had worked for him considered him a most capable engineer and a pleasure to work for.

The members of the executive committee were:

James B. Conant, president of Harvard, was chairman of the S-1 Section and directly responsible for the OSRD atomic program. He would be the "devil's advocate" when optimism exceeded reasonable bounds. He was very smooth and diplomatic, but firm in maintaining and supporting his personal judgments. As I got to know him, I enjoyed his conversation on a multitude of subjects.

Lyman Briggs, director of the Bureau of Standards, was much older than the rest. He was an elder statesman for the project.

Arthur Compton, chairman of the Department of Physics at the University of Chicago and head of the plutonium project, was a tall, handsome man with deep-set, piercing eyes. He soon impressed me with his sincerity of purpose. Prior to the meeting Bush had praised him highly but added, "Don't be surprised if he suddenly decides to make an end run."

Ernest Lawrence, of the University of California and head of the electromagnetic method of separating U-235, was the most outgoing of all, a super salesman, always optimistic, good company, and fun to work with.

Eger Murphree, head of the centrifuge project and engineer for all the projects, was more like the industrial engineers I was accustomed to dealing with. He was stable, conservative, thorough, and precise.

Harold Urey was head of the gaseous diffusion project for separating U-235 and head of the heavy water project. I never could properly appraise him, but my first impression was that he would be difficult.

During and after the committee meeting, I heard more about the prior history of the atomic bomb. Late in 1938, scientists at the Kaiser Wilhelm Institute for Chemistry in Berlin made the basic discovery of fission of uranium. Fortunately for the United States, they sent the information to their coworker Lisa Meitner, in Denmark, where she had fled from Hitler's Germany. She immediately recognized the significance of the discovery and discussed it with Niels Bohr, another

Nobel Laureate. On January 16, 1939, Bohr arrived in the United States, where he informed Enrico Fermi and other scientists of the findings. Recognizing the significance of fission, they confirmed the German experiments and then began to explore the possibilities for atomic power or an atomic explosion.

In October 1939, after conversations with Leo Szilard, Eugene Wigner, and Albert Einstein, one Alexander Sachs, an economist, visited President Roosevelt to discuss the importance of uranium fission and the possibility of using it to create an atomic bomb of extraordinary power. Sachs gave the president a letter from Einstein in which he explained the urgency of doing research in the field. In response Roosevelt created the Advisory Committee on Uranium to look into the matter. By the end of April 1940 scientists had confirmed that the uranium fission caused by neutrons of thermal velocities occurred only in the U-235 isotope, and it was reported that a large section of the Kaiser Wilhelm Institute had been set aside for research on uranium. As a result, the committee concluded that research should be pushed more vigorously.

In June, President Roosevelt announced the organization of the National Defense Research Committee, with Vannevar Bush as its chairman. The NDRC began to contract with several research institutions, using funds that the Army and Navy supplied. On June 28, 1941, President Roosevelt established the OSRD, with its director, Bush, being personally responsible to the president. The OSRD was to serve as a center for mobilizing the scientific resources of the nation and applying the results of research to national defense. Conant replaced Bush as chairman of the NDRC, while the Advisory Committee on Uranium became the OSRD Section on Uranium, soon designated cryptically as the S-1 Section.

The OSRD came into being at the turning point in the United States' atomic energy effort. By July 1941, scientists had recognized some of the problems that had to be solved if an atomic explosion or atomic power were to be attained. They realized that a major part of the effort would have to focus on the problems of separating the fissionable uranium-235 isotope from the uranium-238 isotope.

At the University of California, Glenn Seaborg had discovered plutonium. Some scientists had believed that plutonium-239 might offer a more efficient alternative to uranium-235 in a bomb if plutonium-239 could be produced in sufficient quantity and if a weapon using it could be developed. As a result, work was undertaken to demonstrate the feasibility of a chain reaction in a pile (reactor) as a first step toward developing processes and plants for the production of plutonium in large quantities. Work was also undertaken to determine the quantities of U-235 or plutonium that would be needed to produce an effective weapon. Ultimately, over 90 percent of the cost of the Manhattan Project went into building the plants and producing the fissionable materials, and less than 10 percent was applied to the development and production of the weapons.

Bush reported to President Roosevelt on November 27, 1941, that he was forming an engineering group and accelerating physics research aimed at plant design. In a handwritten response on January 19, 1942, the president approved Bush's decision. By then, of course, the nation had entered World War II, and Bush immediately expedited action aimed at developing a fission weapon. He assumed primary responsibility for the program and followed the president's direction that policy considerations be restricted to him, James Conant, and a "top policy group" composed of Vice President Henry Wallace, Secretary of War Henry Stimson, and Army Chief of Staff George C. Marshall. For recommendations and advice on scientific matters, Bush relied on the S-1 Section.

Bush sent another report to President Roosevelt on March 9, 1942, expressing cautious optimism and indicating that the project might succeed in producing a bomb by 1944. The report also contained the conclusion of Bush and Conant that the Army should take over responsibility for construction. The Army in turn selected General Styer, chief of staff for Lieutenant General Brehon B. Somervell's newly created Services of Supply, as the principal contact with the S-1 committee.

By June, work had reached the point where OSRD was able to recommend that the atomic bomb program was ready for major expansion. In a report to Wallace, Stimson, and Marshall on June 13, Bush and Conant provided detailed plans for the next stage of the project. It dealt with the status of the program as appraised by the senior scientists and contained the recommendations of the program chiefs and the Planning Board as well as recommendations from Bush and Conant. The policy committee approved the report, and on June 17, the president approved it with a handwritten "OK FDR."

The same day, acting on prior discussions with Bush and the recommendations in the June 13 report proposing to have a Corps of Engineers officer responsible for construction of the plants for the project, General Styer requested that Colonel Marshall come to Washington the next day, June 18. Marshall later recalled that after the briefing and reading the June 13 report, "I spent the night without sleep trying to figure out what this was all about. I had never heard of atomic fission, but I did know that you could not build much of a plant, much less four of them, for ninety million dollars. At the moment, among other construction projects in the Syracuse district, I had one for a TNT plant in Pennsylvania estimated to cost one hundred twenty-eight million dollars."[1]

The $90 million represented only seed money for initial construction. The figure proves that no one envisioned that the atomic bomb project would ultimately cost about $2 billion. Colonel Marshall's initial reaction to the figures resulted from the vagueness of the June 13 report, which reflected a compromise between the military and the OSRD. Major General Lucius Clay, the Army representative on the Army and Navy Munitions Board, argued that the scientists should select one method of separating U-235 and see how the costs in money and materials would infringe on the total war effort.

In contrast, Conant, Bush, and the S-1 Section wanted to pursue all four proposed methods of separating uranium. In the June 13 report, they acknowledged, "The program as proposed obviously cannot be carried out rapidly without interfering with other important matters, both from the standpoint of demands on scientific personnel, and demands on critical materials. Judgement must hence be made by balancing the military results which appear attainable against the interference with other programs." The report noted that all four methods of preparing the fissionable material "appear to be feasible. At the present time it is not possible to state definitely that any one of these is superior to the others." As a result, they concluded, "It would be unsafe at this time, in view of the pioneering nature of the entire effort, to concentrate on only one means of obtaining the result." Consequently, the June 13 report requested funds for the design of the gaseous diffusion plant and the building of pilot plants for the other three methods. The total money of $85 million and a $5 million contingency fund were only initial estimates for fiscal 1943.

The report stated: "The scientists in charge of the various phases of this development are now unanimous in believing that the production of bombs of enormous explosive energy is possible, by releasing atomic

---

1. General James Marshall to Jesse A. Remington, January 15, 1968, Office of History, OCE.

energy.'' In conveying the report to President Roosevelt on June 17, Bush acknowledged that estimated costs were higher than his earlier estimates ''principally because it still was considered unsafe to concentrate on one method only.'' In any case, Bush told the president, ''The results to be expected are still extraordinary. Adequate control appears certain. The time schedule contemplates availability early in 1944.''

The chief of engineers selected Colonel Marshall to direct the construction intended to make the bomb a reality because of his fine reputation and his previous construction experience, which included a good track record for starting projects. As he had demonstrated to me in beginning the Rome Air Depot and then the Pennsylvania Ordnance Works, Marshall was a great organizer who delegated both authority and responsibility well. He also was an astute judge of personnel. He would have to summon these skills promptly, since he did not have the luxury of time or the facts available to plan the entire operation before beginning the construction phase. This is where matters stood when he brought me into the picture.

At the June 25, 1942, meeting, the S-1 committee discussion covered all aspects of the June 13 report with particular attention to the status of the separation processes and locations for the several plants under development. At the time, only Lawrence's electromagnetic project appeared to be anywhere near ready for preliminary engineering development. Construction of the centrifuge and diffusion plants were a long way in the future. Supplies of uranium and other vital materials were inadequate even for research purposes, much less for production. The scientists had no clear concept of what they wanted to build. At the same time, with the exception of Lawrence, who wanted a location close to his Berkeley laboratory for the electromagnetic plant, the committee and General Styer wanted to acquire a site somewhere in the middle of the country immediately. Apart from being protected from a possible enemy attack, any suitable location would have to have a large source of electric power close by, plenty of water, and adequate isolation for safety and security reasons. A short distance west of Knoxville, Tenn., the committee had already found such a place, which we were going to have to check out quickly.

By the time we had arrived in Washington for the meeting, Marshall and I had decided that the best way to get the project started without delay was to select an overall contractor. We selected Stone and Webster largely because I had been working with them on the Pennsylvania Ordnance Works and we were familiar with their chief personnel and capabilities.

Following the initial morning meeting, Marshall and I had additional sessions that afternoon and the next day with the OSRD to deal with specific issues, such as funding the project, organization and administration of the district, and the subject of priorities. According to the S-1 committee, apart from obtaining funds, the OSRD's most urgent need was help in obtaining priority items and in procuring scarce materials and equipment required by a multitude of research agencies throughout the country that had been working on the S-1 project.

In a meeting the next day with Lieutenant General Eugene Reybold, chief of engineers, and his deputy, Major General Thomas M. Robins, we received approval of the decision to sign up Stone and Webster. We agreed that we would give the company a letter of intent to do all the work. Later we would parcel out parts of the project to other contractors. At the same time we would give Stone and Webster a commitment that it would share in the construction work.

To confirm the company's willingness to serve as the overall contractor and engineer, I went to New York to visit John Lotz, the president of the company. He knew a little about the project, since the company had been involved with some of the preliminary design work and seemed "greatly pleased" that we had selected them. We drew up a letter of intent containing a limiting cost of $10 million, and Lotz and Marshall signed it that afternoon. Then we discussed arrangements for the site survey trip to Tennessee on July 1.

Very quickly, we were able to secure the cooperation of the Army in obtaining personnel we needed to do the job. The case of a key scientist, John Ruhoff, illustrates the process. On April 17, 1942, Arthur Compton requested the Mallinckrodt Chemical Company in St. Louis to help scale up to industrial output the laboratory process for purifying uranium and reducing it to uranium dioxide as the first step toward producing plutonium. Ruhoff was the top technical man on the job, but he had already received his call-up orders as a reserve officer.

To prevent delays in the uranium processing, Compton asked Marshall to find a way to keep Ruhoff at Mallinckrodt until he had completed his work. Working with the Army personnel office, Marshall had Ruhoff assigned to us, and we assigned him to work with Mallinckrodt. Once the purification process had gone into operation, we assigned Ruhoff to our New York office. Ultimately, Ruhoff was in charge of the design, construction, and operation of all our feed material plants, including Mallinckrodt, needed to supply our main fissionable material production plants.

Acquiring our initial funding for the district also went smoothly. On June 25, Dr. Irvin Stewart, executive secretary of OSRD, had advised Marshall that $15 million was needed by June 30 to continue current OSRD contracts. General Reybold directed him to obtain the money from the Corps of Engineers finance office. When we picked up the $15 million check on June 30, we agreed to draft a letter to the budget officer for the War Department requesting transfer of $85 million to the district as soon as practical.

The district set up its own finance officers and an internal auditing staff under the direction of Colonel Charles Vanden Bulck. We maintained amicable relations with the comptroller general of the United States primarily because we mutually agreed to establish auditors at CEW and HEW to keep the auditing current. As a result the comptroller auditor completed his audit within thirty days from the time we had expended the money. My administrative people involved with the expenditures were very pleased when the comptroller general reported to the Senate Special Committee on Atomic Energy in April 1946, "We have audited, or are auditing, every single penny expended on this project. We audited on the spot and kept it current, and I might say it has been a remarkably clean expenditure. . . . The very fact that our men were there where the agents of the Government could consult with them time after time assured, in my opinion, a proper accountability. . . ."

I believe this achievement occurred because of the way in which Colonel Marshall started the project and organized the administrative aspects of the district. His determination to do things in an orderly manner also was reflected in the way he chose our original headquarters and then went about selecting the site for the production plants. (Unfortunately, this desire for orderly procedures ultimately reacted against him.)

Reybold asked Marshall to locate the new district's office temporarily in the chief of engineers' building and later locate it at our main construction site. Marshall argued for a headquarters in New York to utilize the facilities of the North Atlantic Division and to be close to the Stone and Webster offices in Boston and New York. Marshall ignored Reybold's request and located it in New York City, in the North Atlantic Division headquarters at 270 Broadway. At Colonel Leslie R.

Groves' suggestion, however, we did set up a liaison office in Washington.

With this initial business out of the way, Marshall and I left for Knoxville the evening of June 30 to look at the sites that the S-1 committee had earlier picked out as possible locations for the productions plants. During the three days we spent in Tennessee, we looked at four locations that met our primary requirements: available power and water, proximity to a railroad, and sufficient land for building the four possible plants. We received help in our survey from the Tennessee Valley Authority and a representative from Stone and Webster. Other locations, including sites in the New Jersey-Philadelphia area, near the Shasta Dam, the Grand Coulee Dam, and Boulder (now Hoover) Dam, were considered because they provided some or most of the requirements, but by the end of our three days in Tennessee, we had tentatively selected the Tennessee site on the Clinch River.

Apart from having the TVA power and water, the location had the necessary isolation because the Clinch River provided a boundary on three sides, and a mountain ridge protected the north side. The area also had the necessary number of valleys and ridges to help isolate the separate plants from each other. The S-1 committee approved our choice in a meeting on July 30, but Marshall did not immediately begin acquiring the land because he felt he still knew too little about what was to be built. He, the Stone and Webster people, and I visited the various scientific laboratories, and we quickly confirmed that the project was far from ready to proceed with construction. Marshall felt that since we knew where we would put the facilities, we had started the aerial survey of the site, and we had the authority to obtain the land whenever we were ready, we had no need to begin purchasing property immediately.

Beginning with the trip to Tennessee, the pattern of our method of operation began to emerge. On my part, from July 1942 to the end of the war I was destined to travel over ten thousand miles a month by train, commercial airplane, and bus. Only in April 1945, when the project neared fruition and time became even more critical, did I obtain use of an Air Corps B-25 bomber to facilitate travel. From the start, our philosophy was to go where the work was being done and make decisions as needed on the spot. Our local area staff and contractors then formalized these decisions in writing or in the form of plans and specifications as necessary.

From Knoxville, Marshall returned to Syracuse and then traveled to

Washington for a meeting with Stone and Webster and the OSRD on July 9. Blair and I went to Chicago to meet with Arthur Compton and inspect the proposed site for the experimental pile in the Argonne Forest Preserve. As a result of our two-day visit, we were able to initiate construction of a chemistry laboratory on the University of Chicago campus, arrange to acquire the site in the Argonne Forest Preserve from Illinois' Cook County, and initiate design of a building for the experimental pile.

During my first meeting with Dr. Compton I told him, "If you expect me to do my end of the job, you have to start educating me." So at following meetings, whenever we had time, Dr. Enrico Fermi or other scientists would give me an informal lesson. They all recognized my desire to learn and were eager to teach me what I needed to know. While I never had formal education in physics, I received a superb indoctrination in the subject from renowned leaders in the field.

Having set necessary work in motion, I returned to Washington in time for the July 9 meeting of the S-1 committee and Stone and Webster; the meeting dealt with the requirements for uranium in the coming year, site selection, and priorities. Later in the day, while I continued meeting with the OSRD, Marshall discussed funding and contracts with Groves. Groves informed Marshall that the requested $85 million for the project was not available from the president's funds and so suggested alternative sources for financing the project, at least in its early stage. Groves expressed concern about the indefinite dates on which certain parts of the project would be started, and he urged that we insist upon complete scheduling of the entire project.

On August 11, 1942, Marshall presented to Colonel Groves a draft of a general order forming the new district. They decided to call it the Manhattan Engineer District (MED), since we had our main office in Manhattan, New York City. Giving the project that name would focus attention away from the actual site of the plants. The chief of engineers issued Order No. 33 on August 13, 1942, setting up an engineer district without territorial limits, to be known as the Manhattan Engineer District, to supervise projects assigned to it by the chief of engineers.

3

STRUGGLE FOR PRIORITY, 1942

EARLY JULY JACKIE AND I had vacated our apartment in Williamsport and moved to Alexandria, Virginia. Because of the housing shortage, we settled temporarily with Jackie's sister Marie-Louise and her husband, A. V. Peterson. Pete, a reserve officer, had been called to active duty with the 36th Combat Engineer Regiment before the war began but then was temporarily assigned to the Engineer School at Fort Belvoir, near Alexandria. He knew more about physics and electronics than I did, and during the period the four of us lived together, I probed his knowledge without telling him why. Finally I realized that he could provide a highly effective liaison with the project scientists, so with his consent I arranged for his transfer to the Manhattan District at the end of August. He became a key officer in the district throughout the war.

Pete recalls that, upon reporting to me for duty, "You made clear to me our working relationship on the project in view of our family ties. I well remember that while I would be subject to your orders and authority, any actions on promotions, efficiency reports, etc., would have to come from Colonel Marshall and not from you. And further, if it worked to my disadvantage, that's the way life sometimes is." Actually, it did work to his disadvantage somewhat. On one of our early promotion lists, Marshall discussed with General Groves whether Pete's name should be on it. Groves thought that in view of our relationship he would wait until the next list was prepared. Subsequently, Groves frequently used Pete for special assignments, and I believe Groves ultimately shared my high opinion of Pete's capabilities and superior performance throughout the project.

41

Pete made many friends. In spite of his outstanding stature and impressive physique, he carries his six feet, four inches with modest dignity. He is a soft-spoken, considerate man with a well-ordered mind and superior intelligence. Above all, his loyalty and integrity made him particularly valuable to me and of course to the entire project.

While the overall difficulty of obtaining a top priority for the project continued to be one of our prime concerns, we did early on solve the allocation problem for one scarce resource, the huge requirement of copper for Ernest Lawrence's electromagnetic separation process. Copper was required for electric windings to form the large electromagnets. The pilot unit, originally planned for the Berkeley campus, would need 120 tons of copper, while the full-scale plant to be built in Tennessee would need five thousand tons or more of the metal. Copper was in desperately short supply because of the demands of the war industries. For the electromagnetic process, however, silver could substitute at the ratio of eleven to ten. Since the government would own the plants and the silver could be returned after the war, we decided we should approach the U.S. Treasury to borrow the needed metal from the silver repository.

As a result, on August 3 I visited Assistant Secretary of the Treasury Daniel Bell. He explained the procedure for transferring the silver and asked, "How much do you need?" I replied, "Six thousand tons." "How many troy ounces is that?" he asked. In fact, I did not know how to convert tons to troy ounces, and neither did he. A little impatient, I responded, "I don't know how many troy ounces we need, but I know I need six thousand tons—that is a definite quantity. What difference does it make how we express the quantity?" He replied rather indignantly, "Young man, you may think of silver in tons, but the Treasury will always think of silver in troy ounces."

With our contrasting perspectives expressed, we then settled on a form of agreement that was ultimately used to transfer some 14,700 tons of silver from its storage place at West Point to New Jersey, where it was melted down and cast into large ingots for shipment to Allis-Chalmers in Milwaukee for further processing. Each month during the war, I signed an inventory for the Treasury stating that we had in our possession over four hundred million troy ounces of silver (expressed to hundredths of an ounce). We established very strict procedures to avoid loss, and when the silver was returned to the Treasury after the war "less than one thirty-six-thousandths of 1 percent of the more than 14,700 tons of silver—was missing."[1]

An even more important task in the early months of the project was to ensure that we would have enough uranium to make the fissionable material for the bombs. Initially, the scientists did not even have enough of the element for experimental purposes and had no real idea of how much we would need once the separation plants went into operation. Before the engineers were given the responsibility for constructing the project in June 1942, the University of Chicago researchers, under Arthur Compton, had become involved in trying to locate supplies of uranium ore and worked on reopening the Great Bear Lake mine in Canada. The scientists had contacted Boris Pregel, who at times was known as the radium king of the United States. Through him they made contact with Eldorado Gold Mines Limited, which had a processing plant in Port Hope, Ontario. The president of the company had discovered the Great Bear Lake mine, which was in the far northern part of Canada and was the only uranium mine that had been operated recently in North America.

Once we arrived on the scene, we had a considerable go-round with Mr. Pregel concerning the reopening of the Great Bear Lake mine, requirements for barges, and several other things. Although his story never seemed to hang together, we finally reached a tentative agreement with him to deliver uranium to us. However, Pregel immediately began to have new problems getting mining operations started, and his story became even more confused. Fortunately, we found another and better source of ore right in the New York area.

In late 1940, when Germany appeared in a position to threaten much of Africa, Edgar Sengier, the head of Union Minière, ordered shipment of approximately twelve hundred tons of high-grade uranium ore from the Shinkolobwe mine in the Belgian Congo through Portuguese West Africa to New York, where he had it stored in a warehouse on Staten Island. Early in 1942, while in Washington on other matters, he mentioned this stockpile to Thomas Finletter and Herbert Feis, State Department officials concerned with international economic affairs. Sengier then followed this up with a letter to Finletter dated April 21, 1942, in which he described his supply in detail.

1. Vincent C. Jones, *Manhattan: The Army and the Atomic Bomb,* Center of Military History, United States Army (Washington, D.C.: U.S. Government Printing Office, 1985), p. 133.

I first learned of Sengier's supply of ore on September 7, when Finletter called me to say he had a letter requesting permission to ship to Canada some black uranium oxide an African company had stored in the United States. The letter came from Raphael Rosen of Standard Oil Development Company. Taken by surprise, I immediately began to ask about the stockpile. I could not reach Mr. Rosen at Standard Oil until the eleventh. Rosen quickly filled me in on Sengier's stockpile, explaining that the Belgian wanted to dispose of the ore under conditions satisfactory to the United States. Rosen told me he had advised Sengier that the University of Chicago metallurgical laboratory would be interested in the material and in turn had learned that about two hundred tons already had satisfactory purity. The rest of the uranium did need to be refined. Before leaving for an S-1 executive committee meeting in California, I instructed my assistant, Captain Allan Johnson, to call Finletter about the uranium to arrange for us to make a test of the ore.

Tom Crenshaw, area engineer at Berkeley, and I attended the S-1 meeting at the Bohemian Grove, north of San Francisco, on the thirteenth. I informed the executive committee about the uranium and it recommended that we acquire from Sengier all the ore available. Apart from the uranium question, the meeting dealt with several significant issues on the thirteenth and fourteenth. It decided to proceed immediately with the acquisition of the Tennessee site and locate the electromagnetic pilot plant and the major portion of the plutonium pilot plant there. While recommending that we should start the design and procure materials immediately for the one-hundred-gram electromagnetic plant, the committee reserved the right to change its mind and recommend canceling orders for material for the plant any time up to and including January 1, 1943.

The committee also recommended that we secure a chemical company to assist in developing and operating Compton's plutonium process. Compton had discussed this idea with me in July and had brought in some engineers from du Pont to assist in developing the chemical process for separating out plutonium, but the committee's recommendation gave a major impetus for involving the huge chemical company in the project, a development that had a profound impact on the ultimate success of the whole program.

In regard to priorities, Conant advised me that in his position as head of the S-1 committee, he would recommend that our project be placed ahead of the development of synthetic rubber. He did not want to put this in writing, but he said he would repeat his decision to General Styer or others in corresponding positions in the military. He also told me that Bush had been working very hard in the past week trying to arrange for a higher priority for the entire project.

After returning from the West Coast, I visited Sengier in his New York office on September 18 to discuss purchasing his stockpile. I found the Belgian a courteous gentleman in his sixties. He had a somewhat pallid face, and his light hair was thinning. He was immaculately dressed, and he spoke excellent English in rather curt sentences. Before inviting me to sit, Sengier asked to see my identification: "You say you are from the military. Yet you are wearing civilian clothes." After reading my ID card, he motioned me to a seat and inquired what I wanted. I told him I understood he had some uranium for sale. In turn, he asked, "Are you a contracting officer? Too many people have been around here about this uranium and they just want to talk. Do you have any authority to buy?" Perhaps too flippantly, I answered, "Yes. I have more authority, I'm sure, than you have uranium to sell."

After thinking about it for a moment, he demanded to know, "Will the uranium ore be used for military purposes?" I hesitated. I didn't know what I should say to a foreigner about the country's most secret war project. Sengier rescued me: "You don't need to tell me how you'll use it. I think I know. All I want is your assurance as an Army officer that this uranium ore is definitely going to be used for war purposes."

I could, of course, give him that assurance. He replied, "Good. Then let's make a deal, Colonel. My company, the Union Minière, has twelve hundred tons of uranium ore stored on Staten Island." While I had some inkling from my conversation with Rosen that Sengier had a considerable stockpile, I was surprised at the actual amount, particularly after our desperate effort to find sufficient supplies just to get the research going.

We reached a general agreement in about half an hour. He wrote the details in longhand on a yellow legal pad. I took the original for our lawyers, and Sengier kept the carbon copy for his.

The draft agreement contained only eight sentences:

1. Stock of uranium rich ore in United States (about 1200 tons).
   U.S. Government will purchase entire stock from African Met-

als, Afrimet retaining ownership of radium for use in U.S. or
England.

U.S. Government undertaking the refining contract, direct
with Eldorado.

African Metals making necessary arrangements for radium.
Part of stock (say 300 tons) to go to Canada under U.S. control,
the disposal of the remaining stock to be decided later.

2. Stocks in Africa:

Arrangements to be made to ship to U.S. and store in U.S.
available rich ores now in Belgian Congo.

U.S. to have prior rights to purchase.

U.S. paying for freight and storage and insurance.

Before I left, we arranged for additional meetings to work out all the
details of the purchase, a process that would take almost two weeks.

Back in Washington on September 19, I briefed Finletter on my
agreement with Sengier. In turn, Finletter offered to put pressure on
Eldorado Gold Mines to deal with us directly.

Returning to New York, I met with Sengier in our office on the
twenty-third to discuss four proposals he was suggesting for the handling, sale, and refining of the uranium. We agreed on a price of $1.60
per pound of U-308, $1.00 of which went to his company and 60 cents
to Eldorado for processing the ore. I followed this up the next day by
meeting with representatives of Eldorado. They had no desire to refine
the ore as processor, wanting instead to purchase the ore from Sengier
and sell us the processed uranium, but during the afternoon session, we
reached basic understanding. The company agreed to continue to develop the mine at Great Bear Lake to fulfill our initial contract using
Canadian ore. More importantly, the company would enter into a
refining contract to process the uranium we had bought from Sengier.

The next day, Major Ralph Cornell, the MED lawyer, and I returned to Sengier's office to purchase immediately one hundred tons of
the ore already here and to arrange for its shipment to Canada. We
would place the remaining ore in safekeeping on a government reservation, with first right of purchase. African Metals would inform the
United States in the near future about the quantity, location, and shipping arrangements of the ore aboveground in the Congo. Once shipping had been arranged, we would enter into a contract with the
company for payment of transportation, insurance, and storage charges.
This ore also would be stored on a government reservation after it

arrived, and we would have first right of purchase. A few days later, Sengier arranged for shipping the approximately three thousand tons of ore from Africa and told me that the first shipment of 250 long tons could leave on October 10, the second on October 20, and the third on November 10, with the balance in amounts of four hundred tons a month from December to May 1943. I authorized the first shipment immediately so that Sengier could move the ore from the interior to the port in time to make the October 10 boat and told Sengier I would give him a decision about the balance of the amount the next week. In Washington, some thought was given to having all the ore shipped from Africa by convoy, but ultimately we went along with Sengier's belief that it was safer and less of a security risk to send the uranium in smaller amounts aboard fast freighters that could outrun German U-boats. In this he proved reasonably correct, and by the end of May 1943 all the ore that had been aboveground had reached the United States with the exception of one shipment lost in a torpedoed ship.

In addition to Canadian and Belgian Congo ore we refined uranium as a by-product of vanadium mines in the West. We also initiated an exploration for uranium in the United States and Canada. One other source we located was the waste piles of South African gold mines, which contained uranium at a very low percentage.

Our best source, the Shinkolobwe mine, represented a freak occurrence in nature. It contained a tremendously rich lode of uranium pitchblende. Nothing like it has ever again been found. The ore already in the United States contained 65 percent U-$_3$O$_8$, while the pitchblende aboveground in the Congo amounted to a thousand tons of 65 percent ore, and the waste piles of ore contained two thousand tons of 20 percent U-$_3$O$_8$. To illustrate the uniqueness of Sengier's stockpile, after the war the MED and the AEC considered ore containing three tenths of 1 percent as a good find. Without Sengier's foresight in stockpiling ore in the United States and aboveground in Africa, we simply would not have had the amounts of uranium needed to justify building the large separation plants and the plutonium reactors.

Although we had solved our requirements for silver and uranium, our struggle for priority continued through August and September. General Clay and the ANMB continued to be obstacles. In hindsight, and probably with more objectivity than I possessed at the time, I must concede that Clay was acting correctly based on his responsibilities and the information he had. He simply felt that military combat requirements deserved a higher priority than a nebulous secret weapon in the earliest stages of development. As Clay saw it, without the rubber, the planes, the tanks, and the ships as quickly as possible, the war might end with an Axis victory or stalemate before we could build an atomic bomb. Our disappointment over his repeated refusals to support our requests for higher priorities resulted in part from our belief that the president had not only approved the Manhattan Project as crucial to the war effort but also had cleared the way for a crash program. Instead, we continued to find ourselves wasting valuable time fighting for needed materials with the AA-3 priority that had been assigned.

It was now obvious that only the president could resolve the issue. Accordingly, Marshall and I met with General Styer, who directed us to ask General Clay what data he needed to secure from the ANMB and the War Production Board an AA-1 priority. Styer said we should incorporate the information in a letter that Bush should present to President Roosevelt. I went to see General Clay and outlined to him the status of the four pilot plants and the heavy water project at Trail Canada and reported the decision of the S-1 committee and General Styer that we should try to obtain an AA-1 priority for these five projects. I told him that General Styer had suggested we contact him for his advice concerning the information we should put in a letter from Bush to the president to get an AA-1 priority. Instead of a letter to the president, Clay recommended that General Styer ask the Joint Chiefs of Staff to write a letter to the ANMB stating that AA-2 priority be given the Manhattan Project. He then told me he did not think such a letter should be sent, inasmuch as our project contained so many critical materials. Turning to the directive on priorities, he stated he did not believe that we ever qualified for an AA-1 rating. He contended that the reference to highest priority (in the June 13 report) meant that we would be given the highest priority obtainable to one of the four methods. He remained convinced that our project was not as important as tanks and other munitions of war. Given his unwavering position, it seemed clear that we would have to go over his head to get results.

Meanwhile, Bush had concluded that the AA-3 rating simply would not get the job done. In a strongly worded letter on August 29 to Harvey Bundy, a special assistant to Secretary of War Stimson, Bush pointed out that the difference between AA-3 and the AA-1 ratings would mean at least a three-month delay in completion of the pilot plants. He found this exasperating in light of the small amount of material involved, which had a value of no more than $250,000. He explained that it was his understanding as a result of the president's approval of the project on June 17 that the experimental and pilot plant work would be expedited to the fullest possible extent. He felt now that the attitude of the ANMB threatened to delay the entire project badly unless some changes took place. He therefore concluded: "From my own point of view, faced as I am with the unanimous opinion of a group of men that I consider to be among the greatest scientists in the world, joined by highly competent engineers, I am prepared to recommend that nothing should stand in the way of putting this whole affair through to conclusion, on a reasonable scale, but at the maximum speed possible, even if it does cause moderate interference with other war efforts."

On September 1 Bundy showed Bush's memo to Stimson, but the secretary of war took no immediate action. Bush then took further steps to solve our problems. He went directly to President Roosevelt to pressure chairman of the War Production Board Donald Nelson and the ANMB for a higher priority rating and to raise various issues that bothered Bush. As a result, things were changing by the time I returned to New York on September 16 from the OSRD meeting at the Bohemian Grove in California. At 5:00 P.M. I received a call from General Styer requesting that Colonel Marshall or I be in Washington the next day. With Marshall out of town, I told the general I would be there. He also said to bring Colonel Leslie Groves with me.

Consequently, I called Groves and relayed Styer's message. When Groves asked what Styer wanted, I told him I was not accustomed to asking a general what he wanted to discuss with a colonel. Groves sputtered a little bit and then asked if I would have a car. When I told him I would have one, he said, "Well, pick me up."

When we arrived at his office, Styer told us that Bush had complained to the president about the way the project was progressing and had cited four areas in which he was dissatisfied with Colonel Marshall. He had set up the main headquarters in New York instead of Washington, leaving me to do much of the contact work in the capital. Bush also was displeased with the priority situation and had been working on ways to get the problem corrected. He was unhappy that Marshall had not yet acquired the Tennessee site. And he wanted a higher-ranking officer in Washington—someone who knew the ins and outs of the wartime system and who had more rank and so would be better able to contend with the other high-ranking officers in getting his mission accomplished.

Styer went on to say that Bush had asked for General Somervell;

this was, of course, unreasonable. Bush then had asked for Styer, whom I think he really wanted because Styer had served as Bush's initial contact with the Army. Styer told us that Stimson had concurred in Somervell's decision that neither he nor Styer should take over the Manhattan Project. Styer added that before Bush could come back to him with another suggestion to head the bomb program, he and Somervell had selected Groves. At that point, Groves obviously was upset. It was the first and only time I had ever seen him at all disrespectful to a senior officer. He said, "That's impossible. I won't take it. I was assured that I could go overseas where I have been promised a good job."

Styer looked at me and said, "Out, Nichols." So I left and sat in the outer office. After a few minutes, Styer came out and invited me back in. As I sat down he said, "Now, let's resume where we were." He went on to explain that the president had approved a new setup. Groves was immediately to acquire the Tennessee site and sew up the higher priority. He also indicated that Bush wanted better scientific representation at the higher policy levels. This was to lead to the formation of the Military Policy Committee, which became the policy controlling body for the project. On the key matter of priorities, Bush's efforts to get the project moving during the first half of September had apparently broken the logjam. During the meeting, Styer gave me a letter from Bush to him in regard to higher priorities and told me to return it to Bush and explain the steps we would take to secure the new rating. To accomplish this, Styer instructed me to draft a letter for Groves that was to be addressed to Donald Nelson and signed by General Marshall assigning the Manhattan Project an AA-1 rating.

Styer also touched directly on one area of development that ultimately proved helpful to the success of the project. Apparently at the suggestion of Bush, Styer indicated the desirability of finding out what the Navy was doing in atomic research. When I responded that because I was curious about what the Navy might be doing I had already made an appointment to visit the Navy Research Laboratory at Anacostia in Washington on the twenty-first, Styer said, "Take Groves along." At a meeting later that day between Groves and Styer, they agreed upon the wording of the order for Somervell's signature directing the chief of engineers to release Groves from his current assignment and put him in charge of the atomic bomb project. The memo included a list of instructions to be undertaken at once. Groves was to arrange for the necessary priorities, for a working committee on the "application of the project," and for the procurement of the Tennessee tract. Then he was to transfer all activities to the site; initiate the preparation of a bill of materials needed for construction; and draw up plans for the organization, construction, operation, and security of the project. While not all of these directives were carried out, they did set forth the framework that gave a new sense of urgency to the atomic bomb project.

General Groves has written that he had received an inkling of his fate earlier in the morning, before our meeting with Styer, and of his initial disappointment, which helps to explain his reaction. He recalled that after testifying at the congressional hearing on a military housing bill, he met General Somervell in the hall and asked permission to accept an "extremely attractive assignment" overseas. Somervell told Groves he could not leave Washington because the secretary of war had selected him "for a very important assignment and the president has approved the selection." When Somervell told him the assignment was in Washington and if he did the job right "it will win the war," Groves said his spirits fell because of his premonition it had to do with the Manhattan Project, which he did not consider as big a job as his current post of deputy chief of construction of the Corps of Engineers.[2]

Despite his initial unhappiness with the assignment, once Groves fully took over the reins of the project, on September 23, 1942, I never again heard him voice a single word of regret. Once in charge, he quickly moved to assume de facto control not only of construction but also control of all research and development, operation of the plants, development of the bomb, and the military aspects of the project. The implementation and operation of the reorganized bomb program did not officially begin until Groves actually received his brigadier general's star the next week. In the first meeting with Styer, he had pointed out that his position with the scientists would be stronger "if they thought of me from the first as a general instead of a promoted colonel." He later claimed that this decision proved a wise move. From my experiences during the Manhattan Project, I would agree with his comment about the scientists that "the prerogatives of rank were more important in the academic world than they are among soldiers."[3]

2. Groves, *Now It Can Be Told* (New York: Harper Brothers, 1962), pp. 3–4.
3. Ibid., pp. 4–5.

Even before he was formally in charge, however, Groves began to act on his mandate. In regard to the priority issue, he decided to leave General Marshall out of it, and he changed my draft letter into a letter from Nelson to him stating that the project should be given authority to issue AAA priorities, the highest available. When Groves visited Nelson on the nineteenth and set forth his position that we must have the highest priority, the head of the War Production Board first gave a completely negative response. But when Groves then said he would have to recommend to the president to abandon the project because of the intransigence of the WPB, Nelson quickly reversed himself and signed the letter Groves had drawn up, which stated: "I am in full accord with the prompt delegation of power by the Army and Navy Munitions Board through you, to the District Engineer, Manhattan District, to assign an AAA rating, or whatever lesser rating will be sufficient, to those items the delivery of which, in his opinion, cannot otherwise be secured in time for the successful prosecution of the work under his charge."

Finally the Manhattan Project had the priorities and standing to enable it to secure anything it needed, if available. We had found the uranium and the silver required, and we had orders to acquire the site on which to build our plants. However, we did not yet know about the many special materials that had to be developed. Nor were we aware of new scientific findings that would threaten our success and make the task far more difficult. In presenting the assignment to Groves, Styer had said, "The basic research and development are done. You just have to take the rough designs, put them into final shape, build some plants, and organize an operating force, and your job will be finished and the war will be over." This proved to be a wildly optimistic assessment of the situation in September 1942. But at least we now felt we had overcome some of the bureaucratic and organizational obstacles that had hindered our early efforts.

The big lesson we learned from the struggle for priority was that if we were to succeed we could not rely on normal channels to accomplish our mission. The main objective in the Pentagon was to win the war with conventional weapons or other means readily available. We were a questionable major distraction from this effort. We had to maintain a direct command channel to the chief of staff, the secretary of war, and the president. Everyone else was too occupied with his own primary mission to devote the necessary time and effort to participate in decisions pertaining to our project. We needed a separate command structure with adequate capabilities to perform all functions, independent of existing organizations. A corollary to this is to accept cooperation only when it is available without any infringement on our independence. We definitely needed cooperation, but only on our terms, and fortunately we got it.

4

TAKEOFF AND LANDING
IN THE NEW WORLD,
1942

EVEN THOUGH GROVES did not take command of the Manhattan Project until his promotion became official, he began his direct involvement immediately. Up to then, he had concerned himself with our work only in construction matters. He had not questioned us about scientific difficulties. Now, leaving Styer's office, he wanted to determine to what extent our work would be based on real knowledge, on plausible theory, or on the unproven dreams of research scientists. He also asked about the available supplies of raw materials and the situation with the priorities.

According to Groves, he was "horrified" at the information I gave him: "It seemed as if the whole endeavor was founded on possibilities rather than probabilities. Of theory there was a great deal, of proven knowledge not much. Even if the theories were correct, the engineering difficulties would be unprecedented."[1]

That afternoon of the seventeenth, following our discussion, Groves and I called on Bush to return his letter to Styer and to discuss Groves' new assignment. We soon realized that Bush did not know about Groves' appointment. Bush seemed quite mystified about just where Groves fitted into the project and what right he had suddenly to be asking the questions he was now asking. As a result, Bush was reluctant to discuss the issues Groves was raising, and as soon as we left the office, he contacted Styer to find out what was happening.

Bush was quite disturbed. He told Styer that he felt Groves dealt

1. Groves, *Now It Can Be Told* (New York: Harper Brothers, 1962), p. 19.

too aggressively with people and might have difficulty with the scientists. After his conversation with Styer, Bush immediately sent a note to Stimson's assistant Harvey Bundy. Recounting his conversation with Styer, he explained he had wanted to establish the Military Policy Committee before selecting a man to carry out its policies, and he voiced his doubts that Groves "had sufficient tact for such a job." Noting that Somervell apparently had seen General Marshall that day about the appointment, Bush closed: "I fear we are in the soup."[2]

Bush was not alone in discovering Groves' new position after the fact. Colonel Marshall did not learn the full details of the new organization until he and Groves visited Styer on the nineteenth. Styer assured Marshall that the district's status and particularly his job as district engineer remained the same, except that he now reported to Groves instead of the chief of engineers.

Colonel Marshall then met with General Robins, who informed him that the chief of engineers and Robins were both relieved of any further responsibility for the Manhattan Project.

These changes meant that Groves would be handling most of the contacts in Washington and effectively removed the need for my being stationed there. As a result of both my experiences with Groves in Nicaragua and having gotten to know him better in the early months of the project, I did not relish being under his direct control by being in the same office. Consequently, in early October I arranged for orders transferring my permanent station to New York City.

After the appointment of Groves, our immediate task became to assist him in discovering what he had gotten into. To this end we had another meeting with Bush, on September 21. Marshall explained Groves' relationship to the project. In turn, Bush acknowledged that after talking with General Styer, he understood the situation thoroughly. He admitted he had been somewhat confused during his meeting on the seventeenth with Groves, and in a way apologized for his failure to talk freely in their initial meeting. Bush then provided Groves with a brief history of the project and explained that at the explicit direction of the president the Navy had not been involved in the main effort to develop the bomb. Bush felt that this had been a mistake and that the Navy effort should be kept much more fully coordinated with ours than during recent months. He also told Groves, in answer to a

2. Ibid, p. 20.

question, that he felt the Tennessee site provided adequate room for safety. However, for the first time, he did raise the possible need for a second site to accommodate the plutonium reactors. Finally, when Groves told Bush that he and I were on our way to see Admiral H. D. Bowen, the director, and Dr. Ross Gunn at the Naval Research Laboratory, Bush warned of previous difficulties with the admiral but said he hoped we could get some information on what the Navy was doing in its own atomic research.[3]

At the laboratory, Bowen and Gunn explained that the Navy was working on the same problem of separating out U-235 from natural uranium. They were developing a liquid thermal-diffusion method that then consisted of a one-pipe experimental apparatus that was operating. Also, they were building a fourteen-stage pilot plant. According to Gunn, their isolation from the mainstream of atomic research stemmed in part from Urey's belief that the liquid thermal-diffusion method was not practicable. Gunn also criticized the involvement in the project of certain key scientists of "questionable political beliefs." Nevertheless, both Bowen and Gunn indicated they would like to cooperate and coordinate their efforts with the Army and were quite open in disclosing the results of their tests and the procedures they were using. However, we did not volunteer any information about our work except that we were connected with the S-1 project as Army representatives.[4]

During the meeting, I received a call from General Styer to advise me that President Roosevelt had signed Groves' promotion. Styer commented, "Maybe that will make him feel better." Following his instructions, I waited until we were back in our car before I broke the news to Groves. Nevertheless, all the way to the office he continued to grumble about the assignment, making it perfectly clear that he did not like it.

Receiving his star on the twenty-third, Groves officially took charge of the project. That afternoon, he attended a meeting in the secretary of war's office with Stimson, General Marshall, Bush, Conant, Somervell, Styer, and Harvey Bundy to decide on the form and makeup of the policy supervision of America's atomic bomb effort.

3. Diary, September 21, 1942. "Chronology of District X" is referred to as "Diary." This record contains entries made by Marshall, Nichols, and Blair for the purposes of keeping each other informed during the period June 17, 1942, to June 16, 1943. Copies of the diary are in the National Archives, Historical Division of the Corps of Engineers, and in K. D. Nichols's personal files. There are several other copies in existence.
4. Ibid, Sept. 21, 1942.

During the discussion, it became clear that one man should not shoulder the whole responsibility for making atomic bomb policy. At the same time, the group recognized that each member of any committee selected for a policymaking role must be able to devote whatever time the project demanded. Stimson proposed that the Military Policy Committee consist of seven or possibly nine men drawn from the OSRD, the Army, and the Navy. Groves promptly objected, and pushed for a committee of three. After some discussion, Groves prevailed. One problem with a three-man committee immediately became apparent, however. Everyone recognized that both Bush and Conant should be members of the committee. Stimson resolved that issue when he finally proposed that the General Policy Group appoint a Military Policy Committee consisting of Bush as chairman with Conant as his alternate (the two having one vote), with Admiral W. R. E. Purnell in the office of chief of naval operations representing the Navy, and General Styer, the Army. Stimson also proposed that General Groves sit with the committee and "act as executive officer to carry out the policies that may be determined."

The new committee had the responsibility "to consider and plan military policy relating to the project, such planning to cover production, strategic, and tactical problems and research and development relating thereto." The Military Policy Committee was directed to "report the progress of the project to the General Policy Committee at suitable intervals and further report to them for determination any matters of such importance as to require such determination."[5]

Although Colonel Marshall had delayed purchasing the Tennessee site, we had continued to prepare for the acquisition of the land and construction of the plants. The Corps of Engineers' real-estate office had advised us on July 29 that the total cost of purchasing eighty thousand acres would come to about $4,130,000, with the land itself running $30 an acre. During August, we worked with the TVA to make sure sufficient power would be available to supply our anticipated needs. On the nineteenth, I had accompanied Marshall and Groves to the Corps of Engineers' real-estate office, where Groves gave Colonel John J. O'Brien the directive he had signed directing the engineers to acquire some seventy square miles of land. On the twenty-first, Cap-

5. Bundy Record of Meeting, September 23, 1942, General Policy Committee Memorandum "A" signed by the General Policy Group, September 23, 1942, in Harrison-Bundy Files, Folder 6, National Archives.

tain Johnson visited the War Production Board's office in charge of power allocation to advise it that we were acquiring the Tennessee site. The office reiterated that the TVA could supply sufficient power and indicated that in fact our estimated power needs would not use up all of the TVA generating capacity.

Groves, Marshall, and Blair made a complete inspection of the site on the twenty-fourth. Groves had few overall concerns about the site except for the proposed housing. During earlier discussions about the site, Marshall had argued for exception to the cost limits placed on expenditures for housing. Groves, however, expressed the belief that houses should be quite small and extremely simple in the interests of economy. In turn, Marshall argued that primitive housing could not be expected to meet family requirements of the class of personnel con-templated for the project and that comfortable and modern quarters would have to be provided, at least to the maximum extent that could be reached under the money limitations expected. Groves' concept of the quality of housing needed remained a bone of contention between him and the district engineer both at Oak Ridge, Tenn., site of our facilities (now called the Clinton Engineer Works [CEW]) and at Los Alamos in New Mexico, the site of the development of the first atomic bomb. However, at Los Alamos, Groves assumed more direct control, and the quality of housing suffered because of his concept. Groves considered our houses at Oak Ridge as considerably above average. Certainly they were not luxurious, but the floor plans and sylvan set-ting gave them considerable charm. At our plutonium production plants at Hanford in Washington State, du Pont had a major influence in achieving more suitable housing.[6] With the primary site approved and with the formation of the Military Policy Committee, the Manhattan Project acquired the main aspects of the organization that lasted throughout the war. As head of the project, Groves reported to the Military Policy Committee, which in turn reported directly to the sec-retary of war, or in the case of military matters, through General Marshall. The secretary of war, of course, had direct access to the president, as did Bush. On some major policy and program matters, the General Policy Committee (Stimson, General Marshall, and Vice President Wallace) reviewed the matter prior to referring it to the president for his decision. General Styer and Admiral Purnell were occupying positions in which they could assure the full cooperation of

6. Diary, September 24, 1942.

the Army and the Navy, while Bush and Conant, through the OSRD, could assure cooperation from the scientists. Starting in July 1943, Conant also served directly as a scientific adviser to Groves along with Dr. Richard C. Tolman, dean of the California Institute of Technology.

The reorganization accomplished all Bush's objectives. He wanted a higher-ranking officer, and a more direct chain of command to the president. Previously, with the divided responsibility, he had feared the Army might take over all control from the OSRD. Although initially the Army received responsibility for construction only, Groves now had overall responsibility for practically the entire project. However, Bush had interjected himself into the military chain of command between Groves and the chief of staff and the secretary of war by obtaining the position of chairman of the Military Policy Committee (see the 1945 organizational chart, page 15).

The S-1 committee continued to operate under the OSRD, while the Manhattan Engineer District retained its structure, expanded purpose, and organization. The district engineer reported directly to General Groves on all matters instead of to the chief of engineers. We maintained contact with General Styer and other components of the Services of Supply, U.S. Army as well as other Washington government offices. We kept the chief of engineers and his deputy fully informed at all times of our progress and sought assistance when needed. We recruited most of our officers from the Corps of Engineers. The chief of engineers' office continued to be extremely cooperative.

Throughout the war, in addition to reporting to the Military Policy Committee for policy matters, Groves maintained direct access to both General Marshall and the secretary of war whenever he saw fit. Likewise, Groves would regularly contact directly for information, or to issue instructions, any officer in the Manhattan Engineer District or any individual working for our various contractors for research, engineering, construction, or operation whenever he wanted to. Organizational channels meant nothing to Groves. In addition, he maintained direct control of construction and operations at Los Alamos except for administration and auditing of the University of California contract and some support he expected the district to do. He also assumed control over many of the international contacts, all intelligence efforts, and at the request of General Marshall, military planning for the use of the bomb. An organization purist would say that this hybrid organization would generate friction, confusion, and might not be able to function smoothly. However, having served at the "doing" level, I found that Groves' unusual organizational concepts had more advantages than disadvantages and expedited decisions and results.

Groves decided to establish his headquarters in Washington, at least initially, believing he could move later if necessary. He soon found, however, that the capital was the ideal place from which to work. He was able to stay in touch with the War Department and other government agencies while also readily reaching out into the field. He established his office rooms in the new War Department Building (now the State Department Building). He had only seven rooms and very few staff assistants until shortly before the dropping of the bombs, when additional space was acquired for the project's public information operations. The district's liaison office occupied two of these rooms. This office always remained under the control of the district engineer and allowed him to pursue such things as priorities for obtaining needed materials. At the same time, Johnson did handle Groves' personnel problems and frequently turned to Groves when he felt he needed help in dealing with Washington problems.

On the twenty-sixth, however, we met with Mr. Lotz and other Stone and Webster officials in Washington to review the program to date and discuss current work preparatory to meeting with the S-1 committee. In addition, we decided to approach du Pont to see if the company had men available for the development and operation of the chemical plant that would separate plutonium from the product of the reactors ("gunk," as the scientists called it). We had already brought du Pont into the project to work on the production of uranium metal and at the metallurgical lab in Chicago to furnish men to help develop the chemical process.

Later that day Stone and Webster, Marshall, Groves, and I met with the S-1 committee, which concurred with the several major actions we proposed to take or had already taken: acquisition of all available uranium ore, including ore in Africa; designating Ruhoff to be in charge of all phases of full-scale uranium metal production; and contacting du Pont as the possible contractor to develop and operate the plutonium chemical separation plant.

Groves set up a meeting with du Pont for October 2. Before the du Pont people arrived, Groves, Colonel Marshall, Blair, August C. Klein of Stone and Webster, Compton, and I met to prepare for the conference. Compton emphasized his desire to have du Pont available for consultations on questions other than those strictly related to the separation plant. He then presented us with a schedule for completion of the four planned piles, beginning with the experimental pile at the metallurgical lab; he expected the pile to become operational by the end of 1942.

Compton briefed the du Pont officials, and I explained that the contract we had under consideration would be similar to the one du Pont had for the TNT plant I had been building in Pennsylvania. It was agreed that we would send a letter of intent to du Pont covering design, consultation, and procurement of equipment for the plutonium facility. On October 5, Groves paid his first visit to the metallurgical lab to inspect the facilities and discuss the work with the scientists. Apart from wanting to know the status of the work and to meet the key scientists, Groves wanted to find out exactly how much plutonium and how much U-235 would be needed to fuel a reasonably effective bomb. Hearing the estimates, he asked how accurate they considered them to be. For the first time, he encountered the wide range of many of the estimates for key items. Upon receiving an assessment that the estimates were probably correct within a factor of ten, Groves compared it to a caterer being told to prepare for anywhere from ten to a thousand guests. (Fermi's early estimates were as little as twenty kilograms, but possibly as high as one or two tons.) As a result of his visiting Chicago, Groves was aware for the first time of the size and difficulty of the project. He also was impressed with the prospects that the plutonium project might be achievable earlier than any of the others, provided we could solve some very difficult technical problems. Additional meetings during October convinced us that Stone and Webster could not carry the full load of all the engineering work even with subcontractor assistance, let alone handle all the construction. As a result, we concluded that we needed du Pont to take on the entire plutonium project.

At a meeting with Vice President Willis Harrington and Dr. Charles Stine, also a company vice president and distinguished chemist, Groves proposed that du Pont take over the entire plutonium project, not just working on the chemical process for separating out the plutonium. Harrington and Stine both protested vigorously that du Pont had no experience in or knowledge of physics and that they were incompetent to render any opinion except that the entire project seemed beyond human capability.

Returning to Wilmington, Harrington and Stine discussed the request with the du Pont executive committee. Primarily because Stine and Harrington had great doubts about the specifications of the proposed chemical process for plutonium separation, the company advised Groves that before it could give him an answer, its personnel would have to inspect the research at the metallurgical laboratory.[7]

Representing some of the best engineering and scientific talent in the company, the du Pont investigators, including Stine, Roger Williams, Thomas Gary, and Thomas Chilton, arrived in Chicago on November 4 and spent several days observing every aspect of the activities at the lab. After the group returned to the company's headquarters in Wilmington, Del., du Pont scheduled a meeting to discuss their findings, and on the tenth I accompanied Groves to the conference in Wilmington.

Prior to the meeting, Groves talked privately with Walter S. Carpenter, Jr., the du Pont president, to emphasize the importance of the project. At the conference itself, Gary, Chief Engineer Everett G. Ackart, Stine, Williams, Harrington, Crawford H. Greenewalt, and Thomas H. Chilton represented du Pont, and Drs. Compton and Norman Hilberry came from the metallurgical lab. I found the discussion thorough and very enlightening.

The company began by saying that the plutonium process was feasible, but emphasized the lack of data to make firm engineering decisions, the uncertainty about the best pile design, the need for heavy water as an alternative to graphite, the many chemical problems confronting the designers, the problem of getting rid of waste material, the need to expedite metal production, the need to protect employees against hazards, and the need to protect du Pont against liability. They also indicated that if du Pont were to become involved, they would need approval to drop some of their other military work. They also wanted to control all aspects of the work—that is, to design, construct, and operate the plants. Their proposed schedule was to produce a few grams of plutonium sometime in 1943 and a production rate of one kilogram a day sometime in 1945.[8]

When Groves tried to pressure them into a faster schedule, Stine countered that we probably should add six months to all schedules. He further stated that he thought there would be only a fifty-fifty chance of

7. Groves, Record of Preliminary Negotiations, undated; du Pont, Design and Procurement History, Manhattan Project Records, Office of the Commanding General File, National Archives, p. 14.
8. Diary, November 10, 1942.

success and a much lesser chance of completing the work before the war ended. However, he added that the United States must make an effort to do it and that du Pont would have a better prospect for success than any other company. Two days later, Carpenter advised Groves that he and the executive committee would recommend to the board of directors that the company undertake the project in spite of the scientific and economic hazards du Pont would face.

The plutonium project was not, of course, our only area of concern. On the eleventh, Groves came up to the Manhattan District's New York office to discuss the status of the uranium acquisition program, the electromagnetic process, and the decision to substitute du Pont for Stone and Webster on the plutonium process. He told me that he and Conant had decided to bypass the electromagnetic pilot plant in favor of the immediate development of at least a portion of the full-scale plant. He indicated that they would be taking a similar approach on the gaseous diffusion project, and he approved the recommendation that Conant, Klein, and I had made to eliminate the centrifuge method. On the twelfth, the Military Policy Committee ratified these decisions with du Pont to build the plutonium facilities and Stone and Webster the electromagnetic plant. Depending on a future report of M. W. Kellogg Co., the MPC indicated it might authorize a six-hundred-stage gaseous diffusion plant.[9]

On the fourteenth, I attended the S-1 executive committee meeting, which routinely ratified the MPC's decisions. Although it was not mentioned at the meeting, a new crisis emerged during the day, which was to have a profound impact on the entire Manhattan Project. During lunch, Wallace Akers, the British technical chief on its atomic bomb project, informed Conant that James Chadwick, one of the key British scientists, had recently concluded that plutonium might not be a practical fissionable material for weapons because of impurities.

That evening, after the S-1 meeting, Conant consulted Lawrence and Compton about the unexpected information. They acknowledged that scientists at Chicago and Berkeley had known about the problem since October but could offer no ready solution. Conant then advised Groves of the development, and he immediately assembled a special investigating committee comprised of Lawrence; Compton; J. Robert Oppen-

9. Diary, November 11, 1942; minutes, Military Policy Committee, November 12, 1942; the Harrison-Bundy Files, National Archives.

heimer; and Oppenheimer's associate, Edwin McMillan, newly re-cruited to the atomic bomb project after working on the development of radar.

Four days later, Groves presented to du Pont representatives the committee's conclusion that any problems with the plutonium process could be overcome with higher purity requirements. Nevertheless, du Pont expressed strong doubts about the solution. Earlier, Chilton and Ackart had accused Compton and me of misleading them about the difficulties of the chemical separation of plutonium. I remember Chilton stating, "The material doesn't even exist." In fact, the only knowl-edge about plutonium at that time was Seaborg's microchemistry re-search involving micrograms of plutonium produced in a cyclotron. Chilton and Ackart had wanted to review the entire project, but at the time Groves approved review of the plutonium project only.

This new information from Chadwick reinforced du Pont's recom-mendation in its report on the metallurgical laboratory that a survey of the entire Manhattan Project was in order. In light of the questions du Pont had raised and the possible problem with the plutonium process, Conant and Groves decided on the eighteenth to initiate an overall review of all aspects of the bomb research and development. To head the investigating committee, they selected Dr. Warren K. Lewis, pro-fessor of chemical engineering at M.I.T.

Groves called me down to Washington to inform me of the recent events and the decision to review the entire project. The committee included Dr. Eger Murphree of Standard Oil, and Tom Gary, Roger Williams, and Crawford Greenewalt of du Pont. They attended a pre-liminary meeting in Washington on Saturday, November 21, at which Groves outlined the problem. I then spent Monday and Tuesday in New York with the committee, attending briefings at Columbia Uni-versity on the gaseous diffusion project by Harold C. Urey, John R. Dunning, associate professor of physics at Columbia, and Dobie (Percival C.) Keith, vice president of Kellogg Co. The group then went on its own to Chicago and California.

On its way back East, the committee stopped again at the metal-lurgical lab on December 2, the day the pile went critical for the first time. We had not been able to complete the pile building in the Argonne National Forest by the target date due to labor problems and delays in getting agreement on the size of the building. Anxious to begin as-sembling the pile, Fermi convinced Compton that his calculations ruled out the possibility of a runaway reaction or explosion, and without seeking approval from us or the university, Compton gave permission to construct the pile under the stands at Stagg Field on the University of Chicago campus.

When Conant learned about the decision from Compton on November 14, he is reported to have turned white. Nevertheless, no one chose to stop them because of their urgency to begin testing and our confidence in Fermi's calculations. The dramatic test on the second provided the crucial and convincing proof that it was possible to create and control a self-sustaining nuclear chain reaction. This successful test was the single most important evidence that the plutonium project was feasible.

Of the visiting Lewis Committee members, Compton invited only Greenewalt to witness the historical event, explaining later that as he was the youngest, he would be able to talk about it for the most years. When the committee left, Compton called Conant. "Jim," he said, "you'll be interested to know that the Italian Navigator has just landed in the new world." Conant excitedly responded, "Is that so. Were the natives friendly?" Compton answered, "Everyone landed safe and happy."[10]

I was very disappointed that I had not been there for the test. Having visited Chicago a week earlier, I had not thought it possible to complete the work in time for the visit of the review committee. However, to impress the Lewis Committee, Fermi had expedited the work; also, the pile went critical with fewer layers than he calculated.

When I did return to Chicago on December 14, Walter Zinn, a met lab physicist, helped assuage my disappointment over missing the birth of nuclear power by giving me a private demonstration of the pile in operation. In the midst of the one-person show, however, Fermi came in, started reading the instruments, and then suggested Zinn shut down the reactor because it was attaining too high a power level. At the time I was less than appreciative of Fermi's concern. We were standing at the controls on a balcony not far from the reactor, which was completely unshielded. We were all aware we faced risks working with so many unknowns. Earlier Compton and Fermi had briefed me about the physics of the pile and their belief that they could proceed safely under the Chicago stadium. I also knew about the suicide squad that had the job in the initial test of pouring boron onto the pile if the theory and calculations proved wrong and a positive temperature coefficient instead of the expected negative one resulted

10. Compton, *Atomic Quest* (New York: Oxford University Press, 1956), pp. 141–44.

in the power level rising too rapidly to be controlled by the control rods. However, with the initial criticality test safely behind them, there was little immediate concern that something would go wrong, although we were very anxious to rebuild the pile at the more isolated Argonne site as soon as possible.

I did not feel in potential danger as I stood looking at the pile. On many other occasions while inspecting construction work at Hanford or Oak Ridge, I had far more qualms about my personal safety (for example, when a contractor or one of my officers would lead me across a six-inch steel beam, probably to see if I had the nerve to follow him). In recalling how Zinn put the pile in operation, I remember only that I thoroughly enjoyed looking at the dials, and I was exhilarated by the experience and the wonderful feeling of success.

At that point, I was not yet aware of how much the demands on me would increase during the next two and a half years as pressures were applied to complete the bomb at the earliest possible date. The autumn of 1942, which included the assignment of Groves and the selection of du Pont and other major contractors, coupled with the first controlled chain reaction, was undoubtedly the most crucial period for the ultimate success of our efforts. As the momentum increased, the sheer volume of my work accelerated, and the problems I faced became more and more complex. Moreover, the work required more frequent travel from one project location to another to expedite decisions and the work.

At that time I regularly commuted between New York and Washington and often visited Wilmington, Boston, Cleveland, Oak Ridge, Chicago, and occasionally San Francisco. As the pace of our efforts picked up, however, my itinerary (usually by train or commercial airline) added such places as Hanford; Trail; Ottawa; Ames, Ia.; Toronto; St. Louis; Decatur, Ill.; Morgantown, W. Va.; Charlottesville, Va.; Milwaukee; and occasionally Los Alamos.

Initially, Marshall, Blair, and I operated independently on separate tasks that needed emphasis while coordinating our efforts in regard to the overall project. I devoted most of my attention to ore procurement, feed materials, and the plutonium project. At the same time, I kept informed about all other activities. Marshall and Blair spent much of their time in the first several months of the project creating the organization, coordinating design of the electromagnetic plant, the heavy-water plant at Trail, and planning the central facilities and the town of Oak Ridge at the CEW. Vanden Bulck did a tremendous job setting up administrative procedures for each area and each project. (See the

April 1943 MED organizational chart, pages 16–17, for the many others who were working hard to accomplish our mission.) After September 23, Groves was driving everyone to get things done faster.

We coordinated our efforts by telephone, short meetings, and by entries in the project diary that Marshall had started the day Styer appointed him district engineer. I soon found I had two bosses and carried out the requests of both men. Marshall proved far more pleasant to work for and Groves more difficult and demanding. Both were extremely capable.

The preparation of the December 1942 progress report for President Roosevelt provides an example of how we obtained policy decisions and presidential authorizations. During a November 19 meeting with Groves in which he briefed me on the plutonium crisis and the creation of the Lewis Committee, he directed me to write a draft of a report to the president. Although I knew we would have to revise the draft after we received the report of the Lewis Committee, I outlined the existing situation within the project and a proposed program for the future, including an estimate of cost.

The Lewis Committee submitted its report dated December 4 to Groves on December 7. Unexpectedly, it strongly supported the gaseous diffusion process, contending that "of all three methods, the diffusion process is believed to have the best overall chance of success, and produces the more certainly usable material." Consequently it recommended that we design and construct the entire, full-scale gaseous diffusion plant. While agreeing that the electromagnetic process probably was the most immediately feasible of all the ones under consideration, the committee concluded that it seemed least likely to produce U-235 in sufficient quantity. Therefore it recommended we build only a small plant of 110 units to produce one hundred grams of U-235 for physical measurements.

The report recommended continuing the plutonium project despite the many problems that remained and indicated the process might even provide "the possibility of earliest achievement of the desired result." Finally, the committee suggested that "experienced operating organizations" be hired and given "full responsibility and corresponding authority" for each of these projects. Informally, the committee suggested industrial organizations suitable for various parts of the project. In effect, this furnished us with a blueprint for the complete industrial organization of the project, and Groves followed most of their recommendations. Most important, the Lewis Committee report gave us more confidence concerning the feasibility of producing sufficient quantities of fissionable material.

With the Lewis Committee's recommendations in hand, Groves and I met with the S-1 committee on the ninth to discuss it and also the draft of the progress report to the president. Conant began by reading a letter from Dr. Murphree, who had not been able to accompany the Lewis Committee. While agreeing in general with the report, Murphree dissented from the recommendation concerning the electromagnetic process, believing it was more important and should be pushed to a greater extent than the committee indicated. Groves then read my draft of the progress report. Using the comments from the S-1 committee and after further discussions with the Military Policy Committee on the tenth, Groves completed the final draft of the progress report and submitted it to the Military Policy Committee, which submitted the final report to the General Policy Committee.

The report, titled "Report on Present Status and Future Program on Atomic Fission Bombs," dated December 15, 1942, recommended a full-scale gaseous diffusion plant costing $150 million, plutonium plants costing $100 million, and additional heavy-water plants producing two and a half tons per month and costing $20 million. Modifying the Lewis Committee's recommendation that we should build only a small electromagnetic plant that would produce a total of one hundred grams of uranium, the report to the president called for building an electromagnetic plant to produce one hundred grams of uranium a day and costing $10 million; the plant might later be expanded to full scale. Conant ardently supported this change because he felt the gaseous diffusion plant would take too long to build and to achieve equilibrium for producing bomb-strength material. As a result, he felt the electromagnetic plant essential for early delivery of a uranium bomb. Events proved him correct. The electromagnetic plant was absolutely essential. Without it, the U-235 bomb would not have dropped on Hiroshima on August 6, 1945, or within months of that day. Responding to du Pont's concern about liability and hazards, the report recommended that "in view of the unusual and unpredictable hazards involved in carrying out the work under this project, the President authorize the incorporation of suitable provisions in the contracts where such action is deemed advisable by the Chief of Engineers, providing that all work under such a contract is to be performed at the expense of the Government, and that the Government shall indemnify and hold the Contractor harmless against any loss.''

The report provided estimates for beginning bomb production as possibly June 1, 1944, a better chance before January 1, 1945, and a good chance during the first half of 1945. The total cost was estimated as about $400 million. The report included the opinion, based on what intelligence we had, that it was highly improbable that the Germans would have a weapon in 1943, but it was possible that they might be six months to a year ahead of the United States in research and development.

Bush wrote a cover letter summarizing the report and transmitted it to the president on December 16, 1942. In his letter, Bush explained, "The situation has changed notably since the report of June 13, 1942. There can no longer be any question that atomic energy may be released under controlled conditions and used as power. Furthermore, there is a very high probability that the same energy may be released under suitable conditions in such a small interval of time as to make a super-explosive of overwhelming military might.'' Bush acknowledged that no one could "be absolutely certain" about this until the first bomb was built, and this would require "a considerable amount of material.'' He then said that the estimate of the actual amount of material needed for each bomb "has unfortunately been increased by the scientists since last June.'' According to the current time schedule, Bush contemplated "availability during the first half of 1945 of a sufficient quantity of material to produce six super-bombs. This can only be attained if the recommendations herewith transmitted are approved and the highest priority assigned to this project.''

We did not wait for the president's approval of the report to proceed with its recommendations. Three days before the report was completed, Groves, Marshall, and I met with Keith and J. H. Arnold of Kellogg to inform the company that we wanted it to proceed with the design of the gaseous diffusion plant and that we would issue a letter of intent for that purpose. On the same day, we met with John Lotz, the president of Stone and Webster, to inform him that we would be moving the plutonium production facility from Oak Ridge because of space and safety considerations. We also advised him that we had decided to reduce the size of the electromagnetic method to five hundred units and that Kellogg would be designing the diffusion plant.

*   *   *

Lieutenant Colonel Franklin Matthias and I went to Wilmington on December 14 to discuss with du Pont officials the site requirements for the relocated plutonium process. Given the needs for a much larger area than the Tennessee site and a great deal of electric power and water, relatively few suitable locations existed. The area in central Washington State near the Grand Coulee Dam seemed to offer the most promising site, and on the fifteenth, Matthias and A. E. S. Hall and G. P. Church of du Pont left to survey that area, and other locations in California and Nevada. Although they examined several sites, probably no better location existed anywhere than the Hanford area in Washington, on the Columbia River. Matthias reported this to Groves on December 31.

While making progress with du Pont and with the site selection for the plutonium facility, we found we had our first crisis with the scientists at the metallurgical lab. All during the period of our negotiations with du Pont, Compton was having serious problems with his key scientists. Although Compton had first recommended du Pont and had fully participated in most of our discussions, he reversed his position at one point because of the company's conservative approach and its skepticism about early success. His key scientists not only thought du Pont acted too conservatively but also believed they could do the job better themselves. As a result, they became highly disappointed as they saw the overall responsibility for design, construction, and operation of the plants slipping from their control. This problem continued throughout the war.

I understood the reasons for the scientists' opposition to du Pont and to government control. The situation has been compared to mothers facing the need to surrender control of their children as they grow up. Nevertheless, I did not agree that Compton's scientists had the knowledge or skills to do the design, construction, and operation jobs faster or better. In fact, I believe chaos would have developed if we had assigned the scientists the responsibility to design, build, and operate the plutonium facilities. For the duration of the project, many of the scientists continued to complain about du Pont. I found I had to devote a lot of my time dealing with flare-ups between the scientists and du Pont, which occurred regularly. I have no question that we tested a plutonium bomb as early as we did only because of du Pont's role in the Manhattan Project. Du Pont's engineering, construction, and production experience was just as essential for success as were the scientific capabilities of the met lab. Teamwork is essential for a difficult project of such magnitude, and a great variety of experience and talent has to be available on the team.

Another major decision during the autumn of 1942 was the selection of J. Robert Oppenheimer to head the scientific effort to develop and fabricate the bombs. Prior to the time we came into the project, Compton had selected Oppenheimer, the leading light of theoretical physics at Berkeley, to head the theoretical group for weapons design. As a result, he was as well informed as anyone about the project and the theoretical possibilities of an atomic explosion. I first met Oppenheimer when he briefed me in July during one of my visits to the met lab. Groves met him initially on October 8, during his first visit to Berkeley.

One of Groves' priorities had been to decide on the establishment of a laboratory to do the design and development of the bomb as well as to select the person to head the effort. Some of the key scientists greatly underestimated the time and effort that would be required to develop the weapon. They saw no urgency in starting it because they thought it could be done in a few months with less than a hundred men. Waiting would have the advantage that they would have U-235 and plutonium for experiments. Groves had discussed the subject with Oppenheimer during their first meeting and had been impressed very much by the physicist's ideas. Nevertheless, Groves recognized that he would encounter opposition both in the scientific community and in Army security if he were to select Oppenheimer to direct the bomb lab.

Before the war, Oppenheimer had become involved with left-wing or Communist-leaning groups and had enjoyed friendships with known or suspected Communists, including his brother, Frank; his girlfriend; and his wife. While he had broken contacts with the questionable organizations once he began work on the bomb project, he retained many of his friendships. Even though the USSR was on our side, these contacts were considered to be security risks. In addition, Oppenheimer had not won a Nobel Prize, which contributed to the scientific prestige of the other project scientific leaders—Lawrence, Fermi, Urey, and Compton.

Discussions with scientists at Berkeley and at the met lab on his way back East after his first meeting with Oppenheimer failed to provide Groves with a suitable alternative to head the bomb research and development. With plans already made for him to return to Chicago on

October 15 to meet with Compton, Marshall, and me to discuss the status of the plutonium effort and du Pont's involvement, Groves asked Oppenheimer to join us there to continue the discussion about the location and organization of the bomb lab. With many issues still remaining as the departure time for the *20th-Century Limited* approached, Groves invited Oppenheimer to join us on our way back to New York.

After dinner on the train, the conversation continued in my small roomette, with Groves, Marshall, Oppenheimer, and myself crowded into the small space with no elbow room. I cannot remember how we all found a place to sit (a roomette designed for one person was about forty by eighty inches). Despite the close quarters and the late hour, the discussion proved most fascinating. It covered all aspects of setting up an organization, building the needed facilities for laboratories and housing, procuring scientific equipment, and dealing with such expected issues as recruiting scientists and confining them in a laboratory in a remote area, and whether the scientists should be commissioned as officers in the Corps of Engineers. When Oppenheimer left the train in Buffalo, there remained no doubts in my mind that he should direct the new lab despite the difficulties we would have in clearing him.

Oppenheimer went to Washington to continue planning for the bomb lab with Groves. The most immediate need was to find a suitable site. Colonel Marshall had already initiated the search in early October when we borrowed Captain John Dudley from the Syracuse District to make a survey "for an installation of unnamed purpose." The initial criteria included a location more than two hundred miles from any border or ocean, in a sparsely settled area with a rather mild climate, in a natural bowl surrounded by hills and suitable for an ultimate population of 450 people, including children and guards. A site with some buildings already there would facilitate an early start, while an isolated area would simplify the security problems. Nevertheless, access by road, rail, and air must not be too distant.[11]

In late October, after meeting with scientists to ascertain their needs, Dudley spent several weeks traveling, part of it on horseback, through the Southwest. By the beginning of November, he had settled on Jemez Springs in New Mexico as coming closest to the stated criteria.

11. Letter from Brigadier General John Dudley to Jesse Remington, May 5, 1968; John Dudley, "Ranch School to Secret City" in Badash, Hirschfelder, and Broida, *Reminiscences of Los Alamos, 1943–1945* (Hingham, Mass.: Kluwer Academic, 1980), pp. 3–4.

To confirm the choice, Dudley arranged for Oppenheimer and Edwin McMillan to meet him and Groves in New Mexico on November 16. Before Groves arrived by air, Dudley toured Jemez Springs with the two physicists. Despite his earlier enthusiasm, Oppenheimer, seeing the site for himself, found it unacceptable because "the surrounding cliffs would give his people claustrophobia and he refused to ask them to live in a house that had been previously occupied by a Mexican or an Indian."[12]

In any case, Oppenheimer received support for his position from Groves when he arrived in midafternoon. In response to Groves' inquiry, Oppenheimer was ready to offer an alternative site, the Los Alamos Ranch School, located on a mesa about forty miles from Santa Fe. In fact, as early as August 29, the district engineer's office in Albuquerque had made inquiries about the school, and Dudley had visited it during his own search. Using the initial set of criteria, however, he had rated it a poor second to Jemez Springs. Now, with Oppenheimer having recommended the location and with a new set of criteria, Los Alamos became the prime candidate as the site for the atomic bomb lab.

Los Alamos did in fact fit the basic requirements except that the road system would require much work and the water supply was adequate for only about five hundred people. Consequently, after visiting the school and surrounding area, Groves agreed with Oppenheimer that the site was acceptable and immediately set into motion the process to acquire the school buildings and property.[13]

With the selection of Los Alamos for the bomb development laboratory and the Hanford site for the plutonium production plants and with the acquisition of the CEW site, the Manhattan Project had the locations for its major facilities. On December 28, the president had written to Bush that "in regard to the special project, I will approve the recommendations." That gave us the firm presidential support we needed to move our effort from a vigorous research program under OSRD with some preliminary construction, procurement, and production of feed materials into an all-out design, construction, and production effort, including the formation of a scientific team to develop the weapon itself. On that same day, Groves came up to the district office for an end-of-the-year review of the project with Colonel Marshall and

12. Dudley to Remington, May 5, 1968.
13. Dudley to Remington, May 5, 1968; Dudley, "Ranch School to Secret City," in Badash et al., pp. 3–5.

me. After our meeting, we visited Union Carbide to discuss the company's willingness to become the operator for the gaseous diffusion plant.

The next day, Groves and I went to Wilmington, where we approved a series of decisions with du Pont on the production of heavy water. Back in New York the next day, Marshall and I met with Kellogg and Union Carbide to go over Kellogg's design of the gaseous diffusion facility. Thus, by the end of the year, work was under way on three methods of producing fissionable material, with the goal of producing at the earliest possible date an atomic bomb that could fit into the bomb bay of a B-29. Considerable progress had been made, and the rate of the progress was accelerating. We all knew we were in a race against time and that a single day of unnecessary delay would cost dearly in lives and dollars in the normal course of the war. Moreover, we still had no accurate intelligence on the progress Germany might be making toward developing her own atomic weapons.

5

ORGANIZING FOR CONSTRUCTION, 1943

BLAIR AND MARSHALL WERE BUSY in early January, redefining Stone and Webster's responsibilities. Stone and Webster retained responsibility for design and construction of the electromagnetic plant (Y-12) and for the design and construction of central facilities such as roads, railroads, water and sewage systems, offices, laboratories, etc., for the Clinton Engineer Works (CEW) and the town of Oak Ridge.

At the same time, I learned that du Pont and Compton disagreed on the location of the plutonium semiworks (X-10). The resolution of this particular problem and others dealing with the start of construction for the plutonium production facilities required a series of meetings during the next several weeks.

On January 6, I went to Wilmington and met with E. B. Yancey, Roger Williams, Ackart, Gary, and others and confirmed that because of its concerns with safety, du Pont would not recommend building the semiworks at the Argonne National Forest, as Compton had expected. Williams told me that the company had tentatively decided to construct the semiworks at CEW on the site originally intended for the full-scale plutonium plant. After the meeting I went on to Chicago to discuss with Compton and Hilberry du Pont's proposal for shifting the pilot plant to CEW to provide greater safety.

Compton argued that the Argonne National Forest was safe enough and maintained that the worst hazard would not affect anyone for a distance greater than five hundred meters. In addition, he expressed his belief that he would have a serious morale problem and difficulty in staffing the semiworks if we located it in Tennessee. Finally, he maintained that he did not have sufficient personnel to staff adequately three locations simultaneously.

When I got back to New York the next day, I called Groves to advise him of the conflicting views and recommended that Roger Williams of du Pont and Compton should discuss all aspects of safety requirements. I would then get Compton and du Pont together again at Wilmington to settle the site and several other outstanding issues. Williams, Compton, Marshall, and Groves did meet in Chicago on the eleventh and agreed on CEW as the site for the semiworks. We then held a conference in Wilmington on the sixteenth. Yancey, Evans, Williams, Gary, and Greenewalt represented du Pont. Compton, Hilberry, and Martin D. Whitaker came from the met lab, and Peterson and I represented the Army.

Williams introduced a new consideration when he stated that the company's executive committee considered it inadvisable for du Pont to operate the semiworks, since the research groups usually operate such plants. In addition, he said the company did not have the qualified personnel available at that time to assume the responsibility for such an operation. Du Pont therefore recommended that the met lab run the semiworks, with the goal of separating one-tenth gram per ton of output from a pile with the objective of obtaining one gram as soon as possible. The pile itself was to be simple and air-cooled.

After discussing the responsibilities of all the parties involved, we agreed to a program that called for du Pont to design and construct the semiworks. Final approval for the du Pont design was to be given after Peterson had secured Compton's opinion concerning its workability. Following additional discussion, I referred to the location of the semiworks at the CEW and authorized du Pont to proceed with the design and construction. However, I deferred approval of the met lab as the operator of the plant until Compton had more time to consider it. Compton said he would discuss with Conant the advisability of the met lab operating the semiworks.

The series of meetings that resolved the semiworks problem was typical of how many of the decisions were made. My job was not to tell competent organizations what to do or how to do it, but rather to see that they promptly eliminated areas of disagreement through discussions. Once an agreement was reached, I would authorize proceeding with the work. If an agreement could not be reached, I would decide which party to support or if necessary take it up with Groves. Once the decisions had been made, the district organization had the responsibility to ensure that everything proceeded as fast as possible and in accordance with the overall policy decision and with our administrative procedures for reimbursing the contractors. Time always was of the essence.

In addition to du Pont's work on the plutonium project, I met with their people in Wilmington regularly to discuss all aspects of their involvement in our effort. On January 21 I visited Wilmington with Blair and Cornell (head of our legal division) to confer on the heavy-water plants (P-9) and the semiworks. Then, on the twenty-third, Groves and I returned to Wilmington and met with F. A. C. Wardenburg, head of du Pont's Ammonia Division, concerning the heavy-water plants and their possible expansion. We also met with F. M. Yancey, head of du Pont's Explosives Division, and Roger Williams, assistant general manager of the Ammonia Division, in regard to the safety aspects of the Hanford site to ascertain how much land we should purchase outright and how much to lease.

The following day, I went to Tennessee with J. O. Wilson, who was to be the company's project manager for the semiworks. We selected and approved the specific site for the semiworks plant and arranged for du Pont temporary office space in a school near the site. I then went to Chicago to arrange for construction of additional facilities at the Argonne lab. At Peterson's suggestion, I also conferred with William B. Harrell, the University of Chicago's business manager, in regard to various administrative problems connected with the university's understanding of the project.

Groves, Marshall, and I again visited Wilmington on February 8 to discuss safety matters at the Hanford site. Summarizing the situation, Yancey expressed the opinion that although there was some gamble attached to the site, he knew of no better one. Satisfied that Hanford appeared to be the best site in general and, more specifically, best in regard to safety, Groves stated that he would see Under Secretary of War Patterson the next day to obtain approval for acquiring the site. We also discussed the question of who should operate the semiworks at Clinton. I made an entry in the Manhattan Project diary that "du Pont appears to be immovable in its decision not to operate" the semiworks. At the same time, the company assured us that the met lab, assisted by some of du Pont's best employees on the met lab payroll, would be capable of operating the facility. They assigned S. W. Pratt, one of their very capable plant managers, to be an assistant to Martin Whitaker, who was designated director.[1]

On February 10 and 11, Groves, Marshall, Peterson, and I met with Conant, Urey, Lawrence, Compton, Murphree, Briggs, and Harry Wensel of OSRD staff. While the S-1 committee had given up its policymaking role to the Military Policy Committee, it still served to provide us with scientific advice. We discussed all projects in considerable detail, with key personnel making presentations. As usual, many conflicting opinions were expressed. In regard to plutonium, we spent considerable time on the subject of heavy water as an alternative to graphite as a moderator. Speaking for du Pont, Greenewalt emphasized that heavy water was necessary only as a second solution for the plutonium project. Although it had advantages over graphite, the design could not possibly be completed as early as for graphite. Dr. Urey pushed for research on an alternative sulphide method to produce heavy water if expansion was necessary.

Lawrence advocated doubling the plutonium project and proceeding with a full heavy-water project. However, Compton thought the present program for heavy water about right, and any decision about heavy water was put off until the next month. Greenewalt then discussed the doubtful factors in the present helium-cooled pile design for the plutonium project. He felt that no further discussion could be made in regard to heavy water until we decided which pile design was in the approved program.

The peacetime application of reactors for power was discussed and it was decided that this was not the time for an independent group to study the matter; it should not be allowed to interfere with the war effort. There also was considerable discussion concerning how far the operating contractors such as du Pont should carry the metallurgical processing of plutonium before shipping it to Oppenheimer at Los Alamos.

Concerning radiation, health, and safety, Groves stated that the district should set up a health group and that each research group and operator should set up its own group to work with the district group. Radiation as a health hazard was recognized very early to be a major concern, and Compton already had set up at the met lab a health division under Dr. Robert Stone of the University of California.

Shortly thereafter, Compton solved the problem concerning du

1. Diary, February 8, 1943.

Pont's proposal that the met lab operate the semiworks at Oak Ridge. He found that Harrell, the University of Chicago's business manager, would support him if Compton believed it was essential to the war effort that the met lab take on that responsibility. However, they both had many serious doubts about the wisdom of accepting the responsibility. Conant was not at all helpful, telling Harrell that he would not accept such responsibility at Harvard and expressed serious reservations about any university assuming such a responsibility. Nevertheless, Compton finally made the decision and Robert M. Hutchins, the university's president, backed him. Once the Board of Trustees approved the decision, we negotiated a contract with the met lab to operate the semiworks.

Concurrently with this decision, the met lab had been considering changing the design of the plutonium reactors. In the December 1942 report to the president, the plans called for helium-cooled production reactors. However, by January 29, 1943, the thinking at the met lab had shifted to a water-cooled design, and the scientists proposed the change to du Pont. After considerable study on its own, the company concurred with the proposed change. We approved it. Meanwhile, procedures and the organization for the plutonium project gradually evolved. Although du Pont was absolutely dependent on the met lab for the basic research and preliminary design, the company assumed the responsibility for the final design. A few individuals from the met lab, hired by du Pont, assisted du Pont directly at Wilmington. Greenewalt provided the liaison between du Pont and the met lab. He relayed the company's requests for research data and basic design concepts for pile design to the physicists working with Fermi and Eugene Wigner, and to Glenn Seaborg and James Franck for research and process data for the chemical plant. Radiation was the big new factor that complicated the design. As du Pont engineering designs were completed, Greenewalt would take the design drawings back to the met lab for checks and approval. This procedure relegated the met lab scientists to a secondary role, which they did not like, and their discontent continued to grow.

The scientists bitterly protested being in the role of blueprint checkers. For the district, our coordination mission involved easing the friction and resolving design problems between du Pont and the met lab. Once designs were completed and approved, du Pont alone had the responsibility to construct and operate. The district had Major William Sapper at Wilmington as area engineer for administrative support,

Peterson as Chicago area engineer for the met lab, and Matthias at the HEW. During 1943, I visited Wilmington almost every week, Chicago twice a month, and Hanford about every three months. These regular visits were essential to keep abreast of our work and to ensure maximum progress. In addition to checking on design progress at Wilmington, Arthur Levin, our attorney negotiating the du Pont contract, and I met with Rittenhouse, the du Pont attorney, and Roger Williams, the top official of the company who devoted full time to the project. The attorneys would try out their legal language on Williams and me to see if it expressed what we thought we were agreeing to. Williams' superior F. M. Yancey would join us frequently.

The contract with du Pont proved the most complicated of any the district had to negotiate. The crucial issues included the uncertainty of the project's success, the hazards involved, the dependence of du Pont on the met lab, determination of the company to operate the plutonium plants only during the war, responsibility for the liability for radiation health problems that might arise twenty or more years later, and the conservatism of du Pont itself. From time to time, Williams or Yancey would insist that Groves meet with us to be sure that he also understood and agreed to the scope and terms of the contract.

The president had approved holding du Pont harmless from liability due to hazards. This combined with the moral considerations that du Pont possibly would be exposing employees to radiation hazards without revealing the full risk to them (due to secrecy) led du Pont to insist on a form of insurance for each employee. They contrasted our situation to a TNT plant where they could explain the hazards to an employee, pay him hazard pay, provide a safety chute to slide down in case of incipient danger of explosion, and hope that he made a fast enough exit. They were understandably worried about effects of radiation being delayed for twenty or thirty years and then being confronted with a government decision to absolve itself from responsibility, leaving the company holding the bag against damage suits. They proposed and insisted that a trust fund of $20 million be set up in a bank to cover this future liability. This was a large sum in those days and was based on the then-current $10,000 life insurance policy issued to soldiers.

Groves outlined the special provisions in the du Pont contract to the comptroller general of the United States. Groves reported to the Military Policy Committee on May 5, 1943, that the comptroller replied he "would not be forced to object." We considered this the equivalent of approval. In addition, in view of the government holding du Pont harmless against such hazards, du Pont proposed that instead of the usual fixed-fee contract (wherein the fixed fee represented profit), their contract provide for no profit and likewise provide that all actual costs be covered. We finally settled on a $1 fee, which our legal advisers claimed necessary to make it legal, and went into considerable detail about what costs were covered and how they would be audited.

In spite of the unusual requirements, du Pont was most cooperative in negotiating the contract. They were very considerate of all the government legal requirements that Levin presented and were experienced with government procedures due to the many war contracts they had. They wanted to be sure that any departures from routine procedures could be justified and would stand up against subsequent audit by the General Accounting Office. In addition to providing a good postgraduate course to supplement my West Point law course, the attention to detail that du Pont exhibited and their insistence on precautionary provisions served as constant reminders that success remained uncertain and that unfortunate accidents might occur.

In addition to developing safety methods, the company ardently supported the health research at the met lab under Dr. Stone. They cooperated with Dr. Stafford Warren, whom we had selected to head the district's medical section because of his radiation experience. The present nuclear power industry has profited from the many provisions for safety that du Pont initiated during the Manhattan Project. As a result of this wartime effort, our nuclear industry today probably knows more about the hazards of radiation and the potential dangers of reactors, radioactive materials, and plutonium than most other major industries knew about the hazards of their materials and processes during their early development. Thanks in large part to du Pont and the met lab, the U.S. nuclear power industry still enjoys an amazingly superior safety record, one that contradicts the misinformation provided by the antinuclear lobby. (See Chapter 15 for discussion of the Three Mile Island and Chernobyl accidents.)

My visits to the met lab were mainly to keep informed and to attempt to prevent the discontent among the scientists from jeopardizing full cooperation with and support of du Pont. Arthur Compton and Norman Hilberry always were most pleasant to work with. I soon developed tremendous admiration for Compton. I have never dealt with anyone else who had the mental capacity, the high ideals and moral standards, and the determination to achieve an objective, combined with such great consideration of others, as he had. He always was extremely cooperative and had a keen understanding of the Army's problem of dealing with the variety of individuals and organizations involved in the project. He reorganized the met lab to assist better in the overall effort to produce atomic weapons at the earliest possible date. He also cooperated in our efforts to administer the project to account for expenditures and avoid waste or corruption. He was very patient in trying to help me understand the physics and the technical problems. He encouraged me to attend occasionally the round-table discussions of Fermi and the other key scientists.

On one occasion, Compton asked me to attend a meeting to hear the severe criticisms of du Pont directly from the scientists. The main complaint was the conservatism and slowness of du Pont. Some of the scientists felt that du Pont was carrying the requirements for safety to a ridiculous point, that it took them too long to make design decisions, that they were incompetent, and on and on. Finally, Enrico Fermi, in all seriousness, strongly recommended to me that we eliminte du Pont from all work except that of the chemical separation plant. He said, "Well, we don't need this great organization. They are too conservative." He claimed that the met lab could design the reactors in half the time du Pont was taking, and concluded, "If you people will just hire for me the laborers and supply them with brick, I'll tell them where to lay them."

Needless to say, I was surprised to hear such a claim coming from a man of Fermi's stature. However, my admiration for him was not diminished; instead, his statements made me realize how serious the conflict had become. In response, I tried to explain to them the major talents du Pont brought to the project and how important it was for Fermi, Wigner, Franck, and the others to continue to cooperate with the company.

When the meeting ended, Compton, Hilberry, and I went to Compton's office to discuss how to deal with the concerns of the scientists. To my comment that this was all "unwarranted," Compton responded, "Well, sometimes, you know, I'm inclined to agree with Fermi. If we just had somebody to design the waterworks and the roads, I think I'd almost be willing to back Fermi and take on the technical design." Stifling a chuckle, I said, "Well, A.H., that's the one thing I can do. I'm a hydraulics expert and I have built a lot of roads and runways. I can design the waterworks and the roads. Let's do it."

Compton started to laugh and admitted, "I guess we had better keep du Pont." He was a great man and did a magnificent job of maintaining morale at the met lab. But he wanted us to listen to him and try to find some solution to the problems of keeping his tribe of talented, precocious people reasonably happy. The scientists all worked their heads off, but they always thought they knew how to do everything. As the research phase of the plutonium project declined, it became necessary to keep them occupied. By then, heavy water from Trail was becoming available, and interest in a heavy-water pile increased. As a result, the met lab began to devote more effort to basic research and to the design of a heavy-water pile as an alternative to graphite for the second pile at Hanford. A review committee headed by Warren Lewis supported a heavy-water effort and the building of an experimental heavy-water pile at the Argonne lab. The committee also listened patiently to the continuing complaints about du Pont. The S-1 committee approved the heavy-water-pile development program.

The overall result was that the scientists had a heavy-water pile as a new objective. More important, they continued to cooperate and to furnish timely support to the du Pont design effort so necessary for our success. When plant operations began at Hanford, it proved necessary to have available the talents of Fermi and others to solve new, unexpected problems. Fortunately, the district and Groves had available, in addition to Compton, the advice and help of Conant and Tolman to assist in the continuing requirement to reconcile the differences among the scientists, industrial engineers, and Army officers, three different breeds of cat.

Ernest Lawrence provided outstanding scientific leadership for the design, construction, and operation of the electromagnetics plant (Y-12). In my opinion, Lawrence probably was the most dynamic of all the physicists involved in producing the atomic bomb. A Nobel Prize winner, he had distinguished himself as a great experimenter and the leader in the development of the cyclotron. He also was a super promoter and salesman. Before the war, he had demonstrated that he could raise the money required to fund his expensive scientific ambitions. His enthusiasm and optimism combined with a most captivating personality won him the support of many affluent friends. He was a doer who attracted scientists and engineers to carry out his projects, including the building of a 184-inch giant cyclotron. Its magnet now provided a facility for developing the electromagnetic process.

At our first S-1 meeting, Lawrence had explained to me the theory of the electromagnetic process. The theory was relatively simple. The experimental production unit called a calutron consisted of a U-shaped vacuum tank placed within the magnetic field of the huge magnet. At one end of the U, a power source would ionize uranium chloride. The path of the ion beam would, due to the magnetic field, be an arc and end in box collectors at the other end of the calutron. Both U-235 and U-238 ions would be similarly charged and be bent by the magnetic field. The curvature would depend on their masses. The U-238, being heavier than the U-235, would follow a larger arc than the U-235, and the beam would be collected in two separate boxes, with the inner box collecting the more enriched U-235. Initially the enrichment attained was very low. But Lawrence believed that through experimentation he could perfect the process.

During the first year, Marshall and Blair spent more time than I did on the electromagnetic process; however, I visited Berkeley in August 1942 and was fascinated with Lawrence, his organization, and his process. The scientists were devoted to him, and I found it inspiring to see the results of their long hours of experimental work. My Ph.D. was in engineering and experimental hydraulics, and all my activities at the U.S. Waterways Experiment Station involved experimental work. Consequently, the Lawrence approach appealed to me. With the experimental facilities available and operating, it was relatively easy for Lawrence to convince us that his method was nearly ready for engineering design for a pilot plant or even for a full-scale production plant. Once a single unit could be perfected, it need only be reproduced several hundred times to be a useful production plant.

Lawrence hoped to be able to produce weapon-concentration U-235 in one stage. For a while I was in his doghouse because I pointed out on several occasions that his experimental results suggested that at least two stages would be necessary. As an optimist and promoter, he apparently felt that one stage would be easier to sell. I am sure that he knew that two stages would be necessary, but he was not about to admit it at that time.

The electromagnetics plant also appealed to Stone and Webster. It was essentially an electrical engineering project, and they felt competent to handle it. Westinghouse was selected to design and produce the calutron tanks, liners, sources, and collectors. Allis-Chalmers received the contract for the magnets and General Electric the complicated high-voltage electrical control units. Other suppliers received assignments for valves, vacuum pumps, and other components.

Lawrence's enthusiasm and drive helped expedite design, construction, and production. He was most cooperative at all times, and I enjoyed his friendship long after the project ended. I shall never forget how after a long day at the laboratory, I would spend a pleasant evening with him and his key staff at Trader Vic's followed by a late return trip up the hill to see how things were going. Later, when he was spending considerable time at Oak Ridge, he'd stop by for a visit with Jackie and me. Although I do not like to single out one individual, Lawrence, without doubt, was more responsible than anyone else for our success in producing the U-235 necessary for the Hiroshima weapon. He provided inspiration to the whole team.

On my first visit to Berkeley, Stone and Webster was already at work tackling the engineering design and procurement problems. Vacuum pumps and magnets of unprecedented size had to be developed and produced. The calutrons, sources, collectors, and tanks were in a very early stage of development and would require precise engineering, and already it was obvious that they would be difficult to maintain. The controls would require many new developments. Most important, there was considerable question concerning operators. Would it take hundreds or even thousands of Ph.D.s to operate the plant? Lawrence recognized that there was a chemical processing problem, but like most other physicists he believed it would require only a few chemists and chemical engineers to develop a chemical process and to design and operate the necessary chemicals plant. To handle the chemicals problem, Groves decided that Tennessee Eastman Corporation, a division of Eastman Kodak, should be the operating contractor. This followed the informal suggestions of the Lewis Committee. Tennessee Eastman already was involved in war work and was responsible for the Holston Ordnance Works, an explosives plant near Kingston, Tennessee. They knew industrial chemical operations, the labor situation in Tennessee, and had the full technical, managerial, and financial backing of Eastman Kodak. Tennessee Eastman worked closely with Stone and Webster and took full responsibility for developing the chemical process based on work done at Berkeley.

By the middle of March 1943, the scope of the Y-12 plant was determined. There would be five racetracks (the magnets were arranged in an oval form like a racetrack) of ninety-six tanks each for the

Alpha stage. Lawrence finally had acknowledged that a second or Beta stage would be needed. The Beta tanks were smaller, arranged in the form of a rectangle, with the plant consisting of two units of thirty-six tanks, two sources each. For the Beta plant, everything had to be more precise because this plant would be handling 10 to 36 percent U-235 enriched material as compared to only 0.7 percent U-235 material in the Alpha stage, and losses could not be tolerated.

Tennessee Eastman also started plans for training personnel and decided not to use Ph.D.s to operate the control units. Instead, they planned to use girls from Tennessee having only a high-school education or less. In Berkeley, many of the scientists were skeptical about this decision.

As the teams of designers and manufacturers set to work to design and procure the necessary equipment, they were handicapped by the lack of specifications. Berkeley had not yet completed its experimental work to develop a satisfactory unit, and concepts were constantly changing as the experimental equipment and tests were obtaining better separative results. There was a great need to freeze design. Nevertheless, everyone proceeded to do what he could. Stone and Webster had broken ground for the first Alpha building on February 18, 1943.

Foundation problems were encountered and solved. Railroad sidings and roads were built into the site to handle the tremendous quantities of materials that would be needed. Labor was recruited with the help and cooperation of the U.S. Employment Service, the Building Trades Council, and various national craft unions. Soon we encountered shortages of many types of materials—electronics tubes, transformers, rectifiers, switch gears, and insulators.

We had few problems with the scientists at Berkeley. Stone and Webster and the equipment manufacturers worked closely with them. They all enthusiastically entered into the spirit of the project to perfect the Alpha calutron unit as soon as possible so final engineering could proceed and the vast amount of equipment could be procured. They also welcomed the need to design the new Beta units. The chemists set new objectives for the chemical process required. Berkeley cooperated with Tennessee Eastman, training senior operating personnel. The major difficulty we had was pressing for necessary compromises with the scientists, who desired to continue improving each component, whereas meeting construction goals required freezing designs before procuring equipment and building the plants. We recognized that a better design would produce more material but that it might come too late to be of any value in World War II.

*   *   *

Lawrence toured the entire electromagnetics project in May 1943 and was amazed at the scope and magnitude of the effort. As a result of designing and building cyclotrons, Lawrence had far more experience with large equipment than most scientists. But even so, he had difficulty comprehending what was involved in building 552 calutrons and accompanying magnets and controls. However, his amazement at the progress increased his enthusiasm and his dedication to final success. Moreover, his enthusiasm permeated the entire project.

The district organization leaders for the electromagnetics project included Harold A. Fidler, area engineer at Berkeley; Benjamin K. Hough at Boston with Stone and Webster; Warren George for construction at Clinton Engineer Works; and Major W. E. Kelley, unit chief to coordinate the entire effort. Charles Vanden Bulck arranged for administrative support at each area site, and Allan Johnson supported solution of all priority problems, which quickly arose, particularly for electronic components. John Ruhoff arranged for feed material to supply Tennessee Eastman. Stone and Webster was responsible for design and construction. (See April 1943 MED organizational chart, pages 16–17.)

Our objective was to complete the authorized electromagnetic plant by the end of 1943. Hopes were high, but unexpected bad news arrived in the middle of 1943, when Oppenheimer tripled the estimate of the amount of U-235 that would be needed for an effective weapon. In response, Lawrence immediately began a campaign to expand the Y-12 program. He stressed the progress being made at Berkeley on multiple sources and other means of increasing production. He pointed out the difficulties being encountered in barrier development for the gaseous diffusion plant and that delay in reaching full production with the gaseous diffusion method would be inevitable. Therefore he proposed that partially enriched material from the lower stages of the gaseous diffusion plant be used as feed material for the Beta stages of the electromagnetic plant. Others had previously made the same proposal. He believed this combination would achieve our production schedule for weapon-strength material months earlier than production from the upper stages of the gaseous diffusion plant.

On September 2, 1943, finding that all the Y-12 contractors were in accord and supported Lawrence on how to expand the electromagnetics program, Groves approved the expansion. With the backing of Conant, the Military Policy Committee on September 9 approved

Groves' recommendations to build Alpha II, consisting of two build- ings, each containing two rectangular tracks of ninety-six units each. Four sources were to be in each tank, and many other improvements were included to improve production and ease maintenance. Two ad- ditional Beta units also would be required. The estimate for the electromagnetics plant now approached $250 million. As part of this decision, Keith was informed to defer work on the upper stages or top of the diffusion plant and concentrate on sufficient stages to produce 36 percent U-235 product. Believing that a completed gaseous diffusion plant would produce the required amount of weapon-strength material on schedule, Keith, head of Kellex (a newly organized subsidiary of Kellogg), did not agree with the decision. He continued to hope that the top stages would be authorized later.

In contrast to the relative ease in organizing and expediting the Y-12 project, the K-25 (gaseous diffusion) project was by far more difficult to organize, coordinate, and expedite. The process involved more development engineering than science. Again, the principle involved was relatively simple. Uranium hexafluoride gas would be pumped through a bundle of porous tubes contained within a tank called a converter or filter. Passing along the length of the tubes, approximately half the gas would diffuse through the porous wall of the tubes; the other half would flow out the end of the tubes. Due to the difference in molecular weight of U-235 and U-238, the part of the gas that diffused through the porous walls of the tubes would contain a slightly higher percentage of U-235 than the gas that flowed through the tubes. This was a proven process for separating different gases of unequal molec- ular weight.

The slightly enriched gas would be passed up the cascade to the next higher converter, where it would mix with gas coming down from the next, even higher converter. Likewise, the slightly depleted gas would be passed down the cascade to the next lower stage, where it would be mixed with gas coming up from the next, even lower stage. The process would consist of thousands of stages. Normal uranium hexafluoride would be introduced at about the one-third point of the cascade. Highly enriched uranium hexafluoride would be drawn off at the top of the cascade, and deleted uranium hexafluoride at the bottom. Again the principle was simple but the implementation was compli- cated, difficult, and frustrating.

The most suitable process gas to be used for separating isotope

U-235 from U-238 is uranium hexafluoride. This gas is highly corrosive, and nickel is one of the few metals that will resist its corrosive effects. A production plant would consist of miles of barrier tubes; miles of interconnecting pipes; thousands of large converters confining the tubes in each stage; thousands of motors, pumps, and pump seals; and instrumentation piping and instruments to control pressures and volume of flow throughout the plant. The entire system had to be leakproof to a degree never before achieved by such a large system involving flowing gas. The statement was frequently made that the leakage into the entire process could not exceed the equivalent of one pinhole. If air leaked into the system it would react with the uranium hexafluoride, and the barriers would become plugged and useless.

At first it was thought that essentially every part of the process would have to be made of nickel. If that had been the final conclusion, the answer was easy. Sufficient nickel was not available, and the process would have been dropped. The main problems to be solved involved expert metallurgical, chemical, and mechanical engineering as well as scientific research. The hardest problem was developing and producing a suitable barrier material. Everything depended upon the separation or enrichment rate of the barrier. It would determine the number of stages and the size of the plants.

Due to corrosion, it soon became apparent that a nickel barrier of some type would be necessary. In the spring of 1943, no one knew how to develop a satisfactory barrier or even estimate when any such development might be successful. However, theoretically, the gaseous diffusion process offered the best potential means to produce U-235 in the quantities necessary for a reasonable rate of weapon production. It would provide a continuous and relatively maintenance-free method and would require far fewer operators than the electromagnetic process. To proceed with the plant without good prospects for a suitable barrier involved the biggest gamble in the whole Manhattan Project. Groves persistently stuck by the decision to proceed with the plant. With the additional requirement to design special pumps and all the other difficulties that rapidly became apparent, that took guts.

Under the OSRD Urey, a Nobel Prize winner, received the responsibility not only for heavy water but also for the gaseous diffusion process. The OSRD made this decision partly to avoid Columbia University's having two members on the S-1 committee and partly because John Dunning, an engineer, was not considered to have sufficient prestige among the scientists involved.

In early April 1943, the Manhattan District assumed responsibility for the OSRD research contracts that pertained to the atomic bomb project. This included the Columbia University contract. On April 17, the day after the S-1 committee had approved the met lab proceeding with the heavy-water pile as an alternative to the graphite pile, I informed Urey that all heavy-water-pile work would be performed at the met lab under Compton and that further research work on new methods to produce heavy water would be dropped. These were Urey's main interests at the time.

Urey never was an ardent advocate of the gaseous diffusion method. I encouraged him to take a more active part in the work on that process. In turn, he indicated that he felt he had a very small part to play in research work at Columbia and that we should deal with Dunning in making the contract. In my notes for the diary of April 17, 1943, I wrote, "The problem of where to concentrate Dr. Urey's energy is still unsolved."

At an S-1 meeting on April 29, we had a rather confusing discussion about the gaseous diffusion organization and then made some tentative decisions about the direction of research and development. Everyone agreed that Dunning had done an excellent job up to that time but that his group needed to be strengthened. Conant stressed that Kellex should be given greater responsibility to head the development of the gaseous diffusion process. Briggs stressed that Keith should be held responsible for all research. General opinion seemed to be that we should rely on Keith to control Dunning. It was suggested that Groves and Conant should discuss with Frank D. Fackenthal, provost of Columbia University, the choice of project director, but we made no decision in that regard at the meeting. In the diary for April 29, 1943, I later added, "I have conferred with Mr. Fackenthal and Dr. Urey, and Dr. Urey is being designated as Research Director for Columbia University work. Dr. Urey will take a greater interest in K-25. I will see Keith and emphasize to him the importance of seeing that the research work is adequate and tell him that he is responsible that sufficient research and development work is done to insure the success of the project."

General Groves and Conant had approved the designation of Urey as director of research for the university laboratories and the expanded facilities in the Nash Building of Columbia called the SAM laboratory. Initially Urey took an active part and aggressively attacked the many problems, particularly the barrier development. As difficulties increased, however, he gradually lost confidence in the prospects for success. This caused serious problems for us.

As this happened, Dunning and Keith assumed the roles as key leaders and advocates for the gaseous diffusion process. Dunning's enthusiasm for the gaseous diffusion process almost equaled that of Lawrence's for the electromagnetic process, but Dunning's ability to convince others fell short of that of Lawrence or Compton. There were many other very outstanding scientists and engineers at Columbia working on the project who made major contributions. Nevertheless, the gaseous diffusion process undoubtedly suffered when compared to the plutonium and the electromagnetic projects because its director, Urey, who had the prestige, never was an ardent promoter of the project.

Keith was an experienced engineer at Kellex and had some experience using the diffusion method for separating two different gases. He tackled the job of engineering development and design of the gaseous diffusion plant with vigor, intelligence, and enthusiasm. He, like Compton and Lawrence, was an optimist. Keith had Manson Benedict, a theoretical chemist, as a principal and extremely competent technical assistant. Along with Karl Cohen, a very capable mathematician at Columbia, they tackled the theoretical and statistical problems concerning the control of the cascade consisting of thousands of units. Whether or not the plant could be controlled remained in question for many months, as did the related question of how much time would be required to reach equilibrium in the plant.

Keith recruited his key personnel from Kellogg and various other corporations. We frequently referred to Kellex as an organization of vice presidents, each a leader in his own specialty. Albert L. Baker, experienced in refinery design, was selected by Keith as his principal assistant. Throughout the project, Baker did a masterful job of coordinating the efforts of a rare group of experts and of administering the project in accordance with sound engineering and accounting procedures. Lieutenant Colonel James C. Stowers was selected to be our area engineer for the separate New York area office established for the gaseous diffusion project. In July 1943 Major William P. Cornelius was selected to be the K-25 construction officer at CEW.

The first item to be constructed was the 235,000 KVA steam electric power plant. In spite of plentiful TVA power, we thought it necessary to have our own power plant to assure a more reliable power source and to provide variable frequency to control the plant properly.

We later found that variable frequency was not necessary. However, extreme reliability of power was a must because a sudden power failure could seriously damage the seals. (We had such a failure during the 1953–55 period when I was general manager of the Atomic Energy Commission. A rat shorted out a transformer at CEW. A switching error followed, which resulted in a complete loss of our tremendous power load, and electrical surges were felt as far away as St. Louis. The plant suffered several million dollars in damages, and several weeks of production were lost.)

For design of the power plant, Keith borrowed Ludwig Skog, a partner of Sargant and Lundy, a leading engineering firm in the power plant field. Our plant was to be the largest single block of steam power ever built at one time.

With our high priority, we diverted turbines and generators on order for other projects and being manufactured by Allis-Chalmers, General Electric, and Westinghouse. The variety and size of the units afforded ample flexibility to have five different frequencies. Boilers already being fabricated were located, and our high priority shifted the orders to Kellex.

J. A. Jones Construction Company was selected to construct the power plant and the gaseous diffusion plant. We negotiated the contract with Edwin Jones, one of the principal owners of the firm, and throughout the project he devoted a large portion of his time to see that the company performed miracles in getting construction done on time. The site for the power plant was cleared and piles driven early in June 1943. Early in March 1944, nine months later, steam was available from one boiler, and early in April, 15,000 kva were available on the first turbine generator. The plant was completed in January 1945. I have yet to learn of any coal electric power plant of comparable size that has been built that fast.

Such miracles of design, procurement, and construction were absolutely necessary if we were to get an atomic bomb at the earliest possible date. On conventional items, such as the power plant involving little new development except for variable frequency, we needed only the right men, the right organizations, and our top priority. However, for the main gaseous diffusion plant we were confronted with a different situation.

We needed to develop practically every component. New methods of manufacture, new methods of construction, and precise quality control had to be established. Allis-Chalmers was selected to develop and to manufacture the centrifugal pumps required, and a new factory was constructed. Houdaille-Hershey was chosen to manufacture the barrier tubes and started construction on a new factory building. Columbia was responsible for barrier development, testing procedures, and manufacturing techniques. Columbia eventually was assisted by Bell Telephone Laboratories, Kellex, Union Carbide, and others. The Norris-Adler barrier appeared to be the most promising but could not be considered satisfactory from the standpoint of separation qualities, corrosion resistance, durability, uniformity, and production techniques. Much additional development was necessary.

Groves selected Chrysler to manufacture the converters, and the company converted the Lynch Road plant in Detroit. Stowers found K. T. Keller, president of Chrysler, very difficult to deal with. On April 22, Stowers and Baker of Kellex met with Keller and his attorney to negotiate the contract for manufacturing the converters. After his initial meeting with Groves, Keller had the idea that he could manufacture the converters and submit a bill to the government, which would pay it without an audit. When Stowers presented the type of contract we were using, Keller hit the ceiling. He claimed he had been misled, and he proceeded to deliver quite an exposition of the many faults of government contracting. Stowers later wrote, "This exposition was the most arrogant, vulgar, and blasphemous that I have ever heard from a businessman."[2] Nevertheless, after Keller cooled off a bit, Stowers tried again, only to have Keller insist on phoning Groves. Groves gave him some assurance "that he and Nichols would look into" some of the provisions of the contract that disturbed him and would revise them for his consideration. This reassured Keller and he told Groves that Chrysler would continue to press the work without a contract in hand. He then turned the negotiations over to his attorney and left the meeting. The discussions then proceeded without too much difficulty. Chrysler had large contracts with Army ordnance for tanks and guns, and the attorney was more familiar with government contract requirements. I later met with Keller and explained our procedures.

After this initial flurry, Keller and Chrysler made great contributions to the success of the project. In addition to manufacturing the converters, Keller would come to CEW and advise us on assembly of the units in the diffusion plant and also, more important, help us with the maintenance of the electromagnetic calutrons.

2. Diary, April 23, 1943, summary of negotiations with Chrysler Corporation.

Keller's assignment was to manufacture the converter units, large horizontal cylindrical tanks, and drill the millions of holes in the spacers to hold the barrier tubes. Keller assured us that he could nickel-plate the interiors of the converters and spacers so they could resist the corrosive effect of the uranium hexafluoride. Urey remained skeptical and would not recommend that Chrysler be authorized to spend $50,000 to demonstrate their process.

Keller asked me if we could reimburse Chrysler if the company demonstrated it could do the job. I agreed. Later I went out to Detroit to observe the process and took back small samples for Urey to test. They met the requirements, but Urey still remained skeptical that Chrysler could do the job on a large-scale basis. However, the company soon was providing larger samples that not only met the Columbia tests but also were tested by other laboratories. These results satisfied Keith that Chrysler had solved the plating problem thanks to the work of Dr. Carl Heussner, director of the Chrysler's plating laboratory.

On June 16, Groves, Stowers, and I met with Keller, James Rafferty, Lyman Bliss, and George Felbeck of the Union Carbide, and Keith and Baker of Kellex, so Keller could outline the scope of his work. Originally we thought that Chrysler would assemble and test the whole converter unit and ship the units to CEW ready to install. Keller now proposed to assemble just a few units in Detroit to see that everything fit. However, for the thousands of converters to be produced when the production lines went into operation, the company would ship the units to CEW, where they would be assembled with the barrier tubes, the pumps, and seals, fluorinated, hydrogenated, and then tested. Keller recommended this division of work to avoid storing and utilizing fluorine in large quantities in a populated area.

Having watched a demonstration of the startling effects and hazards of fluorine at the Columbia laboratory, I supported Keller's recommendation. Like many of the chemicals we used, fluorine required great care in handling. As an extremely reactive element, it was used for conditioning all components that would be exposed to the uranium hexafluoride. In the demonstration I witnessed, elemental fluorine was squirted on a ham bone, which immediately ignited. It made me fully appreciate the hazardous material with which we were dealing and the need to develop means to handle it safely. Ford Bacon and Davis Company would use a similar demonstration as part of its training program for the employees who would be engaged in the assembly and fluorination of the converter units. Hooker Chemical Company developed the method to produce the fluorine in quantities, and they built a production plant at Oak Ridge.

By July we were making progress on the organization required and the plants necessary to produce equipment for K-25. But we had too little in the line of specifications for too many of the components. The situation was not reassuring, and we still had to discover the solutions to many key problems. Nevertheless, we were determined to proceed full speed ahead with all we could do.

My other activities during the first half of 1943 included a visit to Los Alamos; two to Hanford to see how Matthias and du Pont were progressing on construction; visits to each of the heavy-water plants; and visits to feed material plants in Cleveland, Port Hope, Boston, and St. Louis. To reach Trail, I traveled by bus with part of the route following the upper Columbia River. I was impressed by the beautiful scenery that provided a welcome period of relaxation. However, as we approached Trail I saw for the first time how emissions can devastate a beautiful valley and mountain slopes. I had neither the time nor the mission to determine who was at fault, but it was obvious that natural beauty had lost its struggle with man's need for industrial products.

Progress on the Consolidated Mining and Smelting (CMS) plant for heavy water was reasonably satisfactory, but I learned that they were having some difficulty with the Canadian government about the contract. Sometime later, in June 1943, while talking to secretary-treasurer Carl B. French of Eldorado, I learned that he also was having some vague difficulties with the Canadian government about our contract. French advised that we should contact a C. D. Howe in Ottawa. I checked with the district's administrative staff and found there were other reports about Mr. Howe. I asked my secretary to call him. When she got him on the line, I identified myself as deputy district engineer of the Manhattan District and said I had heard that he had questions about our contracts with CMS and Eldorado.

In a pleasant voice, he responded that he was happy I had called. He did have many questions and had been patiently waiting for someone in the U.S. government to contact him. He asked how soon I could come up to Ottawa to discuss the matters. I told him I would be there the next day, June 14. I took the overnight train. Arriving in the Canadian capital, I went to the address Howe had provided and discovered to my surprise that he was the minister of munitions and supply.

Howe turned out to be a most friendly person and I soon realized that he knew what we were doing in the Manhattan Project. He confirmed that the Canadian government had purchased Eldorado and turned it into a Crown company. He laughed and chided me about how long it had taken us to contact him. Now that we had acknowledged that the Canadian government existed, he said he would designate Leslie Thompson as a liaison with us and I could expect no further opposition to the contracts.

He then told me he had heard we were initiating a survey for uranium in Canada and said he felt that we should ask Canada to do that. He confirmed rumors that we had picked up that the United Kingdom was trying to buy an interest in the Congo uranium mine. After a very pleasant lunch, he expressed the hope that Great Britain, Canada, and the United States would agree on resuming cooperation and exchange of information on atomic matters. He assured me that until then our informal arrangement would take care of our heavy-water and uranium-ore refining contracts.

On July 21, with Leslie Thompson I visited the Port Hope refinery to discuss increasing production. In addition to learning about refining uranium ore and the status of our contracts, I found out a little about refining radium from uranium. Dr. Marcel Poshon, Eldorado's technical director, had just completed a batch of radium and held it in a beaker under his jacket close to his chest so I could peek in and see the radium glow. I suspected then and our safety division confirmed later that this is not a recommended practice.

Earlier in July, Groves had told me that he and Conant had met with C. J. Mackenzie of the Canadian National Research Council, who had suggested that Groves meet with Howe. Groves asked me if any of our Canadian contacts could help arrange such a meeting. I told him that I had a liaison man, Leslie Thompson, in Howe's office, and I was sure he could arrange a meeting. When I told Groves how I had placed a call to Howe to meet with him without knowing that he was the minister of munitions and supply, Groves commented, "It is wonderful to be young and not bothered by protocol."

On July 15, Groves availed himself of my contact, asking me to set up a meeting with Howe and Mackenzie in Ottawa for Marshall and himself for July 26. By then, however, new developments had occurred within the Manhattan Project. As a result, I instead of Marshall accompanied Groves.

6

GETTING ALONG
WITH GROVES

MY FIRST YEAR WITH the Manhattan Project sped by very quickly. By mid-July 1943 I began to wonder if General Styer had forgotten that he had told Marshall I was to remain with the project for one year only. Would I be leaving soon?

Late in the afternoon of July 19 I received a phone call from Groves: "Nichols, go to an outside phone where we cannot be overheard and call me immediately." I reached him from the corner drugstore. His message was brief and to the point: "I want to see you early tomorrow morning. Tell no one in New York that you are leaving."

The next morning, after a preliminary meeting with Groves and others, I met with Groves alone. By that time I was rather anxious about the urgency and secrecy of my visit. Groves informed me that the chief of engineers had selected Marshall for an overseas assignment and promotion to brigadier general and that Groves had agreed to it.

I was quite surprised and very pleased when Groves added that General Styer had approved me to be district engineer. At the same time, I was very sorry that Marshall was leaving. I liked working for him and was happy to have him as a buffer between Groves and myself. Groves did explain that he ordered the unusual secrecy to avoid any word getting to district personnel before Marshall received the news of the change.

Groves next asked me if I knew Marshall's whereabouts. I told him that Marshall had gone to Nashville to apologize to the Tennessee governor for a mistake we had made in the way we had informed him about declaring CEW a military district. We had done this to resolve

99

certain jurisdictional questions about CEW, HEW and Los Alamos, and with Groves' help we had obtained a presidential proclamation that the three areas were military districts and not subject to state control. When we received the proclamation, we found that it should be posted in public locations and that certain officials, including the governor, were to be informed.

Marshall had sent the proclamation to Major Thomas Crenshaw, the new commanding officer at Oak Ridge, with a simple note to notify the proper officials. Somehow, a young, inexperienced lieutenant in security received the mission to go to Nashville to notify state authorities. He had met with the Tennessee secretary of state and insisted on personally serving the proclamation on Governor Prentiss Cooper. This was our first contact with the governor, and he was indignant. He tore up the proclamation and ordered the lieutenant out of his office. As a result, Marshall had gone to the State Capitol at Nashville to try to make amends for the *faux pas*.

Groves asked me to get Marshall on the telephone. It took only a few minutes. However, just as I got him, Groves said, "I have to leave. You talk to him." Groves walked out. This placed me in an awkward position. Fortunately, I had a very close relationship with Marshall, and this lessened the embarrassment. Furthermore, we both knew Groves.

When I told Marshall what I had learned from Groves, he simply commented, "That is fine with me, Nick. I have been sitting on my butt outside Governor Cooper's office. He refuses to see me. Now you can come and sit here. Good luck." (Ironically, Marshall had forgotten about his conversation with Groves requesting a new assignment and later recalled that despite his prospects for promotion, he felt disappointed when he heard about his reassignment. To him, it sounded like he was "getting fired."[1])

I have always attributed Marshall's leaving to a combination of factors beginning with Groves' appointment to overall command. Marshall had believed he was to have charge of the project, and he did outrank Groves on the permanent promotion list. Consequently, when Groves received his star first and then began consolidating his control of the project, an awkward situation began to develop. Moreover, Marshall and Groves disagreed in a major way on how to handle personnel. This increased tensions once Groves took charge. Hence the

---

1. Marshall interview, April 19, 1968, Office of History, OCE.

opportunity to promote Marshall and transfer him to a desirable overseas assignment provided the means for Groves to resolve the clash of
personalities to the benefit of both men.

My own advancement, while indeed welcomed, ended for all practical purposes my hopes for an overseas assignment as a combat engineer. By this time, however, I realized the importance, urgency, and
challenge of my work, so I did not mind giving up my dream of
commanding engineers in combat.

In fact, I had little time to think about any alternative. Nor was I
asked. The next day, July 20, 1943, I received my orders from the
office of the chief of engineers designating me District Engineer, MED,
effective upon the departure of Marshall. Because the official orders
forming the Manhattan Engineer District had been dated August 13,
1942, Marshall decided to make it a full year of officially being district
engineer and leave on August 13, 1943.

When Marshall left, he said he had only one request to make.
Would I transfer Virginia Olsson to Tennessee when we moved the
district office there and make her my secretary and keep Anne Phillips
as my secretary in the New York office? This concern for personnel
was typical of Marshall. Complying with this request proved to be
far more rewarding than I anticipated. The two young ladies cooperated extremely well. They maintained close communication and
anticipated many problems. They made all arrangements for my appointments, travel, and hotels as well as accounting for the classified
documents needed on trips. Most important, they provided superior
secretarial and office management service at the two offices. All
through the war, they devoted long hours to the task with never a
complaint.

Marshall made a second request concerning personnel. He asked
that any officer personnel whom he had selected and whom Groves
later might insist on firing or who otherwise desired or needed a transfer should be sent directly to him in the Pacific Theater. He made
arrangements for this with Colonel E. A. Brown in the chief of engineers' personnel office. Most of the officers so transferred earned promotions in their new assignments.

Marshall has always claimed that he got along fairly well with
Groves: "We never got to any impasse or anything like that. He was
pretty sly. He would find some way of doing it his way even though I
thought it ought to be done in another way, because he was down in
Washington and I was up in New York." However, during the year I

served as Marshall's deputy, I did witness several confrontations between the men.

Groves was abrasive and often very critical. He seldom indulged in casual conversations, and in fact he usually appeared aloof. As a result, many people considered him unfriendly. On one occasion, Groves, Marshall, and I spent the entire day inspecting all construction at CEW, and when we were finally alone, Marshall remarked to Groves, "Didn't you find anything anywhere on the project that pleased you? I heard only criticism of every aspect of the work. Don't you ever praise anyone for a job well done, or a good idea for improving things?" Groves only commented, "I don't believe in it. No matter how well something is being done, it can always be done better and faster."

Whenever things turned out to be wrong or problems were not promptly solved, Groves, in a quiet, low-keyed voice could be most cutting and sarcastic in his remarks and was likely to conclude, "How can anyone be so dumb?" He never used profanity or foul language, nor did he ever raise his voice in anger. And his personal ethics were beyond reproach.

His treatment of individuals varied from man to man and the particular need for a person's talent at the moment. Initially, when everything depended on the key scientists, he handled them with kid gloves. Frequently, however, he would criticize certain scientists to me and urge me to make them more responsive to his desire for greater speed and for less vacillation in their recommendations. During the later stages of the project, when research had become less important, he tended to ignore many of the key scientists who had already accomplished their assignments.

Groves prided himself on being an excellent judge of the capabilities of individuals and claimed he could size up a man the first time he met him. If the estimate was unfavorable, the individual could do nothing right and would become the target of endless criticism or would be replaced. On the other hand, if Groves' estimate was favorable, the criticism was less personal and less harsh. In my case, Groves always gave me superior efficiency reports and noted that this "officer works best under pressure." And Groves certainly knew how to apply plenty of pressure.

When we had serious problems, Groves always became involved and we would have to contend with the reaction of the people he ruthlessly criticized or blamed for the poor results. Following such occasions, both Marshall and I felt it necessary to restore confidence and smooth ruffled feathers. Marshall showed skill in this area, and I learned from his example. After I had succeeded Marshall, one particular incident irritated me, and as soon as we were alone I told Groves that I felt he had been unnecessarily harsh in his comments to the individual. I then told him that if he ever criticized me in public in that manner, I would immediately respond in such terms that he would have no recourse except to fire me. I added that I didn't mind his being critical when we were alone because I could take it. I also told him that at any time he didn't like my performance, he could have me reassigned overseas. Thereafter both Groves and I endeavored to avoid any public confrontations, and I must admit that he treated me with more tact and consideration than I ever expected.

I could predict what Groves' reaction would be under most circumstances. In cases when I knew he would disagree on a course of action I felt necessary, I would wait until just a few minutes before I had to catch a train or plane; then I would bring up the matter. I would tell him what I was doing or going to do and explain why. As he started to respond, I would say that if he wanted me to do it differently, or wanted to discuss it further, he would have to phone me because I had to leave. The procedure usually worked, since Groves seldom would phone me.

On many matters, he used a less aggressive approach when he knew I personally had taken some action he didn't like. For example, at Oak Ridge, with a population of seventy-five thousand, we ran into many situations where we needed money to do things for which it would be improper to spend government funds. To obtain adequate funds, we set up a "Recreation Association" and among other things gave them the beer and other concessions to generate revenue. These funds were generally used to support recreational facilities, libraries, civic associations, and some welfare. Groves disliked the system and told me it might be illegal. When I asked him if he could provide a better solution, he only commented, "No, but I have serious questions about your way. You had better make certain that everyone involved remains completely honest."

Another method he used at times when he felt I might disagree would be to have Mrs. O'Leary, his secretary, phone me and say that General Groves wanted me to take a certain action. If I agreed and wanted to do it, I would say, "Okay." Otherwise I would tell her that I would not comply unless Groves himself called me. I would then add the reasons why I disagreed. In most cases, he never called. However, in some instances, Groves would break into the conversation to discuss the matter with me, confirming my belief that he frequently listened to the call on the other phone to get my reaction. Mrs. O'Leary and I were on the best of terms. She was Groves' only executive officer and in my opinion the most capable of his assistants. On major design, construction, and operation questions, I generally had plenty of opportunity to influence his decision on important matters. After hearing the opinions of the involved project leaders as well as mine, Groves would indicate what course of action he preferred. I then had the responsibility for seeing that it was done. I never had any doubt that Groves had the final responsibility for major decisions.

We both traveled extensively to gather our facts directly from the organizations and individuals involved at the work sites. Generally each of us established his own schedule. But we did meet at least once a week at a place where we had a problem in which either of us thought we both should be involved. At times, a contractor would request that we both meet with him. Our separate visits posed the possibility of conflicting decisions or instructions that bothered some of my area engineers, who were my representatives concerning contract provisions. My instructions to these officers were: Groves is in overall charge of the project. When he tells you to do something, if you are in agreement, do it, and then inform me about it. If you disagree or think I would disagree, call me and I will tell you how to handle it. I also instructed them that Groves' intervention would of necessity be intermittent, and when he left the ball in the air, as he frequently did, it was their job to catch it and to carry on as usual.

On parts of the work that were progressing on time and in which no one had complaints about delays or the need for coordination, Groves paid little attention, except to ask that he be kept informed.

In the case of Los Alamos, Groves made it clear that he personally would do all the direct supervision of the work. However, he indicated that I should keep myself informed by visiting Los Alamos or by meeting with Oppenheimer elsewhere concerning progress and coordinating technical specifications for U-235 and plutonium. In addition, I was to work out with Oppenheimer the means to determine the percent of enrichment of U-235 that would be the optimum compromise between possible production rates, which was my responsibility, and bomb efficiency, which was Oppenheimer's field.

At the same time, the district was to support Los Alamos in acquiring many unusual and rare materials, and see that all expenditures under the University of California contract for operating the laboratory were properly audited. These duties required only occasional visits to Los Alamos—once in March 1943, once in April 1945, and then more frequently after the war. Contrary to the 1982 BBC television series on Oppenheimer that portrayed me serving as a personal aide to Groves on frequent visits to Los Alamos, the only time I ever accompanied him to Los Alamos was during my March 1943 visit. Most of my meetings with Oppenheimer took place at Oak Ridge, Berkeley, New York, Chicago, or Washington, when problems arose.

Many times, people have asked me how the various contractors and scientists responded to Groves' concept of organization and direction of the work and how I, the district engineer, fitted into the scheme of things. Arthur Compton, in his book *Atomic Quest,* helps answer that questions, describing my working relationship with Groves probably better than I can.

Having explained the transfer of responsibility for bomb design from him and the met lab to Oppenheimer, with Oppenheimer now reporting directly to Groves, Compton wrote:

> The division of authority was part of General Groves' administrative strategy. He did not want any one man under him to have so much responsibility that he would become indispensable. The Army knows that men may die or become disabled and that their places must be taken by others. Besides, the responsibility for seeing that atomic weapons were built was his. He felt that he must not overburden anyone in the project. Thus the General made use of several scientific advisers. James Conant and Richard Tolman especially discussed with him both scientific and policy matters. In this way he avoided the troubles that he feared if some single scientist had been in a dominating position. Groves preferred to talk over his problems informally with each of the score of men responsible directly to him and then to form his own decisions. With his exceptionally comprehensive mind and tireless energy, this procedure brought satisfactory results.
>
> More than any other person, Groves depended on Colonel K. D. Nichols.—My own planning of general policy and major lines of action was authorized in discussion with Groves. As to budgets, personnel matters, and technical relations with other branches of the project, Nichols was my prime contact. Groves kept direct

responsibility for the work at Los Alamos, while Nichols took complete responsibility for the operations at Oak Ridge. At Hanford, Colonel Franklin T. Matthias, the officer in charge, took instructions as I did from both Groves and Nichols. Groves left in Nichols' hands full responsibility for procurement and treatment of ore, except for the later negotiations with foreign suppliers. The dividing line between the responsibilities of the two men was thus somewhat hazy, but they worked together effectively and with complete understanding.[2]

Nowhere was this blurring of effort more visible than in the security for the project. Groves assumed control of the intelligence effort, which focused on ascertaining the status of the German atomic research and development. I had the responsibility for internal security at the various facilities. In fact, both operations overlapped and our men in charge, John Lansdale for Groves and first Tony Calvert and then Colonel W. Budd Parsons for me, generally worked well together, but on occasion they clashed over jurisdiction.

Sometimes the final determination of personal security clearance would rest on the importance of the contribution the person was making, sometimes on other factors. In one instance Parsons informed me that Ernest Lawrence had two Communists working for him at Berkeley. When I called Lawrence and told him that he had to fire the men, he did not want to believe the information and also argued that they were essential to his work. I finally agreed to send him the raw surveillance reports and wiretaps so that he could decide for himself. In a few days he called to say that he was firing the men. He told me that the material had not convinced him that they were Communists, but he had learned from our recordings that they were wife-swappers and he "wouldn't have any of that in his laboratory."

As much as possible, Groves and I cut red tape to avoid delays. To save paperwork and time as well as to preserve secrecy, very little written material passed between Groves and the district. After I became the district engineer, he asked me if I desired to have more in writing or continue with just verbal instructions and authorizations. I told him that so long as I participated in the preparation of progress reports to the president or saw the approved reports, verbal instructions were adequate. I did not see how it would be possible to cover in

2. Compton, *Atomic Quest*, pp. 130–31.

writing all the informal decisions that were made at the numerous meetings involved. Formal procedures would be time-consuming for both of us and definitely would slow the work.

Practically all aspects of the Manhattan Project were directed by the most outstanding leaders in their field, and it was a great team. Having the best talent allowed us to issue mission-type orders. Supervision was required primarily to coordinate the various activities of many organizations toward achieving our common objective. Moreover, we had the advantage of having a clear-cut objective, producing an atomic bomb that could be delivered by a B-29 at the earliest possible date, to end the war and save hundreds of thousands of lives.

It was necessary to appraise, from time to time, the prospects for success of specific parts of the effort and change direction or method if necessary. Frequently we started an alternative approach to a problem to ensure that a backup would be available in case the initial approach failed. Cost in dollars never became the governing consideration, but costs in time, manpower, and critical materials were major factors. In addition, we concentrated on eliminating all wasted effort such as unnecessary refinement or expending effort on developments that could not be achieved for use during the war.

Working with Groves was not always easy. At times I became extremely irritated with him. However, this never lasted very long because I came to realize that his leadership and distinctive characteristics would achieve our goal in the shortest possible time. Generally I refrained from expressing my opinion of him to anyone. However, when Groves and I were invited to lecture at the National War College, I did discuss the matter.

The first day, I had listened to his talk covering the accomplishments of the entire project. My assignment at the next day's meeting, which Groves did not attend, was to discuss the organization we had used to get the job done. In the question period that followed, a young Navy officer asked the first question: ''We heard General Groves yesterday and you today. I understand that you probably know him better than anyone else. I have heard that he is pretty rough on subordinates. Will you please give us your personal opinion of General Groves?''

The commandant of the War College immediately stood up and announced, ''That is an improper question and you do not need to answer it.'' Having avoided the question many times before, I quickly decided I had an audience that would appreciate a frank answer. I responded that I did not mind answering it and proceeded to do so:

"First, General Groves is the biggest S.O.B. I have ever worked for. He is most demanding. He is most critical. He is always a driver, never a praiser. He is abrasive and sarcastic. He disregards all normal organizational channels. He is extremely intelligent. He has the guts to make timely, difficult decisions. He is the most egotistical man I know. He knows he is right and so sticks by his decision.

"He abounds with energy and expects everyone to work as hard or even harder than he does. Although he gave me great responsibility and adequate authority to carry out his mission-type orders, he constantly meddled with my subordinates. However, to compensate for that, he had an extremely small staff, which meant that we were not subject to the usual staff-type interferences and heckling. He ruthlessly protected the overall project from other government agency interference, which made my task easier. He seldom accepted other agency cooperation and then only on his terms. During the war and since, I have had the opportunity to meet many of our most outstanding military leaders in the Army, Navy, and Air Force as well as many of our outstanding scientific, engineering, and industrial leaders. And in summary, if I had to do my part of the atomic bomb project over again and had the privilege of picking my boss I would pick General Groves."

Sometime after my comments at the War College, I was serving as a member of the Army's Scientific Advisory Panel along with Gene Vidal, who had been one of the very early pilots in the Army Air Corps. After a long, pleasant evening of drinking with Gene following one of our meetings, he asked my opinion of Groves. I repeated what I had said at the War College. Gene enjoyed the story. At our next advisory panel meeting, he told me, "I know you probably did not intend for me to tell your story to Groves, but he is my classmate and lives near me in Connecticut. Recently I saw him out working on his lawn and dropped by for a chat. I embellished your story a bit. After the first sentence, Groves, in a shocked voice, asked, 'Did Nichols say that?' After each sentence, he repeated, 'Did Nichols say that?' Groves continued to express astonishment until I got to the punch line and then he broke out into a big grin and beamed with pleasure." Later some of my friends from the project told me that Groves inquired whether they agreed with certain other officers in the district who thought he was an S.O.B.

I most likely got along with Groves as well as I did because we shared the single, overriding goal of producing atomic weapons in time to help end the war. Some of the industrial people have suggested that

I contributed more to the success of the project than he did. As an Army officer and someone with a strong ego, I never paid much attention to this type of comment. I always saw Groves as my boss, the commanding officer in charge of the entire Manhattan Project. I had my role, which I fulfilled to the full extent of my abilities. As a Corps of Engineers officer, I found more than enough satisfaction in the ultimate success of the project, the contribution it made to the surrender of Japan, and the saving of so many lives, not only American but Japanese as well. Success was due to teamwork of hundreds of organizations and tens of thousands of individuals. Groves was the strong project leader.

7

NEW RESPONSIBILITIES, 1943

THE MONTH OF AUGUST 1943 marked the beginning of more responsibility for me and an even busier schedule. Colonel Marshall spent
most of the last half of July and the first half of August preparing for
his new assignment.

Even before he left New York on August 15, I had begun to assume
the responsibilities of my new duties. As a result, I now became more
involved with the construction of the electromagnetic plant, the central
facilities at the CEW, and the town of Oak Ridge as well as all the
other aspects of the work.

After I became district engineer, the first new task that confronted me
involved not construction but international relations. On August 21,
Groves asked me to help him write a status report on the project
addressed to the General Policy Committee. We worked late into the
night until Groves was satisfied with the contents.

He had become disturbed about the forthcoming Quebec Conference between President Roosevelt and Prime Minister Churchill.
Groves heard that Lord Cherwell, Churchill's personal scientific adviser, was included in the official party. To Groves, this meant that
Churchill intended to put atomic bomb technical cooperation on the
agenda. Harry Hopkins, Roosevelt's most intimate adviser; Stimson;
Bundy; Bush; and Conant had held several individual conversations
with Lord Cherwell, Sir John Anderson, and Churchill concerning
renewal of interchange of atomic information between the two nations.

Bush realized that the British wanted an agreement that would extend into peacetime and would include commercial applications in addition to weapons development. Both Bush and Stimson felt that any agreement should cover only wartime cooperation. They believed that an agreement regarding postwar problems was beyond the president's war powers authority.

The position of Groves and Conant was to limit cooperation strictly to those items that would assist our wartime atomic bomb effort. Negotiations had reached the point where a Bush-Anderson draft agreement had been put together that included a British commitment to reservations satisfactory to Bush. His only worry was that Roosevelt would not stand firm for the commitments to which Anderson had agreed.

Most of these negotiations were unknown to me when I arrived in Groves' office. I learned about them later. I also suspect that Groves was not as fully informed as he would have liked. He told me that the purpose of the report was to ensure that when Roosevelt met Churchill the president had full information about the status of the atomic bomb project so he would not make any unnecessary concessions.

When we finally finished the report, instead of having Bush sign it as chairman of the MPC (which was the normal procedure), Groves signed it. I then delivered it to Bush the next day. When he finished reading it, he commented to me, "Okay, but it won't do much good." I got the impression that he was a little put out by the president's not inviting him to go to Quebec. In any case, I then took the War Department courier plane to Quebec and briefed General George C. Marshall, chief of staff, U.S. Army, on the content of the report. This was my first meeting with General Marshall. He was friendly but cool. He listened very attentively, asked me a few questions, and then said, "I think you are too late. They are all out on a boat ride now. I will deliver the report to the secretary as soon as he returns. You stay here until I tell you to go back to Washington."

I was glad to have the rest of the afternoon to pursue my interest in military history by visiting the Plains of Abraham, site of the historic battle between Wolfe and Montcalm. The next morning, General Marshall told me that although he had given the report to Secretary Stimson, it was too late, since Roosevelt and Churchill had already discussed the matter. I doubt if the report was ever shown to Vice President Wallace, the third member of the General Policy Committee, and it never reached the president.

The Quebec Agreement to provide for top U.K., U.S., and Canadian cooperation established a Combined Policy Committee (CPC) consisting of Stimson, Bush, and Conant for the United States; Field Marshal Sir John Dill and Colonel J. J. Llewellin for the United Kingdom; and C. D. Howe, the minister of munitions and supply, for Canada. It provided for interchange of information, pledges never to use the weapon against each other, never to exchange information with a third party without mutual accord, and never to use the weapon against a third party without mutual agreement. Information concerning commercial or industrial use passing to the United Kingdom would be limited to what the president considered fair and equitable.

The first meeting of the CPC was held in Washington on September 8, 1943. One of its first tasks was to arrange for British scientists to participate in the U.S. operations and to reach an agreement on the interchange of information. Some twenty-eight British scientists were assigned to the Manhattan District laboratories. A subcommittee consisting of James Chadwick for the United Kingdom, C. J. Mackenzie for Canada, and Tolman for the United States was appointed by Styer (acting for Stimson as chairman of the CPC) to work out arrangements for British scientific participation and rules for exchange of information. Initially, Los Alamos, the centrifuge project, the gaseous diffusion process, the liquid thermal diffusion process, and the heavy-water project were included in the arrangements. Another subcommittee of Chadwick, Mackenzie, and Groves was responsible for establishing the rules and conditions for exchange of information with the technical effort in Canada, later to be centered at Chalk River.

Groves assumed full responsibility for supervising the exchange of information but arranged for Peterson to carry out its operational aspects, including making periodic visits to McGill University in Montreal, where the technical effort was being carried out at the time. Also, visits by U.K. and Canadian technical people to various U.S. sites were led by district personnel, particularly Peterson and Traynor.

Churchill's support and close contact with Roosevelt helped maintain top priority for the Manhattan Project, and some of the British scientists made significant technical contributions to the bomb development at Los Alamos and to the design of the gaseous diffusion plant. Our support of their effort on heavy water at Chalk River was of considerable future value to Canada in their postwar development of heavy-water atomic power plants. But the cooperation also brought Klaus Fuchs into our midst and gave him the opportunity to convey important information to Russia. Groves had wanted to investigate all

British scientists, but the CPC denied him this request. The committee considered that the British should be responsible for any investigation of their own personnel.

After Quebec, I implemented the decision Groves and Marshall had made to transfer the Manhattan District headquarters from New York City to CEW. By early September, we had completed the relocation. Jackie and I drove to Oak Ridge, arriving on October 3. Soon Jackie was immersed in the sea of mud that Oak Ridge often became in its early days. She quickly became involved in the many new and unique experiences and responsibilities that were part and parcel of being the commanding officer's wife in a rapidly growing town.

After the departures of Marshall and Blair I operated without an overall deputy. Lieutenant Colonel Earl H. Marsden was my executive officer. Continuing the principle that Marshall had established, I chose a man for each assignment whom I felt had the ability to make decisions and carry them out expeditiously. Once selected, I trusted him to do his job unless he proved deficient. I believe that giving a man such freedom of operation creates an atmosphere in which he will strive to live up to the confidence vested in him and usually achieve more than even he thought possible. This became even more crucial as we implemented compartmentalization throughout the project.

Groves ardently supported compartmentalization of the atomic bomb project. He did not want too many individuals having complete knowledge of all the work. He believed compartmentalization was the only way to maintain secrecy to the maximum possible extent. Moreover, knowing the inquiring minds of scientists, he felt they would spend too much time prying around into other parts of the project, and he wanted to keep everyone's nose to his own grindstone.

Compartmentalization required that individuals in each major project be restricted to classified information for that project only, and within the separate projects limited to what they needed to know. The scientists did not favor the idea, as it directly contradicted the belief that science thrives on the free exchange of all information. At Los Alamos, they insisted on complete open discussion of all aspects of bomb design and won the support of Oppenheimer. Groves reluctantly approved it. (This undoubtedly aided Fuchs' espionage efforts.)

While they won their local argument, they did not have full access to information on the production of U-235 and plutonium. Compartmentalization was more readily enforced in regard to the production of fissionable and feed materials.

With the MPC establishing policy, the S-1 committee became less important and was eliminated by the middle of 1943. Groves then adopted the policy that only Conant, Tolman, Compton, Lawrence, and Oppenheimer among the scientists should have full access to all parts of the project.

After the war many scientists cited instances where they claimed critical information had been withheld from those who needed it. In practically all the cases I have checked, the author did not know that the knowledge he thought had been withheld had, in fact, been passed on to the organization needing the information. Conant, Tolman, Compton, Lawrence, and Oppenheimer saw to it that interchange of scientific information was provided when necessary. Likewise, Groves and I were alert to arrange for access where beneficial. As a result, in my opinion compartmentalization did accomplish Groves' objectives and certainly did not jeopardize the success of the project. However, it did annoy many scientists.

To carry out the mission of this widely spread and highly complex organization, compartmentalization increased the need to make more personal contacts to coordinate the various offices. As a result I was at CEW, New York, and at least one other office or site every week. In anticipation of the move of the district headquarters to CEW, we had split the varied New York functions into several area offices as part of the compartmentalization policy. The former district headquarters became the Madison Square area office responsible for all uranium feed materials and development of production methods for other special materials not commercially available, such as elemental fluorine, with John Ruhoff, now a lieutenant colonel, directing the operation. I retained an office and conference room there for my use when I visited New York, with Anne Phillips remaining as my secretary.

The New York area, under Lieutenant Colonel James C. Stowers, handled the Kellex contract for design and procurement of all components of the gaseous diffusion plant, K-25. The research on the gaseous diffusion process continued at Columbia University and the SAM laboratory, now all designated the Columbia area, with Major Benjamin K. Hough, Jr., as area engineer. The Murray Hill area, under Major Paul L. Guarin, administered a contract with the U.S. Vanadium Corporation to survey uranium resources in the United States, Canada, and several other countries.

By September 1943, a total of 175 officers were serving at the various MED offices. At the end of the year, the total exceeded three hundred officers and continued to grow. We did acquire a few more regular Army officers, such as Captain Peer de Silva, security officer at Los Alamos, and Captain Mahlon E. Gates, whom I brought to Oak Ridge after he returned from the Burma Theater because he was the sole surviving son of his family. But for the most part, we staffed the Manhattan District with the Army reserve officers as mentioned before, civilians given direct commissions, naval officers, civil service employees, Women's Army Corps personnel, and a Special Engineer Detachment consisting of technically trained enlisted men.

Before I had moved to Tennessee, the most urgent matter at CEW had been to appease Governor Prentiss Cooper of that state. I had to pick up where Marshall had left off on July 20. We needed the governor's help, particularly in dealing with the general problem of access roads. The road net around CEW was inadequate to support the volume of commuter traffic in and out of CEW every day. At our peak of construction over forty thousand workers were commuting from locations as far away as Chattanooga. We were also having an immediate problem with a judge in neighboring Anderson County; the judge was threatening to close the new Solway bridge because it was no longer useful to the county except for access to the CEW. The county's taxpayers were protesting that they were paying off the bonds for a bridge that was now benefiting only the U.S. government. In addition, we needed at least one new road all the way to Knoxville to supplement existing highways.

On July 27 I notified Groves that I would be at the CEW on July 29 and 30 and that I had made an appointment with the governor for 10:00 A.M. on Saturday, July 31, to attempt a reconciliation. Groves sent word that he would come down on Friday for the day to help work out with me what part of the improvements the U.S. Government could pay for and what we hoped the state of Tennessee would finance. Groves returned to Washington on the evening train, leaving me to deal with Governor Cooper.

I arrived a little before 10:00 A.M. on Saturday at the State Capitol in Nashville. In view of Marshall's experience, I was prepared to wait. But Governor Cooper met me at the door as soon as his aide announced that I had arrived. He offered me a chair close to his desk, and I began by explaining that I was succeeding Colonel Marshall as district engineer. I said I wished to apologize for our having failed to contact him much earlier. I told him that the fiasco of the first contact resulted from our failure to instruct properly the very junior officer we had sent over to Nashville.

I emphasized the tremendous importance of the CEW and stressed the magnitude of the project, the expected employment, and the anticipated size of Oak Ridge. I discussed our traffic problems in getting workers in and out of the CEW. I stated that I wished to make amends for the affront to him. I continued, "Governor Cooper, if you could visit the CEW at your earliest convenience, General Groves and I will personally escort you over all parts of the project, including the town. However, I cannot invite you inside any of the production plants under construction because of the extreme security. I hope you understand."

I then outlined how we proposed to get government funds for a new direct road between Knoxville and the CEW. In turn, the governor raised the question of the bridge and the local political problem that had developed. I assured him that I thought I had a solution that would be satisfactory to the judge and the county. I concluded by saying, "Governor Cooper, when you visit Oak Ridge, General Groves and I hope you will allow sufficient time to accept an invitation to my home for a reception in your honor. It will give you an opportunity to meet the chief contractor personnel and local county officials and perhaps deal with some of their local problems caused by our project."

When I had completed my case, the governor let me know in rather frank language what he thought of our negligence in failing to make an earlier and more effective contact with him. He was particularly vitriolic about Lieutenant Colonel Thomas Crenshaw, the commanding officer at the CEW, stating that "he was just another of those damn Yankees who thought they could ride roughshod over Southerners." Cooper went on to say that Crenshaw was not a gentleman or he would have made personal contact earlier and also to discuss the military district order, which Cooper thought was unconstitutional. He said, "The least you should do is fire the commanding officer."

After continuing on in a similar vein for a time, he asked me what university if any Crenshaw had attended. Having done my homework, I had been patiently waiting for this question. I replied, "Princeton," adding that I had always thought the university did a pretty good job of turning out well-educated and cultured graduates. He smiled and said he agreed, acknowledging that he had graduated from Princeton.

Seizing the moment, I replied, "Crenshaw is one of the best men I have, but if your terms for peace are firing the commanding officer, I will do so. Nevertheless, it will be a great loss for the project and I will have great difficulty replacing him with anyone equally competent." Letting the matter ride for the moment, I turned to the problems we had concerning uncertainties about jurisdiction, cooperation with the surrounding counties and nearby Knoxville, the size of the project, and the economic benefits to Tennessee the project would provide.

After some additional conversation, the governor indicated that he would visit CEW and that I should arrange with the secretary of state's staff the date and the plans for meeting him in Knoxville. He added, "I will cooperate and help in any way I can to assure the success of your project." Thanking him, I asked, "Do you still request that I fire the CEW commanding officer?" He replied that he still thought the man should be fired, but if I thought he was essential, I should do what I thought best. I answered, "Governor Cooper, Crenshaw is a valued assistant and it is best that I retain him."

Grateful for the more favorable prospects for cooperation with the state, I turned to other business. Administering Oak Ridge proved a fascinating, demanding, and difficult problem throughout the war. I deliberately put several layers of organization between me and the residents to cushion myself from involvement in the day-to-day problems of town management. I simply could not spare the time from my primary mission. The overall policy I was determined to establish was that Oak Ridge approximate a normal midwestern town—at least as normal as conditions at a secret installation under Army control would allow. I resisted all efforts to introduce novel ideas advanced by some people, including such things as a unified religion, self-government, and experimental educational methods.

Governor Cooper accepted the invitation to visit Oak Ridge for November 3. I had gone to Nashville on October 28 to make the final arrangements for the meeting and to sound out Cooper's attitude concerning alcohol. Anderson County, in which Oak Ridge was located, was dry, and the county judge was expected to be a guest. The Tennessee secretary of state told me that the governor appreciated good bourbon and suggested I serve punch at the reception. "Make it real good and no one will ask about the contents," he said.

Groves had difficulty with the date set and had to postpone a Military Policy Committee meeting, but given the importance of the visit, assured me that he would arrive on the morning train. Together, we met the governor and his party at the Andrew Johnson Hotel and then proceeded to the CEW, accompanied by a police escort.

Just before we reached the CEW gate, our motorcade was passing four huge LeTourneau scrapers bound for the K-25 plant site. At that moment, the lead police car blasted its siren. This startled one of the LeTourneau drivers, and as our car passed his scraper, it toppled off the edge of the road and down the steep bank. The scraper rolled over a couple of times, the driver scrambling to stay on the top side. When it stopped, he scrambled up the hill. Quite a spectacle!

I promptly ordered our driver to stop and went back to see if the man was injured. Miraculously, he wasn't. I invited him to meet the governor. As we headed on our way, the governor commented, "If that was staged for me, don't do it again. I am fully convinced that you need better access roads." I assured him that the incident had not been staged.

We showed the governor all the construction, the workers' camps, the central facilities, and the town of Oak Ridge, ending up at my house, where Jackie had arranged a reception for about thirty people, including the governor's party, the local officials, and key contractor leaders.

Not having a large punch bowl and being unable to borrow one, our cafeteria operator had suggested to Jackie that he carve a punch bowl out of a large block of ice. When I arrived with the guests, she whispered to me, "It looks beautiful. I'd like to be with you when you show it to the governor."

After the necessary introductions all around, I invited the governor to go into the dining room, where the table had been set with the beautiful ice bowl as the centerpiece. Just as we entered, the bowl sprang a leak and the punch began pouring out toward us. The governor stopped, looked at it, and a smile came on his face for the first time. "Now I have seen everything," he said. "You have a most impressive town and fantastic plants, but this caps everything. I've never in my life seen bourbon flowing so freely." The incident literally broke the ice.

Prior to arriving at the house, the governor had been fairly quiet. But from then on, he seemed to be more relaxed and gracious and willing to discuss our mutual problems. As a result of his visit, we had little trouble negotiating with the state for road improvements and a completely new access road. Also, we later came to an understanding with the county judge about his bridge. We rented it for the duration of the war. Groves questioned this solution but offered no better one. He called it blackmail. My only comment when he criticized my decision was to say, "I have to live with these people; you don't." Fortunately, the GAO never took exception to the payments.

As time passed and expenditures grew, we occasionally had to bend our desire for complete secrecy to brief some government officials and top military officers who had reason to know something about what was happening at the CEW. The way we had handled Governor Cooper's visit became our standard operating procedure.

In the case of Senator Kenneth McKellar, however, I did not even have to provide such a VIP tour. We learned that his primary concern seemed to be about his political image. He knew nothing about the project and could not answer questions his political allies asked about it. He requested that I meet with him in private to give him an accounting of what was taking place at the CEW. As a result, I arranged to meet him in Knoxville.

We met at the Andrew Johnson Hotel. He excused all his visitors, and I spent about half an hour explaining that the CEW was a vital war project. I told him that the extreme secrecy was ordered by the president himself. I gave him an estimate of the number of employees and the expected population of the town and expressed my opinion that the end of the war would not terminate all employment there. He seemed to be content with the explanation and asked that I meet with him occasionally in Knoxville to discuss any political problems he might have in the future concerning our operation. I agreed. Before I left, he apologized to his visitors and explained that because of secrecy, it had been necessary to exclude them from a very interesting discussion.

In regard to the town of Oak Ridge, I put Captain Tim (Paul E.) O'Meara in charge of administration. Tim was a big, outgoing, handsome Irishman. I first met him at the Rome Air Depot. He was a cement salesman trying to sell the project on concrete runways. He succeeded. Shortly after Pearl Harbor, he was called to active duty. Later, Marshall selected him for duty with the Manhattan District, and it proved to be an excellent choice.

We decided that the district should not try to staff and manage all the required facilities in the town. Consequently, we contracted with Turner Construction Company to take on town management and operations. George Horr, a vice president of Turner, had been the project manager for construction of the Rome Air Depot, so Marshall and I were well acquainted with both him and the company.

Turner organized a special company called Roane-Anderson (named for the two counties involved in the CEW site). They organized and operated the bus lines, garbage collection, the school system, the hospital, management of all the housing, the hotel, the fire department, central eating facilities, and practically everything else pertaining to town operation, including delivering coal to individual houses. Starting in the spring of 1943, when the first housing was completed, the town grew constantly, until at the end of the war it had become the fifth-largest city in Tennessee, with a population of about seventy-five thousand.

The school system posed a special problem. The school population increased every day. Additional teachers and school buildings had to be made available as needed for children arriving at Oak Ridge from all parts of the United States and all having different backgrounds and educational needs. Dr. Alden M. Blankenship headed the school system, and throughout the war he did a masterful job of recruiting personnel, supervising the teachers, maintaining discipline, and rendering a most satisfactory service. Few complaints ever reached me about the school system. The school system was legally a part of the Anderson County system. It was built and operated primarily with federal funds. It was not eligible for state aid.

For all town problems, the residents were advised to call Roane-Anderson. I really caught hell one day when Jackie phoned my office. Miss Olsson told me that Mrs. Nichols was stuck in the mud and wanted to know what to do about it. Busy with a problem, I adhered to my own orders and simply said, "Tell her to call Roane-Anderson, like everyone else has to do." That was the last I heard about it until I returned home for dinner. Then I heard plenty.

The medical care at Oak Ridge was one component in the overall health program of the Manhattan Project. To provide essential medical services to the permanent populations of Oak Ridge, Richland (at the Hanford facility in Washington State) and Los Alamos as well as to the construction workers and operations personnel, we built hospitals and manned them with Army medical officers, most of whom we recruited directly from civilian positions. We were well aware of dangers of radiation, and consequently we initially recruited many doctors who were involved in radiation research as civilians so as to have expertise readily available in case of emergencies.

Dr. Hymer L. Friedell became the first medical officer in the district. He joined the met lab in August 1942 from the University of

California, where he had been doing radiation research with Dr. Robert Stone. I encouraged Friedell to accept a commission and work directly with the MED. By the time he moved to Oak Ridge in the spring of 1943 to take over direction of our medical requirements, it had become clear that we would be needing a fairly large contingent of medical personnel in the district. We had to consider not only the health care of the workers at the various sites, but we also had to conduct research on the hazards radiation posed, to deal with safety considerations within the plants, and to prepare for the effects of the weapons when they would be tested and used. While Friedell had a knowledge of the new field of isotopes not common to young doctors, he recommended that the Manhattan Project have a more senior person in overall charge of the medical program.

When Dr. Stone indicated that he preferred to remain a civilian, we turned to Dr. Stafford L. Warren, then a professor of radiology at the University of Rochester School of Medicine. In March 1943, Friedell met with Warren in the district office, and without explaining why, asked him questions concerning shielding against radiation, protection against radioactive dusts, safe standards of exposure, and other related matters. Warren replied that no experimental data existed upon which to provide answers to most of the questions. As a result, Friedell proposed that Warren conduct experiments, in secret, at Rochester to provide the necessary information. Once Warren produced an outline of the specific experiments needed, we allocated him the funds required for construction and research. Until October, he served as a civilian consultant to Friedell and the district engineer on the overall health program and on "special materials."

Meanwhile, on August 10, I created the Medical Section, MED, with Major Friedell as executive officer. In early October, Warren reported to General Groves that the organization of the Medical Section was essentially complete and that he could no longer function effectively as a civilian. Consequently, Groves drafted a letter, in his usual style, to the Army surgeon general, Major General James Kirk, directing him to cooperate with the Manhattan District. Signed by Somervell, the letter stated in its first paragraph that the district engineer of the Manhattan District was responsible for the health of his command. Groves, as was his wont, asked me to deliver the letter personally to the surgeon general.

After reading the first sentence, Kirk hit the ceiling, stating, "This is an illegal order. I am responsible for the health of all Army personnel.'' As he continued to read, he continued to become more and more irritated. According to the letter, he was directed to cooperate with the Manhattan District, furnish all medical supplies we needed, provide funds for medical care for military personnel, and commission individuals we selected at the rank we designated. He was also to select one man in his office as liaison officer to facilitate cooperation. The officer—and Kirk could designate himself if he chose—was to be the only one cleared to receive classified information about our project.

Kirk put down the letter and launched into quite a discourse on what he thought of engineers and their high-handed tactics. He again repeated that it was an illegal order. At that point, I managed to stop him and asked if I should take the letter back to General Somervell and report to him that the surgeon general had called the order illegal, or should we try to work out agreeable arrangements for means of cooperation. I told him that our high order of secrecy resulted from a direct order of the president and that except for recruiting our own personnel, we were prepared to cooperate fully with his office and not make any unreasonable demands.

Also, I told him we would request only one doctor to be commissioned a colonel, with the rest being majors or captains. He wanted to know if I could tell him the name of the proposed colonel. I said it was Stafford Warren of the University of Rochester. He responded, ''Why do you want that clap doctor?'' He may have heard a report that Warren was experimenting with radiation to combat venereal disease. In any case, not wanting to explain why we had selected a radiologist, I said only that we had chosen him for good and sufficient reasons. Resigned to the *fait accompli,* Kirk named Colonel Arthur B. Welch, MC, to act as our liaison officer, and I then continued my discussion with Welch. Despite this somewhat explosive beginning, I must add that we subsequently had full cooperation from the surgeon general throughout the war and afterward.

On November 3, Warren received his commission as a colonel and was appointed chief of the medical section, MED, and medical adviser to the commanding general, Manhattan Project. The office moved to Oak Ridge later in November and operated with Major Friedell as deputy chief. Warren and Friedell recruited the highly qualified doctors necessary to staff the organization to ensure radiological safety and to help staff the hospitals at Oak Ridge, Hanford, and Los Alamos. Dr. Charles E. Rea from the Twin Cities headed our CEW hospital and proved to be exactly what we needed. An outstanding surgeon, he had, in addition, administrative ability and above all a personality that appealed to everyone. He not only maintained health at CEW but helped preserve the morale of the families as well.

One of the problems we faced had to do with the use of Army facilities and personnel to care for civilians not part of the service. We knew that we should charge for whatever service we rendered, and in those days it was expected that everyone had to pay any bill rendered by the government. With construction labor, in particular, we expected a large turnover of workers. After much discussion with the doctors, they reluctantly came up with a medical insurance program. A single person would contribute $2.50 a month, and a family would pay $5.00. A board of trustees supervised the program, and by good management and good service, Charlie Rea was able to keep the books balanced and the residents reasonably content. In this, he received help from the staff of superior doctors he had selected, many from his hometown area. Many of them had known each other and so were able to work well together as a team.

Charlie Rea did a tremendous job at the Oak Ridge Hospital. Even when time was scarce, he greeted everyone with a smile. Everyone liked him. He loved children, and he and Mary continued to enlarge their family while at Oak Ridge. Rea encouraged others to do likewise. "They grow up so fast—be gone before you know it." As a result, the obstetrics ward remained very busy. During its first three years of operation, 2,910 babies were born at the hospital. For other than births, statistics for the incidence of illness were relatively low for the size of the population, probably because of the low average age of the residents. Although some indoctrination for recognition of radiation illness was given to the staff, no cases involving radiation were encountered during my years there.

Dr. Warren organized the Medical Section of the MED to cover three main responsibilities. First was the coordination of the biomedical research programs, mainly conducted at the University of California and at the met lab at the University of Chicago. Next was the dissemination of this information, which helped expedite development of safety programs we introduced at various stages of the production project. The Medical Section also developed industrial safety procedures required for individual on-site and off-site contractors in various locations as well as the inspection program that ensured their observation. Finally was the assurance of medical care and public health protection to populations at Oak Ridge, Richland, and Los Alamos.

Throughout the war, Staff Warren did a magnificent job of developing standards and safeguarding all our operations. As Friedell later noted, Oak Ridge was "Clearly pretty Spartan and somewhat substandard, but not bad. It certainly wasn't as bad as being under combat; besides, everyone had his family with him. Nevertheless, the relatively primitive environment could have adversely affected the supervisory contractor personnel and the doctors who had given up their careers to come to Oak Ridge."

Warren and his wife, Vi, were helpful in alleviating any ennui and took pains to assist Jackie in making the social life at Oak Ridge more pleasant. Mrs. Warren became president of the Women's Club, and Mrs. Friedell worked with the Red Cross.

The construction and operation of the headquarters and CEW central facilities and the town of Oak Ridge were major undertakings in themselves. During the war, some seventy-five thousand people were housed in dormitories and family housing. To accommodate this population, we had to construct from scratch and then operate a whole city's infrastructure, a sewage system, water supply, and roads, and create fire and police departments as well as provide theaters, churches, stores, libraries, schools, offices, cafeteria, hotel, laundry service, etc.

From the start of construction in November 1942 until December 31, 1945, construction costs totaled $101,193,000 and operational costs totaled $50,446,000. We employed Skidmore, Owings & Merrill to assist Stone and Webster in the layout of the town and the planning and design of the communal facilities and the houses and apartment buildings. Most of the houses were constructed in lots of a thousand through contracts let by competitive bidding. Captain Edward J. Bloch and Crenshaw played a major role in the central facilities and town planning and in selecting the type of houses constructed.

On my last visit to Oak Ridge, for its fortieth-anniversary celebration, I was pleased to see how attractive the original community set in the hills had become. Many of the homeowners had improved the landscaping and added new siding or brick facings to improve insulation and appearance. Some built additions or garages. I found to my surprise that Oak Ridge had become a comfortable retirement community and was glad that we had laid out the town along the ridges instead of in the lower and flatter areas. Perhaps this cost a bit more than absolutely necessary (certainly more than Groves would have liked), but I know it more than paid off both in morale and in recruitment of professional workers during and after the war. The icing on the cake, as far as I am concerned, is that four decades later, residents live in Oak Ridge by choice, and some who came in the early days still were there forty years later. The "43 Club" invited Jackie and me to attend their dinner celebrating the fortieth anniversary of Oak Ridge. It was fun hearing old acquaintances reminisce about the "old days." Many of the tales I had never heard before.

8

CONSTRUCTION: THE SPECTER OF DELAY, 1943–45

FROM THE SPRING OF 1943 to the end of 1944, a tremendous construction effort ensued at all the Manhattan Project sites. In addition, the first production of plutonium and U-235 began. But I remember the period most for the major crises that arose one after another in all phases of the project, with the exception of the production of feed materials. Through Ruhoff's capable leadership and the competency of his contractors, that program seemed to meet all its requirements on schedule.

At the CEW, the construction of the central facilities, family housing, schools, commercial facilities, dormitories, and barracks or hutments for construction workers became an ever-expanding program. For temporary housing, we rounded up every trailer we could locate from every part of the United States. A separate construction camp, which we hoped would live up to its name, Happy Valley, was located near the gaseous diffusion plant to lessen travel time for the seventeen thousand construction men working there.

The Administration Building had been completed early in 1943, and when I moved the district headquarters there in August, it became the main center for control and administration of district activities. We ultimately completed more than three hundred miles of paved roads throughout the reservation, including a four-lane highway down the center of the main valley connecting the town of Oak Ridge with the gaseous diffusion plant (K-25), within the town site, and branch roads to the electromagnetic plant (Y-12), the plutonium semiworks (X-10), and to each of the five security entrance gates. We also had to build a complete water purification plant and two sewage treatment plants. The electric distribution system, telephone, water, and sewage systems had to be expanded continually to meet the exploding population of Oak Ridge. The railroad connections to the L & N and the Southern Railroad were faced with hauling the millions of tons of supplies needed for the three plants, the central facilities, and the town.

During the month of May we finished 1,288 family units, which brought the total to 7,041 residences, exclusive of 113 efficiency apartments. Eighteen dormitories were finished during the month, completing the dormitory program. General site facilities, utilities, and community buildings were 92 percent complete. The rapid expansion continued with the daily arrival of families. Their children poured into our school system, which was already bulging at the seams but had to absorb the ever-increasing number of students. Every time we expanded the plants, we had to authorize more houses and town facilities of all kinds to provide for the welfare of the new influx of workers. We even built a jail, but then had to face the jurisdictional question of whether we had the authority to lock the door.[1]

Work on the plants themselves provided an ongoing challenge during this period. Construction of the electromagnetic plant generated unexpected and difficult problems. The first alpha building with the equipment for the first racetrack was completed and placed in operation on November 13, 1943, my thirty-sixth birthday. However, as a birthday present it proved to be a real dud, as its operation ran into difficulties immediately. Operators could not sustain a beam in the calutrons. Vacuum could not be maintained because of leaks. Some of the tanks pulled loose due to stress caused by the magnets being built in the form of an oval. However, the worst problem was that the magnets shorted out, causing us to stop operations completely to locate the cause.[2]

Groves was at his best or worst, depending on one's point of view, in pinning down the cause of the trouble. He was extremely disturbed about the situation, and his caustic comments and persistent inquiries irritated practically everyone. Even Ernest Lawrence received some sarcastic comments when he recalled that he had experienced a similar problem with his cyclotron. The stress, long hours of work, frequent meetings, major but honest differences of opinion, and flaring tempers generated hard feelings between certain of my construction officers and

1. Narrative Monthly Report on DSM Project, June 8, 1944; KDN files, OCE; Groves Files, National Archives.
2. Ibid., Monthly Report, January 10, 1944.

Stone and Webster's top personnel. The overall result was that we failed to get the first unit operating on schedule. Furthermore, we realized that more stringent quality control methods were necessary and that closer and more detailed supervision over design and construction was essential.

Groves insisted on the need for a more aggressive construction project manager. Consequently, Groves and I met with Mr. Lotz of Stone and Webster, who agreed to hire Frank C. Creedon, who had been with the national Synthetic Rubber Program. Shortly after Creedon had familiarized himself with the project, he called me to recommend that top construction management on the project—Army and Stone and Webster—be replaced by personnel with more experience. Stone and Webster took action, as did I. I hated to lose two good men, but the move was necessary. However, I gained two new very exceptional men from other districts, Lieutenant Colonel John S. Hodgson and Major Walter J. Williams. The decision to send the magnet coils back to Allis-Chalmers for reconditioning by no means ended our troubles at Y-12. We found that we had to replace most of the cooling system piping, clean, and replace it.

Meanwhile, on January 27, 1944, production started on racetrack 2. Again we had trouble. Operation of the second unit remained very erratic, components failed, and after starting to produce enriched U-235, we soon found that the chemical recovery of the enriched uranium was far below estimates. However, by April 1944 we had four alpha tracks in production and one beta track in operation for personnel training.

Tennessee Eastman had eleven thousand employees, including eleven hundred trainees. Nevertheless, we were faced with additional problems, which contributed to a feeling of uncertainty: electrical failures, low chemical recovery, spare parts problems, cracked insulation, and vacuum leaks. Maintenance men were harassed by the number of minor failures. Morale was sinking. Many doubted that such complicated equipment could ever perform reliably. Maintenance difficulties had been predicted, but no one really anticipated the number and extent of our vexations.

Lawrence, however, was undaunted by any uncertainty. He remained optimistic, pointing out that we had made great accomplishments. He believed that his team of skilled operators from Berkeley would find methods of operation that would solve many of our difficulties. By March 1944, some two hundred grams of 12 percent U-235 had been produced and some of it sent to Los Alamos for experimental use, with the entire experimental order shipped by May 2. In addition, during May the first beta track began production, and two shipments of 60 to 65 percent material were made to Los Alamos in June.

With the gaseous diffusion project still having serious barrier problems, Lawrence started a pitch for another electromagnetic plant expansion. In early March, he told me that he wanted four more alpha tracks. Consequently, I was surprised at a March 30, 1944, Military Policy Committee meeting that Lawrence did not make any recommendations for expansion of his project. I thought that Groves and Oppenheimer were prepared to back an expansion. Later I learned that Lawrence thought he would not be able to get approval. No one else raised the issue, probably because Lawrence had concentrated on the difficulties he was having with the plant and the problems for preparing for using gaseous diffusion plant product as beta feeder material.

Sometime later, Lawrence and Marcus L. Oliphant (a physicist furnished by the United Kingdom from Australia) renewed their campaign for expansion. Lawrence again repeated his sales pitch to me, and I knew he discussed the subject with Groves and Conant. Oliphant wrote to me from California about the British view that the gaseous diffusion plant would be of little significance in 1945, something I had already heard directly from the British. However, Lawrence's plans for expansion changed as Stone and Webster brought General Electric into the planning and G.E. gave greater consideration to the relative time it would take to improve the existing units compared to constructing new, improved units. In our monthly report for September 1944, I included this statement: "Development of the alpha-3 design at Berkeley has been largely discontinued, present indication being that a more timely contribution to production can be made by concentrating on the improvement of the present plant performance."[3]

More problems were yet to come. Chemical recovery of the enriched uranium from the beta plant was disappointing to say the least. At times it seemed that essentially nothing was coming out of the spout. Obviously the system required modifications. At one point, components and piping of the chemical plant were cut out and dissolved in an attempt to determine where the holdup was in the process line.

Ruhoff and Charles A. Thomas, head of research for Monsanto,

3. Ibid.

who was already coordinating metallurgical work at the various laboratories and Los Alamos, reviewed the whole Tennessee Eastman Chemical process. These two and James C. McNally, Tennessee Eastman's head of chemical operations of the plant, agreed on ways to improve the process. However, Dr. Mees of Kodak would not concur. In the autumn of 1944, at a conference in New York with the concerned parties, Conant, Groves, and I decided to proceed without the concurrence of Mees. As part of the decision, we had Kelley and Ruhoff exchange places. Kelley became area engineer for the Madison Square area and Ruhoff became our operations officer for the electromagnetic plant. Tennessee Eastman then made modifications to improve the original chemical process and initiated design and construction of a new chemical building to incorporate an additional improved process.

In spite of all these complications, Tennessee Eastman's proposal to train operators with only a high-school education was working remarkably well. As each new unit was turned over to Tennessee Eastman by Stone and Webster, Ernest Lawrence, with a team of scientists from Berkeley, took over operations to eliminate the bugs in the unit. When they achieved a reasonable operating rate, they transferred the unit to Tennessee Eastman. This procedure not only eliminated normal operating bugs but also gave the scientists firsthand knowledge concerning needed improvements. As a result of my comparing production data for the various units, I pointed out to Lawrence that the young "hillbilly" girl operators were outproducing his scientists.

He claimed that this was because his men were experimenting on ways to improve operations. Thereupon I challenged him to make a production race between his Ph.D.s and the young local girls. He agreed and he lost. The girls won because they were trained like soldiers "to do or not to do—not to reason why." In contrast, the scientists could not refrain from time-consuming investigation of the cause of even minor fluctuations of the dials. This little contest provided a big boost in morale for the Tennessee Eastman workers and supervisors.

Time heals all wounds, it seems. Eventually all aspects of the electromagnetic operations showed a gradual improvement. At the end of 1944, the Y-12 plant was 95.3 percent complete. All the racetracks had been turned over to operations. Additional chemical processing facilities still under construction were expected to be substantially finished by March 31, 1945, with all equipment installations completed by June 1945. With the chemical operations making better progress, the electromagnetic plant was beginning to justify Lawrence's bounding optimism.[4]

Meanwhile, progress on the plutonium project continued at a uniform rate in 1944. But still we did not have a signed contract with du Pont. The company was insisting that we have a complete record of negotiations explaining every issue and the legal basis for any major exception to normal procedures. At long last, Roger Williams called me to say that everything was satisfactory and that du Pont was ready to sign the contract. However, he then requested that I send him the basis for my authority to sign and also General Groves' authority to approve. That request started a real flap in our legal section.

Originally, on August 17, 1942, Colonel J. C. Marshall had written me a letter designating me "as contracting officer with unlimited authority to execute contracts for all classes of procurements within the jurisdiction of the Manhattan District." I later found out that Marshall had based his directive only on Styer's verbal comment to him when he became district engineer of the Manhattan District. When I replaced Marshall as district engineer in August 1943, I received a written directive from the chief of engineers' office giving me the authority of a division engineer. This authority had rather low limits and required the chief of engineers' approval for any major contract.

The magnitude and complexity of the du Pont contract would require the approval of the under secretary of war. Moreover, Groves really had nothing in writing to indicate just what his contract authority might be. As a result, legal counsel from the district right up to the under secretary of war's office got busy and in a memorandum to General Groves on April 17, 1944, Under Secretary of War Patterson delegated to Groves the full contractual powers that the president had delegated to Stimson under the War Powers Act of December 27, 1941, "in connection with the work assigned to and coming within the jurisdiction of the Manhattan District." He then authorized Groves to "exercise such powers either personally or through the district engineer of the Manhattan District." The authority was effective retroactive to September 1, 1942.

In turn, in a memo on June 10, 1944, Groves officially delegated the same contractual authority to me, except that contracts involving

4. Ibid.

$5 million or more would require his approval. Groves and I were grateful that the issue had been called to our attention and resolved in such a comprehensive manner.

Despite this paperwork hassle, construction continued to progress. However, a multitude of major problems and some very disturbing crises loomed before us. An early requirement had been to provide uranium metal of satisfactory purity in the form of billets to be drawn down to the proper diameter. Under the direction of the met lab and Ruhoff and with a major contribution by Professor Frank H. Spedding at Iowa State College at Ames, Iowa, Mallinckrodt and other contractors were producing satisfactory slugs. Research and development for a canning process for these slugs for use both in the semiworks at the CEW and the production piles at the HEW turned out to be much more difficult than anticipated.

For canning the slugs in aluminum, we needed to develop a reliable production process that would be vacuum-tight and would withstand the high temperatures and intense radiation to be encountered in the piles. The met lab, General Electric, du Pont, and Alcoa were working on the process. Fortunately, it was completed just in time to start loading the pile at the semiworks on November 3, 1943. The pile went critical early on November 4 after we had loaded about thirty tons of uranium slugs in its twelve hundred air-cooled channels.

The pile operated beautifully and eventually reached a power level equivalent to 1,800 kilowatts, about twice the design capacity. We were jubilant. In December, 1.54 milligrams of plutonium were separated and shipped to the met lab for experimentation. Major Oswald H. Greager had a major role in the chemical separation of the initial quantities of plutonium. He was a du Pont chemist who had been called to active duty. He was essential to the project, so we had him transferred to the MED and made him available to the University of Chicago to continue activity at the semiworks.

In March there was great excitement when the first gram of plutonium was produced. I felt elated. We were now on the road to success. Compton and Whitaker invited Groves and me down to the semiworks to see it. Compton opened the glass door and reached in under the ventilation hood to pick up a small glass vial containing a single gram of green liquid. He handed it to me. Whitaker hurriedly picked up a stainless-steel pan and held it under my hand, saying, "For God's sake, don't drop it on the floor." As Groves reached for it, Whitaker grabbed it and replaced it safely under the hood.

When the gram of plutonium reached Los Alamos, we had achieved the primary objective of the semiworks. With the first gram and the rest that followed, Los Alamos was able to determine more accurately the properties of plutonium needed for weapon development. This research resulted later in a significant change we had not anticipated in the direction of the work. As for the Clinton laboratory, the name of the semiworks, it now was able to divert more time to basic research in physics and to the production of radioactive isotopes.

The primary mission of the met lab, of course, was to develop the piles and the chemical process for the production of plutonium. In addition, some of the scientists undertook the study of the possible use of radioactive materials for military purposes. Fermi, for one, was studying the question of radioactively poisoned foods. In May 1943, while in Washington, Oppenheimer learned from Conant that General Marshall had asked him for a summary report on the military uses of radioactive materials.

The problems inherent in instituting an offensive radioactivity plan ultimately dissuaded us from pursuing it. However, the preparations for the invasion of Europe forced us to consider whether Germany might be able to use radiation as a defensive weapon. We faced some pertinent questions that needed answers: Was Germany sufficiently advanced in reactor development to spread radioactive materials on the staging ports in England or on the beaches of Normandy? If so, how would our troops react to the threat?

General Groves moved decisively, again borrowing A. V. Peterson and sending him to Eisenhower's headquarters in England to discuss the situation. General Marshall had alerted Eisenhower to the importance of Peterson's mission before his arrival. I was interested to learn that Paul Hawley, my old doctor friend from Nicaragua days, now was serving as chief surgeon on Ike's staff and, of course, would be called in on the question.

Peterson recalls his assignment: "Following a meeting with General Eisenhower, and later with his staff, I was requested to prepare an operating plan for use in case Germany actually dropped radiation materials on the Allied forces. The plan, called Operation Peppermint, was worked out with Brigadier General George S. Eyster, G-3 of the ETO [European Theater of Operations] general staff. The plan provided for the detection of radiation and measurement of its dose rate, assignment of responsibilities and actions to be taken, lines of communication among the military commands, and technical and medical support from the U.S. mainland, including the availability of cadres of expert people for dispatch to Europe upon notice. All of the preparations for which we in the United States were responsible were made immediately upon my return to the United States. In this connection, any association with atomic weapon development was completely avoided in carrying out this assignment.

"The widespread distribution of photographic film was used as a handy means for early detection and location of radiation contamination. Directives for prompt reporting of unexplained fogging of film were issued by the Air Force and medical Commands. Separate medical directives also requested prompt notice of the occurrence of certain medical symptoms among personnel.

"In anticipation of the possible need for special radiation equipment, Colonel Nichols had earlier requested me to have appropriate instruments developed and fabricated for ready availability in Europe as may have been necessary. In Operation Peppermint, these instruments were placed with the Chemical Corps for operation and use, and so were available in England prior to the invasion. On June 6, 1944, they accompanied the invasion forces into Normandy.

"I returned to the United States shortly before D-Day. I carried back with me General Eisenhower's confirmation to General Marshall that preparations had been made by SHAEF to meet the possible emergency." Replying to my statement that Pete must have enjoyed his jaunt overseas, he continued, "I don't recall how thoroughly I enjoyed this assignment because it required an enormous amount of work. But I certainly was taken with the taut and firm organization the U.S. field commands had achieved at that time, and the speed with which decisions could be made and actions taken."

As we had expected, the invasion force, equipped with Geiger counters, encountered no radioactive material on the beaches nor as it advanced inland. The fact remains that military commanders must be aware of any and all eventualities. I felt and still do feel that these precautionary measures were justified, and our secrecy was not compromised.

The major production of plutonium took place at the HEW. The Manhattan Project had acquired the site in early 1943. Immediately, under the supervision of Matthias, construction equipment was mobilized and a central construction workers' camp had been built at the site of the small town of Hanford in Washington State. A large labor force was recruited and the construction was started on the town of Richland. Slim (Granville M.) Reed, an imposing man with wide construction experience and a drive like Groves, was the overall construction manager for du Pont. At Hanford Gilbert P. Church served as the du Pont project manager. Matthias now was deputy district engineer.

Du Pont released the first design drawings for the first pile on October 4, 1943, but construction already had started on the large pumping system needed to provide water to cool it. Construction on the first pile started October 10, just shortly before the first electromagnetic plant racetrack was ready for trial operation.

Initial plans called for six piles to be located on the southern bank of the Columbia River at six-mile intervals. Later the number was reduced to four and then to three. Likewise, original plans called for four (later reduced to three) tremendous concrete chemical separation plants in units of two south of Gable Mountain, about ten miles inland from the Columbia River. Radiation protection required thick concrete walls and development of remote control equipment for operation of the chemical process. These massive structures were called canyons. In addition, there were separate sites for maintenance buildings, radioactive waste storage areas, graphite machining, slug canning, and other operations.

The design of the remotely controlled equipment for operating the chemical plant was tremendously difficult. Great reliability was necessary to decrease maintenance requirements because all of the maintenance had to be done by remote control.

The operation of the piles required that the slugs be pushed out the rear of the reactor into a water basin for shielding and then loaded into a shielded vessel on a railroad car, which then was taken ten miles to the separation canyons. After separation of the plutonium from fission products and other radioactive "gunk," it could be processed further with little shielding. However, all the waste material had to be safely stored in large, underground tanks of special design to prevent leakage and to provide protection against radiation. All this was an enormously complex and difficult undertaking. The met lab and du Pont, aided by many other organizations, proved themselves equal to the task.

Once more, du Pont's practice of isolating their design, construction, and operation of the plants from the research organization caused additional discontent among the scientists at the met lab. Du Pont wanted the scientists to remain available and to come to Hanford only when needed. Again, we compromised in favor of du Pont. Only Compton received permission to go to Hanford whenever he desired. Other members of the met lab were allowed only if du Pont invited them or if Groves or I approved a request from Compton. However, as events transpired, du Pont had considerable need for scientific help, and Fermi in particular had to spend a great deal of time at Hanford. In addition, the met lab fully responded to du Pont's request for additional research to help solve some of the problems that developed.

Both expected and unexpected difficulties and serious crises did indeed arise. Although canning uranium slugs for the semiworks had been successfully developed on time, the slugs required for Hanford had to withstand more rugged operating conditions. The slugs not only had to be in a vacuum-tight aluminum can, they also had to be bonded to the can. The met lab, du Pont's Grasselli Chemicals Department in Cleveland, du Pont in Wilmington, and Harshaw Chemical Company all cooperated and worked on developing a reliable canning production process. Other solutions were tried, with du Pont even considering using unbonded slugs. Finally, in March 1944, Greenewalt concentrated the effort at Hanford. However, little progress was made until June, when I could report to Groves that the canning operations had been "sufficiently favorable to secure us an adequate supply of canned slugs for the initial charge of each pile."[5]

Compared to the canning problem, the procurement of very pure graphite to act as the reactor moderator and machining it to extremely precise dimensions caused us only occasional difficulties. Generally, the graphite procured from National Carbon met du Pont's rigid specifications on purity. Nevertheless, for some unexplained reasons, occasional shipments were contaminated. National Carbon did not give a satisfactory explanation for the variation, nor did the company correct it. Roger Williams of du Pont asked that I arrange for a du Pont representative to inspect the National Carbon plant to try to find the source of the contamination. I went to see James A. Rafferty, the Union Carbide vice president who was in overall charge of all our contracts with Union Carbide and its subsidiaries, including National Carbon.

Rafferty threw up his hands when I relayed du Pont's request to inspect the National Carbon plant: "Nichols, do you mean that you want me to okay a du Pont man entering the production area of one of

5. Ibid.

our plants? When I signed a contract with you, I never expected that I'd have to let our chief competitor enter any one of our plants.'' After I explained the importance of the request, however, we reached a compromise. Rafferty knew one of Ruhoff's chief assistants, Major Hadlock, who had been a du Pont employee before he was called to active duty. Rafferty trusted him and agreed that he would okay a request for Hadlock to enter the plant. Rafferty commented, ''If necessary I can hire him after the war.'' Du Pont also was satisfied with the compromise.

Hadlock solved the problem when he found contamination in one of the storage areas occasionally used. To handle the graphite, du Pont set up machining facilities at both the semiworks and at the HEW. The tolerances for machining the blocks of graphite were extremely precise, and the operations had to be conducted by skilled workers in an air-conditioned building to exclude all impurities.

In early July 1944, I flew into Pendleton, Oregon, on my way to the HEW, anticipating a long motor trip the rest of the way. I was agreeably surprised when Matthias landed in a two-seat Piper Cub and flew me back to Richland, saving both of us the long drive by car. Hanford then was about 60 percent complete, with the first pile expected to be in operation before the middle of August. The completion date for the other two piles had been advanced from March 1 to February 1, 1945.

Richland Village was almost complete. Twenty-seven hundred of the forty-three hundred planned family units and nineteen of twenty-five dormitories were finished. From the peak of 44,900 reached the previous month, the number of construction workers had begun to decline. Overall project design was almost 90 percent finished, and procurement of equipment and materials was about 95 percent complete. Canning operations appeared to be progressing at a satisfactory rate. Overall, from a construction point of view, Hanford was over the hump and slightly ahead of schedule. As a result, this period was marked by a spirit of great optimism.

On July 17, I received an urgent call from Groves to be in Chicago in the late afternoon. Following our usual pattern when a problem arose, we met with the concerned leaders, Conant, Compton, Fermi, Charles Thomas, and Oppenheimer. The news was alarming. Oppenheimer had abandoned all hope that the gun-type weapon could be used for plutonium. That discovery affected not only Los Alamos but also Hanford. The good news was that if the implosion weapon could be developed, less plutonium would be required and the capacity of

Hanford could be reduced. The bad news was that if the implosion weapon could not be developed, the HEW would be a total loss and the plutonium effort would be a failure and have to be abandoned.

At Los Alamos, there always had been some apprehension that because of the possibility of predetonation, plutonium might not be suitable for use in the gun-type weapon. Early in 1943, Seaborg had suggested that undesirable plutonium-240 might be formed in the production reactors. As the Clinton lab plutonium became available at Los Alamos, scientists verified that this would occur and that with the more intense radiation at the HEW, the presence of Pu-240 would be increased to the extent that plutonium could not be used unless an implosion weapon could be developed.

There was no question that the gun method could not be used for plutonium. Evidence was overwhelming. We discussed the possibility of using the electromagnetic method to separate the Pu-240 isotope from the desirable isotope Pu-239. However, plutonium could not be handled like U-235. Plutonium was more radioactive and more poisonous. Also, there was a smaller difference in weight of the isotopes. An extensive redesign of the beta stages would be necessary. It seemed unlikely that this could be accomplished in time to be useful in the war effort.

Development of implosion on a crash basis was the only possibility worth considering. Such a decision was in the bailiwick of Groves, Conant, Fermi, Thomas, and Oppenheimer. My concern was Hanford. What to do about it? We did not know how long it would take to develop an implosion weapon. I thought that we should proceed at HEW until scientists at Los Alamos determined it was absolutely impossible to develop an implosion bomb in time to be useful in this war. Not to continue building Hanford ran the obvious risk of determining that an implosion bomb was feasible, but then not having plutonium to fuel it. On the other hand, continuing to build would make our loss that much greater if implosion proved impossible. Groves again demonstrated guts by approving continued construction at the HEW at the same urgent pace. At the same time, the decision to develop the implosion bomb made it possible to reduce the number of piles and canyons to three.

Our best guess at that time was that sufficient U-235 for one gun-type weapon would be available about August 1, 1945. Groves was determined to drive ahead with the implosion weapon with the objective to beat the U-235 bomb. He still hoped we might have a plutonium weapon ready between March and July 1945. Both Compton and Oppenheimer thought this was possible, but Conant was not so optimistic. At the same time, we continued to plan on the gun-type weapon instead of implosion for U-235, in spite of the larger quantities of U-235 needed because Oppenheimer felt more confident that the gun type would work. We wanted to be certain that we had at least one type of atomic bomb.

On the Western Front, the Allies had landed in Normandy, had successfully broken through at St. Lô, and were headed for central France. We hoped to defeat Germany before August 1945. What intelligence we had concerning the German atomic bomb effort eased our fears that she might win the atomic race. Because of the time factor, prospects for use of the bomb against Germany were now remote, but Japan still remained a furious enemy.

Later Groves revised his timetable. Our revised date, August 1, 1945, still was optimistic and would require an overwhelming effort and considerable good luck to accomplish. As a result of all this, coupled with our problems with the barrier for the gaseous diffusion plant, the summer and fall of 1944 marked the low point in our expectations. I have often reflected that had the Manhattan Project not been protected by supersecrecy, it might well have been annihilated by a host of critics proclaiming it an impossibility. It would have been a plausible claim and hard to disprove as our troubles continued during the third quarter of 1944.

The completion of the first pile was delayed by preoperational testing. The pile was turned over to the operator personnel on September 13, 1944, and they started loading the canned slugs. On September 21, the control rods were partially withdrawn, and the initial power operation began. Power was gradually increased. After about three hours of operation, the power level suddenly started to decrease and continued to decrease until the reactor shut itself down completely. No one had anticipated this, and the cause remained a mystery. Something was poisoning the neutron activity. The next morning, the power level started to rise again until it reached the previous high level and then again declined. Xenon-135 soon was suspected as the culprit by the met lab. Walter H. Zinn soon confirmed in the CP3 heavy-water reactor that Xenon 135 was the poisoning agent. Previously the met lab had not predicted the possibility of such poisoning.

Fortunately, the error did not prove disastrous. In *Atomic Quest*, Compton explains it succinctly on page 192: "At this point, the plutonium program was saved by the careful conservatism of du Pont's design engineers.'' One of the early flaps between the met lab scientists and du Pont had occurred when the company had increased the number of tubes in the reactor by five hundred, about one third over the met lab's design. The scientists thought that du Pont had no confidence in their ability to calculate the proper design. Roger Williams again asked me to visit the met lab and support du Pont's decision to have a safety factor and to explain the difference between a scientist's estimate and an engineer's estimate. I, of course, understood their argument; a good engineer always computes the theoretical design requirements and then adds a safety factor.

As a result, I supported du Pont, and now we had five hundred extra tubes. The solution was to charge more tubes. Upon loading a hundred of the extra tubes, the pile operated successfully. Du Pont's conservative approach had paid off. We were lucky, and the incident showed why we needed many different skills and experiences and various types of organizations all working together as a team to gain success. With operating experience and after checking safety factors, du Pont decided to charge all 2,005 tubes in the second pile when it was completed and ready for operation. This would not only overcome xenon poisoning but also increase production.

The first batch of irradiated slugs from the first pile was dissolved in the chemical separation plant on December 26, 1944, and the first production run completed early in January 1945. Matthias shipped the first plutonium to Los Alamos by convoy, and it arrived on February 2. From then on, production increased steadily. Compton, at an early S-1 meeting, had bet that he would produce the first kilogram of plutonium in January 1945. He promised all of us a champagne dinner if he lost. He came darn close, but we warned him that after the war ended, we expected him to settle the bet.

Los Alamos reported that the product was okay except for some minor impurities that could easily be eliminated. Our February 1945 monthly report stated: ''With the exception of minor cleanup work in Richland Village, Hanford Engineer Works is completed.'' All three piles were in operation. Three of the separation canyons were completed. It was determined that due to a shorter cycle than was thought necessary, only two canyons would be placed in operation, with the third in reserve.

With Hanford producing the plutonium, it was now up to Los Alamos to complete the development of the implosion weapon. If the

Hanford production schedules were met, we would have sufficient plutonium for the first implosion bomb before we had enough U-235 for a gun-type bomb. However, it was considered necessary to test the first plutonium implosion bomb. The race was on.

During the autumn of 1943, we made progress on development of many components for K-25. Keith coordinated research and development with his design activities. He continued to have weekly meetings with Dunning of Columbia; George T. Felbeck for Carbide; and our area engineer, Stowers. The development of pumps and seals under Henry A. Boorse of Columbia, aided by the Elliot Company, the Sharples Corporation, and Allis-Chalmers finally produced results, and the tests were satisfactory. Under Willard F. Libby, chemist, more was learned about the chemistry of uranium hexafluoride and the corrosive rates of various materials as well as about materials such as the fluorocarbons that would resist corrosion.

At the CEW, for outside access to the Sleepy Valley site, the roads and bridges were improved. In September we leveled a large area for the enormous process building. Almost three million cubic yards of earth had to be moved and compacted. I questioned the need for this, but Dobie Keith soon convinced me that the foundation required for heavy equipment and the advantages of a uniform floor level for the entire area were ample justifications for having a large level area prior to starting construction. Moreover, no time was lost because of the status of the design. Foundations were started in late October 1943.

In addition to J. A. Jones Company building the roads, power plant, and main process building, Ford, Bacon and Davis Company was engaged to build several large auxiliary buildings to house assembly and testing of components and for later maintenance of components.

The component manufacturing plants for converters, pumps, and barriers were all under way: by Chrysler in Detroit; by Allis-Chalmers in West Allis, Wis.; and Houdaille-Hershey in Decatur, Ill.

However, the picture was not as rosy as it seemed. A discouraging period lay ahead. We still did not have a suitable barrier that could be produced in quantity of millions of square feet. Houdaille-Hershey was tooling up to produce the Norris-Adler barrier. Columbia and Princeton continued to try to improve it and to find a manufacturing method that could produce the large quantities necessary. Bell Telephone Laboratories was experimenting with nickel powder. Bakelite, a subsidiary of

Union Carbide, also was working on a nickel powder barrier. Hugh S. Taylor, chairman of the department of chemistry at Princeton, was evaluating the various types of barriers.

Clarence Johnson, working for Keith in Kellogg's Jersey City laboratory, was experimenting with an improved barrier that seemed to incorporate ideas from many of the individuals working at the various laboratories. He accomplished a technical breakthrough of a type that Urey had told me was impossible. No one individual could take full credit for Johnson's new barrier. Tolman reviewed the process and told me he thought it looked promising. However, a manufacturing process had to be developed.

Keith advocated that it be developed as an alternative to the Norris-Adler barrier. Urey felt it necessary to stick with the Norris-Adler barrier. Stowers, the New York area engineer, and I backed Keith. On November 5, 1943, Groves and I listened to the proponents of the various courses of action, and he decided to continue with the Norris-Adler barrier and also to continue with the new Johnson nickel powder barrier until we had better information on which to base a decision.

Keith took over space, and Taylor plus forty men in the Nash building and SAM lab set up to work on the new nickel barrier. On November 11, Urey, demoralized by the situation and the low prospects for success, wrote a letter to Groves that in effect recommended abandoning the entire gaseous diffusion effort or basing the project on the British design and technology. Groves obviously was upset when he showed me the correspondence. He commented, "We are really on the spot with this letter in the record." It was the type of letter that could be used to crucify Groves if he continued the project, disregarding the advice of his top scientist, and we failed. We had to do something. It would be a tough decision.

Tolman was assigned the task of comparing our progress with the British effort. In addition, Groves expedited a CPC decision to arrange for a British team to come to New York to review our gaseous diffusion project. Wallace A. Akers, accompanied by fifteen British scientists, arrived and started this review. I missed the initial meeting because I had been traveling, first to Montreal and then to Hanford, returning to Oak Ridge late Christmas Eve. I spent December 30, 1943, in New York to get up to date on the gaseous diffusion plant review. Keith had been having trouble with the British team. He felt they were not tackling the main issues and were more interested in discussing issues that were already resolved.

On January 4, I met with Groves in Washington to discuss what we should do. We agreed on taking preliminary steps to expedite development and production of the new nickel barrier. He told me to procure the nickel powder from International Nickel, who knew how to make the most suitable nickel powder, and that Houdaille-Hershey should convert the Decatur plant to the new process. Next we should discuss the plan with the British and try to get their support.

The meeting with the British lasted a good part of January 5, 1944. I attended with Bush, Conant, Groves, Felbeck, L. A. Bliss, Keith, Dunning, and Akers, who had his key scientists with him. The British agreed with Keith to the extent that the new nickel powder barrier might be a better barrier eventually, but they thought the long period of research and development on the Norris-Adler barrier would result in a suitable barrier at an earlier date. Keith, however, believed that if necessary, we could solve the production by using thousands of workers employing laboratory methods.

The British would not indicate which barrier they thought would turn out better if we relied on the crash program Keith was recommending. They did feel we should continue both development programs until we had more information. However, Keith claimed that with our limited personnel, we could not pursue both programs. Bliss suggested that research could be continued on both methods but production confined to one method. We contemplated stripping the Decatur plant to prepare for the new barrier production. The British considered that reckless and that we could not do it on our schedule. With luck, they thought we might be able to be in production using the nickel barrier by the summer of 1946. The British also thought it would be impossible to produce the pumps and converters on our schedule. During a recess, Bush and Conant told Akers that the British were in no position to judge what American industry could do when they had top priority. In any case, we certainly did have a variety of opinions to consider. We were definitely embarrassed by all the British pessimistic advice.

Groves did not indicate during the meeting any decision favoring any of the various proposals made by Keith, Felbeck, or Bliss. Groves' primary concern was to ascertain that Keith understood that he, Keith, had made a commitment to have a barrier plant in production in May 1944. Both Keith and Felbeck expressed confidence that we could meet our overall schedule for the gaseous diffusion plant. Groves and I were disappointed that we got so little support from the British, but there was no possibility of taking any other course than what we had discussed before the meeting. We recognized that our necks were way out and we had to succeed.

As a result of this series of events during the October-to-January crisis, we had to take action on many items, including a decision that Urey had to go. With the cooperation of George B. Pegram, a physicist and dean of graduate faculties at Columbia, Urey was relieved of all responsibility for gaseous diffusion research and development work, while allowing him to retain an office and a secretary to keep informed about the program.

Sometime after we had changed Urey's status, I unexpectedly met him at Grand Central Station in New York City as we were boarding the *Twentieth-Century Limited* to Chicago. We had a most pleasant dinner together and talked long into the night. I was pleased to learn that Urey actually felt relieved to have shed the responsibility for the gaseous diffusion research. However, he remained vitriolic about his relations with Dunning, even to the extent of accusing him of promoting the gaseous diffusion process just to ruin his, Urey's, reputation. I found Urey very likable when he was optimistic; then he became a bundle of energy and full of ideas. In contrast, when he was pessimistic, he became depressed and bitter.

As his replacement, we borrowed Lauchlin M. Currie from National Carbon and assigned him as associate director of the SAM laboratories. He took responsibility for all research on the gaseous diffusion process except the barrier work being directed by Taylor. Together the two men had the task of redirecting the work on the process in accordance with our new program.

When Groves visited the Houdaille-Hershey plant in Decatur, he looked for indications that the facility was being stripped, and he discussed the new process that would be installed. Union Carbide would be responsible for barrier production. Leon K. Merrill of Bakelite would represent Union Carbide at the plant, and in preparation for a new manufacturing process, Hershey was to send personnel to work at the SAM laboratories with Taylor.

Keith appointed Taylor to be associate director of research in Kellex with the responsibility for developing the process to produce the new nickel powder barrier. Johnson had the task of building a pilot plant in the Nash building on the Columbia campus. Taylor also retained charge of the analysis unit at Princeton.

Previously, I had made an arrangement with International Nickel to finance a new production unit in one of their plants. We put the arrangement into effect, and International Nickel started shipments of British-produced nickel they had stockpiled. Our British and Canadian liaison cooperated fully on these arrangements. Earlier, Urey had recruited Edward Mack, Jr., from Ohio State University to direct work on the Norris-Adler process at the Columbia University laboratory, and he had made considerable progress in improving the barrier and production process.

Keith was unable to make his May 1944 production date, so we still were in a struggle to find a suitable barrier. Other components threatened to pile up. As a result, we had no other course of action but to proceed with all of the components except for the barrier. By June 1944, construction of the entire K-25 project at CEW was 45 percent complete, the electric power plant 86 percent finished, the conditioning plant 87 percent complete, and the main plant 37 percent finished, but no barrier material.[6]

We were already over the hump concerning construction workers. The peak employment of 19,680 was reached in April 1944 and had already begun to decline. Our latest cost estimate had climbed to $281 million for the gaseous diffusion plant. As components arrived at CEW, we could condition each piece and put almost everything in place except for the converters, which could not be permanently installed until the barrier tubes were available, the unit assembled, and conditioned.

We had thousands of pumps being produced by Allis-Chalmers. A. O. Smith Company and Whitlock Manufacturing Company produced thousands of coolers to remove the heat of compression. Mass production became the order of the day. All the tremendous electric energy utilized in the plant to drive the pumps would be converted into heat and had to be removed from the building to cooling structures. When the plant was finally completed, we were using at Oak Ridge almost one seventh of the electric power being generated in the United States.

Midwest Piping had to fabricate and erect three million feet of nickel-plated pipe, which had to be leakproof and resistant to hexafluoride corrosion. Crane Company fabricated about half a million valves. A tremendous number of recording instruments, gauges, pressure indicators, flow meters, and thermometers, miles of instrumenta-

6. Ibid.

tion piping, and other devices manufactured by a variety of companies had to be installed and connected to a central control room. All components in contact with the uranium hexafluoride had to be corrosion-resistant and leakproof. In addition, the temperature of the gas had to be maintained at about 140 degrees.

Every component and pipe had to be conditioned and maintained in an absolutely clean environment while being assembled. A new, high level of quality control for cleanliness had to be maintained. Welding needed to be done perfectly to be absolutely leakproof. Special welding classes had to be conducted to train welders to meet these new high standards. Leak detectors needed to be developed and men trained to operate them. Kellex did an amazing job of setting the specifications and standards, and we had to take unusual measures to ensure compliance. In the absence of sufficient acceptable barrier tubes, thousands of unacceptable tubes were sent to Chrysler so they could train men on the assembly line and see that the tubes would be strong enough for the assembly methods used. Production of acceptable barrier tubes did not come close to a satisfactory rate until the end of 1944. In the interim, all acceptable tubes were installed in converters. At the end of 1944, the main K-25 production plant was 65 percent complete. Sixty stages of the first section of the plant were turned over to the operator on December 29. Another 102 stages were expected to be turned over on January 15, 1945, for testing. The first section was expected to be operable by February 15. At Chrysler, sufficient barrier tubes of satisfactory quality to permit current assembly of sixty-five to seventy converters weekly were being received from Decatur.[7]

We were finally climbing the hill to success with the gaseous diffusion process. By March 1945, a total of 402 stages were operating, and 906 stages were to be transferred to the operator by the middle of April. In March, Allis-Chalmers had completed the order of 5,804 pumps for plant requirements, and 1,113 of 2,892 filters being produced by Chrysler had been conditioned. As fast as units were turned over to Felbeck, he put them into operation, with the objective of cutting down the equilibrium time to reach a product that could be used as feed material for the beta electromagnetic plant.

During the spring of 1944, we had pushed construction of the gaseous diffusion plant on faith that someone would find a way to produce a

7. Monthly Report, January 8, 1945.

suitable barrier. It was obvious that if we failed, the whole U-235 project would have no chance of producing enough U-235 for a weapon until sometime late in 1946. If we succeeded, our date of August 1945 would be attainable.

In this climate of uncertainty about success, Groves received a letter from Oppenheimer concerning Philip H. Abelson's liquid thermal diffusion work (S-50) now at the Philadelphia Navy Yard. Oppenheimer proposed that a liquid thermal diffusion plant might be utilized to produce enriched feed material for the electromagnetic plant. This proposal fell on receptive ears. Groves was ready to consider anything that would improve our chances of success.

As related earlier, I had first heard of the liquid thermal diffusion method of separating uranium isotopes late in the summer of 1942. The Navy had continued to work on the liquid thermal diffusion process through 1942, 1943, and into 1944, with only occasional contact with the Manhattan Project. The Lewis Committee had reviewed the Navy's efforts in December 1942, and S-1 committee members, Groves, and Bush had made visits to Anacostia in 1943. During this time, however, the Navy's goal in separating isotopes of uranium was to use atomic energy to propel ships and more particularly submarines.

A major reason the S-1 committee did not give greater support to thermal diffusion was that the equilibrium period was estimated to be six hundred days and a tremendous amount of high-pressure steam was required. The process had the advantage of being very simple, and there were no moving parts except pumps for circulating cooling water in the process plant. Abelson's experimental results proved to be very promising, and ultimately the S-1 committee recommended continued support. To provide greater secrecy and more space than available at the Anacostia location, the project was moved to the Philadelphia Navy Yard.

In the spring of 1944, Admiral Parsons visited the Philadelphia Navy Yard and found that Abelson was building a small plant, which he expected would be producing five grams of 5 percent U-235 per day by July 1, 1944. Oppenheimer had been kept informed of our barrier problems and our dim prospects for producing U-235 in adequate quantities for weapons. Encouraged by Parsons, Oppenheimer recommended to Groves that the liquid thermal diffusion method be used as an additional approach for supply of enriched feed material to the electromagnetic plant. Lawrence promptly supported the idea.

With the approval of the MPC (Military Policy Committee), Groves arranged for a review of Abelson's plant. After study, Groves decided to build a production plant using the exact design of Abelson's hundred-unit plant under construction in Philadelphia, but to increase the size twenty-one times. The fact that the gaseous diffusion plant was delayed made possible the use of the steam from our electric power plant at the CEW. We hoped that some combination of the electromagnetic, the gaseous diffusion, and the liquid thermal diffusion plants would work out to produce an adequate supply of U-235 for a bomb to be available by our target date of August 1, 1945. In case the gaseous diffusion plant failed, the liquid thermal diffusion would not be adequate to achieve the August 1, 1945, date. Consequently, it was not a complete solution to our problem. However, there was no question that success with the liquid thermal diffusion would improve our prospects and might advance our date if both plants succeeded.

On June 27, 1944, we entered into a letter contract with H. K. Ferguson Company of Cleveland to design, build, and operate a liquid thermal diffusion plant. We broke ground for the plant two weeks later, on July 12, 1944, and had most of the materials on order or in the process of manufacture. The critical items we needed were 2,172 identical forty-eight-foot columns consisting of an inner nickel tube, an outer copper tube, a water-cooled jacket, the necessary valves, and interconnecting pipes. Uniform tubes were essential because the annular space between the copper and the nickel pipes was a critical dimension.

The theory, although not completely understood, appeared to be rather simple. High-pressure steam in the nickel pipe heats the inside wall of the annular space. The outer wall of the annular space is cooled by circulation of water within the water jacket about the copper pipe. The U-235 tends to concentrate near the hot wall and the U-238 near the cold wall. Normal thermal convection moves the enriched uranium to the top of the column and the depleted to the bottom. The longer the column, the greater the enrichment. We built the plant adjacent to the steam power plant. Originally we expected to operate the plant only so long as the gaseous diffusion plant did not need the electricity from the power plant. As a result, the thermal diffusion plant had to be built fast to be of any real value. (Later, when our electric plant was needed for the gaseous diffusion plant, we added surplus Navy oil-fired boilers to provide steam.)

I needed a project engineer who had a reputation for speed. Groves suggested I contact Colonel Robert R. Neyland, Jr. (later a successful University of Tennessee football coach), who was just finishing several construction projects in his district and might have some suggestions. After I outlined our problem, he told me, "Nichols, I have just the man for you, Major Mark Fox. He can get a job done faster than anyone else I know. He will drive the contractor and make sure that everything is done on time. He knows construction, and contractors like him. He will do a job for you." I replied, "Sounds good to me. Send him to see me." After telling me he would have Fox in my office the next day, he added, "Nichols, I forgot to mention one other thing. When the job is almost finished, put a good administrative man on the job to clean up the paperwork."

Groves told Fox he had ninety days to get the plant in operation. This meant we had to have it on line by September 16, 1944. Fox complained to me, but I told him it had to be done. I had just bet Dobie Keith $5 that we would be operating the plant on Groves' schedule. Keith, among others, did not think the project was worth building, but Groves felt that saving even a single day in attaining our first U-235 bomb was worth the expected cost.

I had agreed that the chances were good that the thermal diffusion plant would save time. After the war, we calculated that the thermal diffusion plant had saved nine days in providing sufficient material for the uranium bomb dropped on Hiroshima. The cost was $10,605,000 for construction and $5,067,000 for operation. Hewlett and Anderson in *The New World,* the official history of the Atomic Energy Commission, estimated the acceleration as about a week. Considering the daily cost of the war and the saving in lives of even a few days, I believe it turned out to be a good investment.

To meet our schedule, fast construction was absolutely necessary, and the H. K. Ferguson Company, Abelson, and Fox, with the assistance of several Navy officers, did introduce steam into a portion of the plant on September 15, and the first product was removed on October 30, 1944. I took Dobie Keith into the plant during the initial period, and in the cloud of steam caused by leaking joints, he reached into his wallet and gave me $5 with a caustic comment about the definition of "operation." By December the entire S-50 plant was 97 percent complete. All three groups of equipment had been turned over to operations. Production during the month of December 1944 was much lower than expected due mainly to a shutdown for repair of leaks and for miscellaneous adjustments to units previously operated.

9

PEOPLE, PLACES, AND THINGS

RECRUITING THE CONSTRUCTION WORKERS to build the plants was a most difficult task. Colonel C. D. Barker, in the chief of engineers' office, and Lieutenant Colonel Curtis A. Nelson and Lieutenant John Flaherty (U.S. Navy), in the district, did an amazing job of coordinating this effort. The labor unions cooperated in making craftsmen available to the extent possible. Due to the high turnover, our recruiting effort continued until we were well over the construction peak.

When operations started at the CEW and the expansion for the implosion weapon began at Los Alamos, we had difficulty finding sufficient technical talent for supervisors at the electromagnetic plant, and laboratory technicians and workers in the explosives plant at Los Alamos. Fortunately, the Army Special Training Program, which provided for enlisted men in uniform to continue their studies, mainly at technical colleges, to meet the future needs of the military was being discontinued in 1944, and we seized the opportunity to select qualified men to form a special engineer detachment (SED).

The largest SED units were at the CEW and Los Alamos. At the CEW, most of these men felt fortunate to be assigned to our work instead of being sent overseas. But some were unhappy because they were working side by side with civilians who drew more pay for the same work. In addition, the married men were not allowed housing for their wives. To help compensate for the difference in pay scales, we based promotions to noncommissioned officers on the job being performed and also made life easier by furnishing maid service to clean their barracks, make beds, etc., and they were given a food allowance for the cafeteria. Many of the wives found homes with residents of Oak Ridge, others located places to live outside the CEW, and many accepted jobs in the plants. Some of the men called the assignment to the CEW "GI Heaven." But many cases cited in Studs Terkel's *The Good War* tell a much different story, including tales about the effects of radiation on the workers. These tales in regard to radiation have no justification in fact but may very well have been what some of the people believed at the time.

At a MPC meeting at the CEW during a discussion with James C. White, president of Tennessee Eastman, about the difficulties of finding sufficient operators, White commented on how helpful the Army had been in supplying men from the SED. He turned to Admiral Purnell and asked, "What can the Navy do to help us?" In response, Purnell arranged for about sixty young commissioned officers with technical education to be assigned to the MED. The Navy ordered these men to proceed to Knoxville, Tennessee, and call my number. My office would arrange to transport them to Oak Ridge, and they would be assigned to a separate dormitory they called "Good Ship Never Sink." According to their qualifications, they were assigned to supervisory positions, mainly at the Y-12 plant.

Dr. Conant had suggested that in view of the secrecy and the military nature of the work, we commission the scientists as officers in the Corps of Engineers and operate Los Alamos as a military laboratory. On our train ride with Oppenheimer in October 1942, Marshall, Groves, and I found that he was amenable to this type of organization, and he actually took initial steps toward becoming a commissioned officer.

However, Oppenheimer ran into difficulties with the military concept when he tried to recruit Robert F. Bacher and Isidor I. Rabi, two physicists whom he considered essential and who already were engaged in war work developing radar at the radiation laboratory at M.I.T. They both refused to accept commissions. In February 1943, Conant, Bush, Groves, Marshall, and I met with Rabi and Bacher at the Biltmore Hotel in New York and spent the better part of the day trying to convince them that a military laboratory was essential. However, they were adamant and made it clear that even if the initial research activities were conducted as a civilian organization, they would tender their resignations if it were later militarized, war or no war. After the scientists left, Groves retained his plan that provided for a delayed militarization of the laboratory, but we all realized that he could never implement it.

Because of the existence of such a plan, Groves decided that the military commanding officer be responsible only for providing suitable housing and other services for the scientists and for guarding the establishment. Oppenheimer was given full responsibility for the scientific work and for security of information and would be assisted by the commanding officer in carrying out the latter function. Groves retained direct overall executive responsibility under the Military Policy Committee.

As a result, the University of California contract became primarily a vehicle for employing scientists and procuring materials. The military organization at Los Alamos was to assist Oppenheimer as might be required and still was responsible for administration of the University of California contract. To prevent the district from getting involved in the day-to-day problems of housing and housekeeping, Groves directed that the Albuquerque District be utilized by the commanding officer for the construction required.

I have always considered it unfortunate that we began with the concept of a military laboratory, because it resulted in Los Alamos having such a hybrid organization. The more normal contracting procedures utilized for the met lab, Berkeley, Columbia, and other universities would have avoided getting Oppenheimer involved in so many housekeeping and administrative details, and he still could have been given the same overall responsibility and freedom of action as Lawrence and Compton for getting results.

Even though Groves had obtained the approval of the Military Policy Committee for the appointment of Oppenheimer as the Los Alamos director, the Army security personnel would not clear him because of his questionable record involving association with known or suspected Communists. Groves eliminated that obstacle by assuming responsibility for all security and intelligence pertaining to the Manhattan Project. However, he did not act immediately on the clearance, choosing to obtain one more confirmation of Oppenheimer from the project's scientific leadership.

At the S-1 meeting in July, which Marshall and I attended, Groves asked each member to name an individual who was equally or better qualified to direct Los Alamos than Oppenheimer. One by one they stated their confidence in Oppenheimer and their opinion about his qualifications and emphasized that he was absolutely essential to the project. Groves then announced that Oppenheimer would be cleared. He formalized the decision on July 20, 1943, in a memo to the district engineer: "In accordance with my verbal direction of July 15, it is desired that clearance be issued for the employment of Julius Robert Oppenheimer without delay irrespective of the information which you have concerning Dr. Oppenheimer. He is absolutely essential to the project."

Marshall then dictated a letter to our Los Alamos security officer, clearing Oppenheimer. Marshall's secretary prepared the directive for his signature, but he had already left for home by the time she had completed it. She therefore brought it to me and instead of asking her to retype the letter for my signature, which I should have done, I took it to a light table and using Marshall's signature, traced it onto the letter. We both occasionally did this on less important matters to save time and typing.

After I had become district engineer, the letter came back from the Los Alamos security office with this endorsement: In view of Oppenheimer's record, we have scrutinized the signature very carefully and it obviously is a forgery. At least it was reassuring to see that the security office was alert and doing its job. I had the letter retyped for my signature, signed it, and put the original and its endorsement in the "burn" basket. When I next saw Oppenheimer, I told him that we finally had cleared him. Aware of the situation, he commented, "That must have been difficult." I added, "In the future, please avoid seeing your questionable friends, and remember, whenever you leave Los Alamos, we will be tailing you." Unfortunately, his clearance did not end the security problems concerning him.

On August 25 and 26, 1943, Oppenheimer revealed to our security officers, Lieutenant Lyall Johnson and Colonel Boris T. Pash, an earlier occurrence that later became known as the Chevalier incident. Oppenheimer volunteered information that months earlier a member of the Soviet consulate in San Francisco had utilized George Eltenton, a member of the Federation of Architects, Engineers, Chemists and Technicians (FAECT), a CIO affiliate in the Bay Area, to make contact with one or more individuals to help establish a channel of information concerning work at Los Alamos. Oppenheimer told Pash that he was concerned with possible indiscretions that had taken place and explained, "To put it quite frankly, I would feel friendly to the idea of the commander-in-chief informing the Russians that we were working on this problem. At least, I can see that there might be some arguments for doing that, but I do not feel friendly to the idea of having it moved out the back door. I think that it might not hurt to be on the lookout for it.''[1]

In the interview with Pash, Oppenheimer did not provide the name of the person Eltenton had used to make the contacts. I read the transcript and listened to the recording of the interview. I found the story very confusing, and Groves asked his security man, John Lansdale, to follow up on the initial investigation and try to get more information. When Lansdale could not persuade Oppenheimer to tell him the name of the person who had contacted him, Groves obtained from Oppenheimer the name of Haakon Chevalier, a friend who taught French at Berkeley.

On December 12, I wired the security man at Los Alamos, Captain Peer deSilva, that Oppenheimer had named Chevalier and that he believed Chevalier had engaged in no additional activity other than the three original attempts. Afterward, from time to time Groves and I discussed the matter. We both thought we still had not received the whole truth about what had taken place. We felt Chevalier or Eltenton might have contacted Oppenheimer's brother, Frank, who was working on the project, rather than J. Robert Oppenheimer. At the time, however, we did little except maintain our surveillance of Oppenheimer and others.

The history of the near-miracle wrought by the scientists at Los Alamos has received the attention of writers, historians, and media people. Because my own work focused on the production of the fissionable materials, I can provide relatively little new information on that aspect of the project. Nevertheless, I kept generally informed about progress at Los Alamos because I realized that if it failed in its mission, all our production efforts would be of little military value and the Manhattan Project would be a failure.

Although he lacked experience in administration, Oppenheimer did a masterful job of organizing the laboratory. He borrowed research equipment and moved it to Los Alamos. Top personnel were recruited from leading universities and other war projects. Probably never before had there been the assembly of such brilliant minds dealing with physics, chemistry, cryogenics, metallurgy, mathematics, explosives, en-

1. U.S. Atomic Energy Commission, In the Matter of J. Robert Oppenheimer. Transcript of Hearing Before Personnel Security Board, Washington, D.C., April 12, 1954, through May 6, 1954. (Washington, D.C.: U.S. Government Printing Office, 1954), p. 845.

gineering, and practically every other aspect of science. By means of his outstanding scientific leadership, Oppenheimer welded Rabi, who served part time as his consultant; Hans Bethe; Robert Bacher; John von Neuman; Enrico Fermi; Edward Teller; Captain William S. Parsons, U.S. Navy; George F. Kistiakowsky; Joseph W. Kennedy; Cyril Smith; Emilio Segrè; John H. Manley; David K. Froman, Robert Serber; and many others into a most effective scientific team.

From my point of view, I was anxious to ascertain if both U-235 and plutonium could be used in a fission bomb. Next, I needed to know the specifications and how much of each material would be needed for a useful weapon. This information was needed to determine the rate of production required from our plants.

At one of my first meetings in Chicago with Oppenheimer, he described the ''super'' (hydrogen) bomb to me. But the Military Policy Committee soon put that one on the back burner, and the fission bomb was given priority. Teller never did accept this decision. Also, early in the program, Oppenheimer discussed the possibility that a fission bomb might ignite the atmosphere. But that fear soon was laid to rest. As these issues were resolved by the expanding scientific effort at Los Alamos, optimism about our overall chance of success soared. However, it seemed that whenever we received good news, bad news seemed to follow. As related earlier, we first learned that we needed to increase the quantity of U-235 needed for a gun-type weapon. Next we learned that the gun-type weapon was not suitable for the use of plutonium. As a result of this we embarked on a major expansion of the effort at Los Alamos to develop the implosion weapon. The time for the successful delivery of an implosion-type atomic bomb now depended on the time required to develop and test the implosion principle.

Late in 1944, I felt we probably would produce both of the two types of bombs about August 1, 1945. Each project had its difficulties, and success for each remained about a fifty-fifty proposition. Faint hearts and pessimists had no place in the Manhattan Project. I always had the gut feeling that we would succeed. If I had not, I would have found ways to be transferred overseas.

Oak Ridge remained like Topsy—it grew and grew throughout the war. We finished our last housing expansion three months after the Japanese surrender and a nursery school and a gymnasium by Christmas 1945. At our peak of construction, the construction labor force totaled seventy-five thousand. Our operating force started its growth later and peaked just after the end of the war, with a total of fifty thousand workers. The combined employment peak was eighty thousand.

Although I intended to avoid entanglement in the day-to-day problems of the town itself, I soon found that I could not escape them entirely. The contractors' project managers brought various matters to my attention when they became unhappy in their dealings with Roane-Anderson; Tim O'Meara, my town manager; or John Hodgson, my central facilities manager. Normally, whenever I returned to Oak Ridge, Marsden would have prepared about two hours of material for me to read. I liked to tackle such reading alone, and I arrived early at the office to do it before the normal workday began. After about two hours, he would bring in various other topics that needed my attention. He had instructions to keep my paperwork and conferences in my office to a minimum because I felt it most important that I spend the bulk of my time at the CEW actually in the plants checking construction progress, success we were having starting operation of the plants, or increasing production rates. One morning Marsden barged in and said, "Nick, there's a delegation of ministers who insist on seeing you." Somewhat annoyed, I asked, "What the hell for? I'm terribly busy." "Well, we are housing about five thousand single girls in dormitories." Most were office workers, both government and contract employees. "Apparently some girls have complained to their ministers that some other girls are entertaining male guests in their rooms." I told him, "Marsden, we have rules about such things." "There have been some violations," he explained. "In any case, the 'good' girls are protesting about the 'bad' girls, and many of them have enlisted the support of their religious leaders." This would not have been such a critical issue today. However, one must remember the values, customs, and morals that existed more than forty years ago. The clergy proposed that I should move all the "bad" girls to separate dormitories so that the "good" girls would not be offended or possibly be considered in the "bad" category. I told them I doubted if the situation warranted such a drastic step. The ministers insisted that I take immediate action. Fortunately, Marsden who had been thinking about the problem longer than I had, passed me a note. After a bit more discussion, I agreed with the ministers that something should be done and I told them, "I will segregate the girls. But I will need two lists, one of all the 'good' girls, and more important, one of all the 'bad' girls." I asked them to adjourn to the conference room to decide which of them should be responsible for drawing up the lists. Marsden reported back shortly that the ministers had decided to do nothing for the time being. They wanted more time to think about the problem. I never heard from them again.

Another question that bugged Oak Ridgers, and continued to be a sore point for the duration of the war, was that of the coal furnaces in their houses. We had rejected the use of oil and electricity because of the costs. The furnace became the *bête noire* of the Oak Ridge housewife. When the Comptons moved across the road from us on Olney Lane, Jackie greeted Betty with mop and pail to help clean up the mud the movers had tracked in and left notes on the idiosyncrasies of the furnace. We had encouraged the Comptons to move to Oak Ridge with the first contingent of scientists from the met lab. Compton established his main office at the CEW and left Hilberry in charge of the met lab. Compton was known as A. H. Comas as far as the general population of Oak Ridge were concerned. He and Betty were charming, imaginative people who entered into the spirit of things with verve. Their spirit and example kept up morale and cheered the plight of all of their personnel.

Ernest Lawrence also had an apartment in the town and spent much of his time at Oak Ridge during the critical period for the electromagnetic plant. His family did not move, according to him, because "I have too many kids of school age." However, we did persuade Molly Lawrence to visit for a while, and as part of the inducement, Jackie and I spent a long weekend with them in Gatlinburg, a Tennessee resort. The relaxation, the trip through the Great Smokey Mountains, and the chance to get better acquainted furnished a much-needed break for all four of us. It marked the beginning of a cherished friendship.

With production activities increasing at the CEW, I found it necessary to spend more time there. The CEW now was the main center for coordinating district responsibilities. With Arthur Compton and Ernest Lawrence there, many decisions pertaining to the met lab or Berkeley could be made at the CEW. Coordination of the work of the many contractors engaged on the gaseous diffusion plant continued to take a major portion of my time. I commented about this on one occasion to Lyman Bliss, who was a Union Carbide vice president and who General Groves and I personally consulted from time to time about many problems concerning the overall gaseous diffusion project. I told Bliss that I was spending about 80 percent of my time just coordinating the work of all the various contractors. He responded,

THE ROAD TO TRINITY

"Considering the overall importance of K-twenty-five, the only mistake you are making is that you are not spending a hundred percent of your time on it. If the gaseous diffusion plant doesn't produce soon, the entire CEW project will be a failure." He was probably right, but my time was spread pretty thin, and there were too many other items that demanded my attention. However, to ease the load I assigned Lieutenant Colonel John Hodgson new responsibilities as executive assistant for all CEW operations.

In our efforts to make life at Oak Ridge as normal as possible, it was essential to provide for the spiritual needs of our residents. Gradually numerous denominations were clamoring for facilities for their church services, office space, and housing for their ministers. The two standard Army chapels in our basic plan were used by several denominations. These were supplemented as needed by reconditioned old churches, school auditoriums, the theater, and community buildings. By April 1945 approximately sixty-eight hundred residents attended church services regularly.

Because of the shortage of housing, we had expected the clergy to live outside the town, but in response to the clamor of the various groups, eight houses were allocated to ministers, priority being given to the largest congregations. In addition, some of the part-time ministers had housing assigned on the basis of being full-time employees of a contractor.

The policy concerning religion was perhaps the area in the management of the town in which General Groves made his wishes known in greatest detail. As the son of an Army chaplain, he left no doubt of his desire that our overall policy on religion must be beyond reproach.

In the autumn of 1944, I realized that coordinating production of the three U-235 plants at the CEW would be a much more difficult task than anyone had anticipated. The liquid thermal diffusion plant (S-50) could raise the enrichment of U-235 from .7 to .9 percent; the electromagnetic plant (Y-12), which had two different alpha stages, one more advanced than the other, could raise the enrichment from .7 percent to 15 to 20 percent in the alpha stages, while the beta stage could raise the enrichment from 20 to 36 percent to an enrichment useful for weapons. The gaseous diffusion plant was designed to raise the enrichment from .7 to 36 percent.

Both the alpha and beta stages of Y-12 were in operation. The S-50 plant came on the line in segments, beginning with a trickle in October 1944. Our plans were to start operating the K-25 plant in sections as they were completed and tested. As the number of completed sections increased, the potential for the percent enrichment and production rate increased. For each increment, time was required to reach equilibrium. Also, the higher the enrichment withdrawn, the lower the production rate. It would be months before we would have the base plant completed and operating at 36 percent enrichment. In the meantime, we needed to operate the three plants to give us maximum effective production units (epu) for weapons.

Early in September, I appointed a production control committee consisting of contractor production personnel from each of the production plants. I relieved Peterson of his plutonium project responsibilities and gave him the task of heading the committee and more specifically the responsibility for producing periodic operating plans for each of the production plants. In addition to the contractor representatives, he assembled a highly talented staff of his own (including George Quinn, Harcourt Vernon, Dean Bartky, and Captain Joseph King) as well as a team of men from the Special Engineer Detachment to run a battery of mechanical calculating machines to make all the computations needed to compare the many different possible combinations. We certainly could have used a modern computer at that time, but the computer era was still in the future. As production increased, we restricted information about his results to the absolute minimum number of individuals. Peterson and his team did a superior job; it took long hours to keep up with the constantly changing production capabilities at each of the plants. Jackie complained that I was overworking Pete, that he was never home evenings when she visited her sister, Mrs. Peterson. I told her, "Pete is a perfectionist, but the job I have given him can never be perfected. I admit I give him hard deadlines to meet. But it's his decision how long he and his staff should work to get the last increment of perfection." Jackie did not find out until after the bombs were dropped what Pete's job had entailed. But I must say that Peterson's long hours paid off in unexpected ways.

Initially, he had little coordinating to do. However, Peterson studied many potential combinations that might be tried as sections of the plants went into operation. Soon it became apparent that the production from S-50 should be fed to K-25 rather than Y-12. We put this into effect near the end of April 1945. The initial operation plan for K-25 was to produce 1.1 percent material for feeding into the Y-12 alpha stages. As sufficient sections of K-25 were completed, it was expected that it should produce 20 percent or higher enriched material to feed into Y-12 beta stages. But a considerable equilibrium time was necessary to move from 1.1 to 20 percent.

Peterson made studies for operating the plants when all authorized units were completed as well as studies concerning production for the completed K-25 plant, including the top stages under design. When he showed me the various production charts, it was obvious that adding the top to K-25 and operating the entire plant as a single production unit would not produce as much weapon material as the base plant tied in with the beta units of Y-12 would. Likewise, the charts showed that adding more alpha stages to the Y-12 plant, as Lawrence was recommending, was not the best answer for more production.

I showed Peterson's charts to Groves first; then I asked Keith to have Manson Benedict check the charts to see where they might be in error. Dobie soon called and admitted, "I missed the boat. Peterson's calculations are correct. The top stages should be abandoned." He also suggested that production probably could be greatly augmented by building more base units for K-25 and balancing them by building more Y-12 beta units. He said he would come up with a proposal shortly. Peterson confirmed that "the idea had good possibilities."

Groves quickly accepted the findings and recommendations of Keith and Peterson's control group to construct more base units to the gaseous diffusion plant (K-27) and one more beta stage track for Y-12. The estimated cost of these additions was $100 million, with an estimated completion date of February 1946. After we reviewed the final plans for this expansion with the contractors, Groves, instead of approving the plan as I had expected, said, "Nichols will tell you my decision tomorrow." We then went to dinner at the Commodore Hotel. Before dinner, he ordered a second drink. I was surprised; this was unusual for Groves. It wasn't until after dinner, while we were walking to Pennyslvania Station, that he mentioned the subject of the expansion. He asked me, "What would you do, Nichols?" Without hesitation, I replied, "I believe it absolutely essential to build the expansion if we want an adequate production rate for U-235. Our actual expenditures are now well over a billion dollars and increasing every day. Personally, I would just as soon be hanged for exceeding two billion dollars as spending one and a half billion without planning for an adequate U-235 production rate." Groves' eyes twinkled as he said, "I am glad that you think that way. I agree. Tell them I approve it."

In the first months of 1945, the enrichment of the U-235 produced remained well below the required weapon strength. Nevertheless, as soon as enriched uranium tetrachloride was produced, it was shipped to Los Alamos for experimental use. Peterson informs me, "One shipment of enriched material to Los Alamos was carried in your B-twenty-five by Pete Young. I accompanied it. I've forgotten the amount and how much it weighed, but it was substantial." To transport other shipments we utilized armed couriers traveling by train. The material was contained in special hand-carried luggage, and the courier remained as inconspicuous as possible. As time went on, this simple method of shipping in a suitcase the entire output of what was fast becoming the largest construction project in history led to a local rumor that the secret CEW project must be a failure or a boondoggle. Many of the local residents and workers observed that thousands of railroad cars were carrying supplies into the CEW but no one had ever seen anything shipped out.

Production from the sections of K-25 in operation was exceeding our estimates. We continued to have maintenance problems on Y-12, but I was confident that these problems could be solved shortly and we would meet our August 1 production goal. It was not yet a sure thing.

With Hanford gradually achieving a greater production rate, I spent less time on plutonium production. Matthias and du Pont had production well in hand, actually exceeding expectations. However, problems continued to arise at the met lab. The scientists continued to be unhappy with the situation. As stated before, in the summer of 1943 we had alleviated the scientists' discontent by authorizing work on a heavy-water-reactor design. In October 1943 du Pont released the design drawings for the first water-cooled, graphite-moderated reactor. At that point, the scientists pushed for a different design for the second reactor at Hanford. Roger Williams explained the delays that would be encountered if du Pont had to change designs for each of the next two reactors or even if only one change was made. He summed it all up when he asked me, "Is the Army interested in maximum plutonium production or in developing reactor design?" Of course, the answer was obvious, but it took my efforts as well as those of Groves and Conant to help Compton quell the dissent caused by our decision to back du Pont.

Even though the various crises at Hanford required Fermi and others from Chicago to assist du Pont, it was becoming obvious that the role of the met lab was diminishing. For good reasons, the scientists were becoming concerned about the future of the laboratory. Early in 1944, a rumor reached Chicago that 90 percent of the lab's personnel would be released by June. Compton tried to dispel their concerns, and in a conference with them agreed to recommend long-range plans for the facility's future. I supported Compton's modest proposals for continuing research, but the MPC and Groves questioned the legality of supporting extensive research that had only postwar objectives. However, in August 1944, Bush told Compton that he had authorized Tolman to study postwar needs and give other assurances that there would be no disastrous break at the end of the war.

Nothing was really settled in 1944. Late in the year, Compton and I had a long discussion. We both agreed that judged solely by war need, the contract with the University of Chicago for assistance at Hanford could be terminated in June 1945. At the same time, we realized that the met lab could be of great value in contributing to future national security. I asked Compton for specific proposals to obtain Groves' approval, but the attitude in Washington continued to be that neither the MPC nor the OSRD could support any extensive project not directly related to our war effort. Bush then proposed that Stimson set up a high-level committee concerning postwar problems.

Realizing that it would take a long time to establish policy for postwar research, Groves finally wrote to Compton that the MED must limit itself to ending the present war. He authorized only work supporting Hanford, Los Alamos, and research (not development) to determine potential nuclear energy value of thorium. He also announced that in accordance with Compton's recommendations, he would find a commercial concern to operate the Clinton laboratory (the semiworks) to relieve the University of Chicago of this responsibility. This decision further upset many of the scientific leaders of the met lab.

They lost hope that the present leadership in Washington was being sufficiently farsighted concerning the future. Also, the scientists became more alarmed about the political issues concerning atomic matters, and many decided to act on their own. Once so motivated, their faith in the Washington leadership was not restored, even when President Truman approved appointing an interim committee to study the use of the bomb and postwar problems.

The scientific organization for many other parts of the Manhattan Project did not express these concerns to the same extent as at the met lab. Early in 1945, Los Alamos still was expanding and engaged in a crash program for developing the implosion weapons. For K-25, we still were trying to improve the barrier, and construction of K-27 had been approved. Berkeley was busy with the design of the approved Y-12 beta unit.

Shortly after Groves had informed Secretary of War Stimson that he planned to approve expansion of the CEW U-235 production facilities unless the secretary instructed him "to the contrary," Stimson decided to visit Oak Ridge. Groves notified me to prepare for the visit. Considering Stimson's advanced age and failing strength, we made every effort to eliminate the need for him to walk or stand. We built ramps for cars into the gaseous diffusion plant. This led to the rumor that FDR himself was coming for an inspection.

On April 11, 1945, Stimson, Groves, and Colonel William H. Kyle, Stimson's military aide, arrived for lunch at the guest house, where they were to spend the night. After lunch, we drove around the gaseous diffusion plant and entered one building so Stimson could see its size and its interior. Inside, there was really little to see except for the interconnecting pipes and the large steel plate boxlike structures housing the converters and pumps. To show Stimson the components, we drove into the conditioning and maintenance building, where he could see the barrier tubes, converters, and pumps that were being assembled and conditioned prior to installation in the production building.

We then visited the semiworks to see the pile and the outside of the chemical building. From there, we visited the water filtration plant, primarily to get a view of the electromagnetic plant and the town. Stimson was then left at the guest house to rest. To save him time and energy during the visit, we did not have him meet key personnel in the plants, with the exception of Whitaker at the semiworks.

While Stimson was resting, Colonel Kyle accompanied me to my home to discuss how we should arrange for all the key contractor personnel to meet Stimson at a reception that was to follow. Kyle proposed that Stimson should sit on the sofa and that I should bring up the various individuals one or two at a time to sit by him for a few minutes. Kyle insisted on relocating the living room sofa to facilitate this procedure better in the limited space of our small quarters.

I found it amusing that Stimson had other ideas. When he arrived and after I had introduced him to my wife, I led him to the sofa. However, he refused to sit down, telling me, "I want to meet everyone present. Introduce each individual, give me time to converse with each one, and only after I have met everyone shall I sit down." He seemed to be exhilarated by the opportunity to meet the individuals responsible for building and operating the plants. Kyle tried to intervene, but Stimson waved him away. Everyone was impressed with the interest the secretary showed about each part of the project. Finally I said, "Mr. Secretary, I believe you've met them all." Only then did he agree to sit down, in the rocking chair on the porch, and he continued to talk with those who were around him. Overall, this visit by the Secretary of War proved a pleasant, informal, and for Jackie and me, a memorable occasion. Despite Stimson's advanced age, he was alert, cheerful, and friendly. He was keenly interested in our project and stressed the importance of our work to the war effort. Consequently, to me, this proved to be the happiest "working" day I spent at Oak Ridge.

The next morning, we briefed Stimson about the electromagnetic plant before visiting it. We also showed him more of the town and visited one of our construction camps and then drove him to the airport. En route, Stimson seemed elated about the whole project at CEW. He was very complimentary about what we had accomplished. He also inquired about the many new, unpainted small houses along the highway. I told him that most of these housed native Tennesseans who had obtained employment at CEW and who were for the first time in a financial position to build their own homes. He laughed and said, "Next time I see Franklin, I'll tell him that the Army has been able to do more for Tennessee while fighting a war than the TVA accomplished with all its dams." Unfortunately, he never did get to see the president again.

While at Oak Ridge, Stimson recorded in his diary that he had spent the first afternoon "going over the most wonderful and unique operation that probably has ever existed in the world." When he returned to Washington, he wrote down that the trip had cheered him up: "I was there confronted with the largest and most extraordinary scientific experiment in history and was the first outsider to pierce the secrecy of its barricades and to have explained to me the tremendous development which had been going on not only in scientific experiment but in the creation of an orderly and well-governed city." He concluded that the project had "this unique peculiarity: that, although every prophecy thus far had been fulfilled by the development and we can see that success is 99% assured, yet only by the first actual war trial of the weapon can the actual certainty be fixed."[2]

Stimson never had the opportunity to tell Roosevelt about "the most wonderful and unique operation." On April 12, the president died at Warm Springs, Ga., and Harry S. Truman assumed office. What would this mean to the Manhattan Project? Roosevelt, with Churchill's backing, had been an ardent supporter and throughout the war had maintained our number-one priority. Would there be a change?

It proved fortuitous that Stimson had visited Oak Ridge when he did. He now had firsthand information about our progress. He and Groves briefed the new president on April 25. Stimson had prepared a paper on the political significance of the atomic bomb, and Groves reported on the history of the project, its present status, and a forecast of weapon availability. He told the president that "the gun type, uranium bomb will be available about August 1 and a second about the end of the year, while the implosion weapon will be ready for a test in early July."

In a memo to the files, Groves reported that a great deal of emphasis was placed on foreign relations and particularly on the Russian situation during the meeting. In regard to costs, Groves noted, "The President did not show any concern over the amount of the funds being spent but made it very definite that he was in entire agreement with the necessity for the project." In regard to the purpose of the project itself, Groves recorded that he and the secretary "both emphasized to the President that our present interest was purely military, that while there was unquestionably great prospects for future commercial developments our immediate prospects were military only and it was towards that field that our entire effort was devoted."[3]

Stimson thought the briefing was of great interest to Truman and recorded in his diary that Truman "was very nice about it. He [the president] remembered the time I refused to let him go into this project when he was chairman of the Truman Committee and was investigat-

2. Henry L. Stimson diary (1939–45), Yale University Sterling Memorial Library, New Haven, Conn.

3. Manhattan Project Records, Office of the Commanding General Files, RG 77, Box 8, File 24, National Archives.

ing it and he said that he understood now perfectly why it was inadvisable for me to have taken any other course than I had taken.''[4] Later, Groves told me that Truman would continue to support the Manhattan Project and that he was in entire agreement with the necessity for it to continue.

4. Stimson diary, April 25, 1945.

10

ROAD TO TRINITY, 1944–45

IN THE WINTER OF 1944–45, two organizational changes occurred within the Manhattan Project. In November, a classmate of mine, Colonel Elmer E. Kirkpatrick, who had worked for Groves early in the war, became available. Groves asked me if I would object if he used Kirkpatrick as an inspector general. Groves felt we should follow the Army procedure of having an independent investigation of any situation that appeared to warrant it, and he preferred not having the Army inspector general looking into our work. I had no objection, and Groves had Kirk assigned to his office.

However, the system did not work as Groves contemplated. I have forgotten the specific matter Groves first asked Kirkpatrick to investigate but when Kirk showed me his completed report, he told me, "I don't like the job of checking on a classmate. I would rather just be working for you." After reading it, I said, "I agree with your recommendation about how to correct the situation. Why not add that Nichols has already taken corrective action." Turning to Marsden, I said, "Earl, see that it is done."

Groves did not like this procedure and told me that he should order me to take any corrective action. This did not seem necessary to me, so I suggested that a better solution was just to assign Kirk to the district as deputy district engineer. Groves agreed, but added that Kirkpatrick would be needed in April or May to go to Tinian to build the facility for assembling the atomic bombs. Afterward, he could return to the CEW and continue as my deputy.

Later in December, I suggested to Groves that he should have a

169

deputy. As his work load increased, I was having difficulty getting things done in Washington when he was out of town. Mrs. O'Leary, his other assistants, and Allan Johnson were very capable people, but some of the contacts with higher officials required a general officer. In response, Groves told me, "You are right, but who could get along with you and also get along with me?" I admitted few men could fit the bill but said, "I can name one: Tom Farrell."

My memory of this conversation varies somewhat from what Groves has written in *Now It Can Be Told* in regard to how Thomas F. Farrell became his deputy. According to Groves' account, Secretary of War Stimson had raised the issue of what would happen if either or both of us were killed in an accident, and he suggested that Groves needed a deputy. At the time, Groves told me only that Stimson had given him instructions that the two of us should never fly in the same airplane. However, shortly after he advised me of Stimson's directive, I was on a plane en route to Chicago when Groves boarded it at an intermediate stop. Looking for a seat, he saw me, sat down next to me, and said, "You are violating Stimson's instructions. Why didn't you get off the plane when you saw me come aboard?"

Regardless of whoever did first present the idea of a deputy to Groves, he did recognize its validity, and immediately after discussing it with me, he asked General Styer to have Farrell, then a brigadier general on duty in India, assigned to the Manhattan Project. On February 14, Farrell visited the CEW, and I spent the day showing him all aspects of our activities. He then had dinner with Jackie and me, and we spent a very pleasant evening discussing his experiences overseas.

We could say nothing about our work during such social occasions. Jackie had no idea of the nature of my work. She did, of course, know that Lawrence, Compton, Fermi, and Oppenheimer were leading physicists in the country and guessed that we were working on some sort of secret weapon. She never asked me about my work, concluding on her own that we probably were developing some sort of ray gun. She was forty years ahead of the state of the art. Ultimately Jackie did ask that I manage to let her in on the secret just before it was revealed to the rest of the world.

I had known Tom Farrell for several years. He had served as a regular Army officer before leaving the service in the 1930s. At the time he returned to active duty in 1941, he was the chief engineer for New York State. Farrell was a very capable engineer and an effective executive. He was warm, outgoing, and very cooperative. His method of operating was as different from Groves as night and day. Nevertheless, Groves respected and liked Farrell as much as I did. As a result, he was able to fit into the project quickly and serve his hard taskmaster with little friction with me.

Initially, Farrell's assignment was to become knowledgeable about all aspects of the district's work and then to assume direct responsibility for military planning for the use of the atomic weapons.

Groves had arranged with General Hap Arnold, chief of the Army Air Corps, for the formation of the 509th Composite Group to drop the atomic bombs. Arnold selected Colonel Paul W. Tibbets, Jr., one of the leading bomber pilots in the African Theater, to head the group. He furnished Tibbets with the latest models of the B-29 Superfortress and set up a reserve of additional bombers if more proved necessary.

Brigadier General Lauris Norstad, chief of staff, Army Strategic Air Force, cooperated with Groves on target selection and all aspects of delivery of the weapons on target. A special target selection group, consisting of Air Corps target specialists and Los Alamos scientists, was established and worked under the direction of Farrell. I had no responsibility for this part of the Manhattan Project, but from time to time, when I stopped in Washington, Farrell would show me what they were planning to do. I was particularly interested in the preparations for support of the ground forces invading Japan if strategic bombing of the selected target cities did not result in Japan's surrender and it became necessary to carry out the landings on the empire's home islands.

Farrell never interfered with my responsibilities but occasionally did make some very useful suggestions. In addition, he was most helpful on several occasions in getting approval from Groves on actions that he probably would not have approved for me. My suggestion to Groves that he have a deputy lightened my burden without resulting in my having an additional boss in Washington. Both Farrell and Kirkpatrick went to Tinian and stayed there until after the Nagasaki bomb was delivered.

Late in 1943 I had the pleasure of meeting Lord Cherwell, Churchill's personal scientific adviser. Cherwell, Chadwick, Oliphant, Bush, Conant, and Groves visited the CEW. After a tour of the plants, they came to our home for cocktails, dinner, and an after-dinner meeting. As usual, we had received plenty of restrictions from Groves pertaining to a British visit. I was not to show them certain aspects of the K-25

plant and not to visit the Clinton lab. In addition, because Cherwell
was a vegetarian, mashed potatoes with olive oil along with vegetables
would be a suitable dinner menu.

Groves requested that Jackie and her mother, who was living with
us, not attend the dinner even though it was at our home. He explained,
"The British consider all family members can be trusted and they
persist in talking about classified information in their presence. In
addition, your mother-in-law is French, and Bush and I have been
protesting to the president that the British are not enforcing security
with the French scientists involved with Chalk River. It is a sensitive
subject." In response, I told Groves very clearly that either cocktails,
dinner, and coffee at our home would include my wife, or the dinner
would have to be at the guest house. Jackie would solve the problem
by retiring to our bedroom immediately after dinner, and her mother
would visit the Petersons that evening.

As it turned out, Jackie changed her plans when I informed her
about the need for "mashed potatoes with olive oil" and "vegetables." "Instead of serving at the table, I'll set up a buffet; then anyone
who cares to can pour oil on his potatoes." At dinner, we learned that
Olipant also was a vegetarian. As a result of Groves' warning, the
British behaved very discreetly during dinner. Immediately on leaving
the dining room, however, they brought up the subject of the Lancaster
versus the B-29. While Groves tried to silence them, Jackie expedited
her departure. Actually she felt "rather flattered that Groves thought a
humanities major could possibly understand technical jargon."

Lord Cherwell questioned why we had selected the B-29 instead of
the larger British Lancaster to carry the bombs. He thought the size of
the B-29 bomb bay unduly restricted Los Alamos in designing the
weapons. He pointed out that the Lancaster would provide more lati-
tude in designing the bomb. Groves and Bush responded that the
shorter-range Lancaster might be suitable for the European Theater,
but it would require an island base much closer to Japan from which to
stage the mission. As a result, depending on the progress of the war in
the Pacific, we might not have captured a suitable base of operations
by the time the bomb was ready for use. Groves and Bush were not
about to consider any change in the decision to design the bomb to fit
into the B-29, but it still required a great deal of discussion before
Cherwell and Chadwick agreed to give up.

With that finally settled, we discussed many other subjects, including ones involving the rules of exchange of information and security

regulations. Cherwell thought the rules were too rigid. But despite some disagreements, the evening proved most pleasant.

In May 1945, I became involved in the process of getting money from Congress for the project. Prior to this, my only responsibility was to have Vanden Bulck provide Groves with all the information he needed to procure adequate funds. Initially the War Department allocated funds from money already available. However, in 1944, when increased funding was required, Under Secretary of War Patterson, felt it necessary to formalize the procedure.

Accordingly, in February 1944 Secretary Stimson, General Marshall, and Bush met in the office of the Speaker of the House of Representatives with Speaker Sam Rayburn, majority leader John W. McCormack, and minority leader Joseph W. Martin. They received a briefing on the status of the project and financial situation as well as estimates for future requirements.

The congressmen voiced their approval and agreed that expenditures were justified and that they would do everything possible to have the necessary funds included in the coming appropriation bill. They wanted to know how the request for appropriations would be inserted in the bill and said they would inform a few members of the Appropriations Committee that they had investigated this subject and that the item should not be questioned.

Subsequently, the majority and the minority leaders and the chairman and senior minority members of the Military Subcommittee of the Appropriations Committee were given essentially the same information. Nevertheless, we ran into a problem in February 1945 when it became necessary to transfer additional War Department funds to "expediting production" to make them available to the Manhattan Project. Congressman Albert J. Engel objected vigorously when he saw the request and demanded justification. As a result, the MPC decided that a small number of carefully selected legislators should visit the CEW and Hanford as well, if they desired. President Truman approved the idea, and in early May, with the House leadership concurring, he invited Congressmen Cannon, Snyder, Mahon, Taber, and Engel to visit the CEW.

Stimson gave us authorization to answer the congressmen anything they asked except the date the first weapon might be used and questions about our production schedules. Groves assigned Engel to me. I prepared for Engel by bringing along unit costs of many things, knowing he was interested in such matters because of his prior career as a brick contractor. The effort paid off because the detailed knowledge and our concern with costs impressed him. However, I remember in particular his comments when I explained to him how we had improved the chemical process for Y-12. I also showed him the chemical building we were constructing in case the improved method proved inadequate. He seemed reluctant to hear about it. He told me that now I had told him, if we failed, he could be held responsible by his electorate for the fact that we spent so much money on duplications. In turn I asked him, "What would you do in regard to this new building? If we find we need it and don't have it, we can't possibly deliver a U-235 bomb on schedule. All three U-235 plants here depend on this Y-12 chemical process to turn out the final enriched uranium to ship to Los Alamos." He still regretted very much having been informed but he thought we should go ahead.

Redundancy was at the heart of the Manhattan Project. Each of the uranium processes we built at the CEW served as a backup for the others. In fact, all the CEW U-235 enrichment plants were backups for the plutonium effort at Hanford or vice versa. Redundancy unquestionably increased the cost of the Manhattan Project, but we did not feel we dared take a chance concentrating on only one production plant, or even one type of bomb.

In any event, the hearing at Oak Ridge ended very pleasantly when Congressman Cannon, chairman of the committee, asked Congressman Taber whether he had any more questions to ask. (A month or so earlier, Taber and Cannon had come close to blows in a hall in Congress over expenditures.) Taber said he wanted to ask General Groves and Colonel Nichols one more question: "Are you sure you're asking for enough money?" Cannon commented, "Well, I never expected to hear that from you, John. Meeting adjourned."

Ultimately, the Manhattan Project received allocations of about $2.4 billion. Actual expenditures to October 1, 1945, totaled $1.845 billion. By the time the Atomic Energy Commission assumed control on January 1, 1947, we had spent $2.191 billion. Under today's conditions, it would be difficult if not impossible to accomplish the Manhattan Project in four times the time, and the cost would be at least thirty times more.

With production of U-235 and plutonium increasing, I decided to visit Los Alamos on May 1, 1945, to become better informed about their progress. In particular, I wanted to be certain we had the best means to determine the optimum percentage of U-235 we should produce at the CEW. This involved a new unit called "effective production unit" or EPU. Oppenheimer established his requirement in EPUs, and Peterson's task was to get the requirements for the first U-235 bomb at the earliest possible date.

This single day, at a crucial period in the development of the weapons at Los Alamos, was one of the most fascinating days I have ever spent. Oppenheimer had arranged that in addition to discussions on my primary problem concerning percent of enrichment, I should be informed about all aspects of their development work and plans for testing the implosion weapon. It proved to be an intensive education, and it emphasized the problems the scientists were encountering with the implosion weapon and the plans for testing it.

I now realized more fully from the firsthand briefing that we in the production field were not the only ones surmounting difficult obstacles. I particularly enjoyed seeing Fermi again and spent considerable time watching him conduct an experiment involving the passing of a cylinder of enriched uranium through a hollow cylinder of the same material to determine more accurately the critical mass.

Other discussions pertained to expected radiation effects and magnitude of heat and blast effect of the bomb. The decision had been made that an air burst would be utilized not only to achieve maximum blast effects but also to diminish residual radiation. The air burst would have three lethal effects: radiation, heat, and blast. As Conant once had commented to me: "If you were to ride down on the bomb, it would be impossible to determine which effect killed you. They will all be lethal."

While I was at Los Alamos, several scientists attempted to start a discussion with me about the housing or other living conditions at the laboratory. I had a set comment: "That is Groves' problem, not mine."

Three days after I visited Los Alamos, Joseph W. Kennedy from the laboratory met with Matthias, Thomas, and du Pont personnel at Chicago and agreed on plutonium purity specifications. It was now up to the CEW and the HEW to produce U-235 and plutonium as fast as possible and for Los Alamos to complete the development and then the fabrication of the weapons. Oppenheimer expected to be able to test the plutonium bomb in early July.

My spirits were soaring. I was almost certain that the CEW would produce the necessary amount of U-235 for one weapon before August 1 and a second one sometime in December. This assumes the second

U-235 weapon would be a gun type. Alternate planning contemplated designing an implosion bomb for U-235 if the Alamogordo test was successful. This would increase the rate of producing U-235 bombs.

After my return, Groves asked me to review production at the CEW and give him a firm date on which we would ship to Los Alamos the last of the U-235 needed for the first atomic bomb. President Truman was scheduled to go to the Potsdam Conference in July, and the final decisions would be made there for both the invasion of Japan and the use of atomic weapons. Groves had to furnish the information to Marshall and Stimson, and he warned me, "You had better be right, Nichols."

After Peterson and his production group had reviewed all the data and made their production schedules through July, Pete and I reviewed the data. We then went to the Y-12 plant and met with F. R. Conklin and James C. McNally, the two Tennessee Eastman officials in charge of the plant. To them, we revealed the amounts we needed, and I asked McNally when he would have it available for our courier to take to Los Alamos. I told him to give me the earliest possible date and suggested he should literally scrape the bottom of the barrel to advance the date ahead of normal production procedures.

When I later told them that Groves doubted their proposed target date, McNally offered to bet six bottles of Scotch that he would make it. I accepted the bet and later told Groves that I had made the bet in his name but would be happy to pay for half of it because I expected to enjoy drinking some of it. He asked, "Why in my name?" I explained, "I am sure that they would rather drink your Scotch than mine."

I received the following letter from McNally dated July 28, 1945: "Dear Colonel Nichols, Many thanks for your note of the 24th. All of us had great satisfaction in being able to accomplish our part of the program on time and I assure you of the continuance of our best efforts until the job is done. I know that this was one bet General Groves was pleased to lose. The prize will sit aside until the Japs are done for when I will look forward to the pleasure of celebrating the occasion with you at the General's expense." The CEW had completed its initial assignment ahead of schedule. Now it was up to Los Alamos and the 509th Composite Group.

*   *   *

From the very beginning of the atomic bomb development, the president of the United States made or approved all major policy decisions. President Roosevelt as well as Churchill, throughout the project, showed unusual personal interest in the secrecy, the urgency, and the priority that the atomic bomb project should have. Without this personal interest and the backing for top priority, the military never would have given the project the priority necessary to achieve success in time to help end the war.

Stimson, Bush, and General Marshall were the three individuals whom President Roosevelt had depended on most to recommend policy for the project and to see that it was carried out. I believe that Stimson, Bush, and Marshall expected that President Roosevelt would approve use of atomic weapons when they did become available. Only then would they finalize plans for their use.

Roosevelt always was aware of both the wartime and the postwar issues that development of the atomic bomb was creating. On September 17 and 18, 1944, he entertained Winston Churchill at Hyde Park, New York. On the second day of the visit, the men discussed both the military and the industrial applications of nuclear energy. Rejecting Niels Bohr's suggestion that the two countries inform the world about the Manhattan Project, they concurred that the project should continue to have the utmost secrecy. They then agreed that when a bomb finally became available, "It might perhaps, after mature consideration, be used against the Japanese, who should be warned that this bombardment will be repeated until they surrender." Both men then agreed to what was in essence an Anglo-American approach to the postwar world: "Full collaboration between the United States and the British Government in developing Tube Alloys (the British code name for the bomb project) for military and commercial purposes should continue after the defeat of Japan unless and until terminated by joint agreement."[1] This agreement was improperly filed at Hyde Park under "Tube Alloys" and so did not become known to Stimson or General Marshall until after the war, when the United Kingdom furnished a copy. Even then, Groves questioned the document's authenticity until the United States' copy was located.

I always believed that the atomic bomb should and would be used as soon as available. In addition, I had the impression that most of the

1. Hewlett and Anderson, *The New World, 1939/1946, A History of the United States Atomic Energy Commission.* (Vol. 1. University Park, Pa.: The Pennsylvania State University Press 1962), pp. 326–27.

key scientists with whom I had close contact believed likewise. Our objective was to shorten the war drastically by the surprise use of the atomic bomb and thereby save tens or hundreds of thousands of lives. That concept justified the urgency and the project's priority.

As related in the previous chapter, many scientists, particularly at the met lab, deliberated about how best to utilize the atomic bomb. Discussions also increased on the postwar need to control nuclear energy, both domestically and internationally; the need for some form of United Nations; and the political implications of using the bomb. Some of the scientists, including Franck, Niels Bohr, and Compton, wrote letters or made personal contact with higher-level authorities, including Vice President Wallace and members of the White House, in an effort to ensure that their concerns about these issues were being addressed.

Meanwhile, the U.S. Joint Chiefs of Staff were finalizing plans for the invasion of Japan, with the full recognition that such an invasion would be a fierce and bloody struggle. The plans were based on one assumption stated in their introductions: "That the Japanese will continue the war to the utmost extent of their capabilities and will prepare to defend the main islands of Japan utilizing all available means. That the operation will be opposed not only by the available organized military forces of the Empire, but also by a fanatically hostile population."[2] The invasion date was tentatively set for November 1, 1945, and it was expected that the war would end sometime in 1946.

I believe that Stimson realized that after the death of President Roosevelt, greater responsibility for making the recommendations or decisions concerning the use of the weapon as well as the postwar policy for domestic and international control of atomic energy fell on him. In my opinion, Stimson was the best-qualified individual in the United States to undertake this responsibility. He had the experience of having been secretary of state as well as the wartime experience of being secretary of war. He held the respect of the nation. Perhaps most important, he had direct knowledge of the atomic bomb project from the very beginning and so was completely familiar with the unique aspects of atomic weapons. He believed their use should require a presidential decision.

On May 2 Stimson received approval from President Truman to

2. RG 165, Records of the WD General and Special Staffs, Operations Division, Boxes 1842, 1843, National Archives.

appoint an "interim committee" to advise the president on the use of the atomic bomb. The committee consisted of Stimson as chairman; Bush; Conant; Karl T. Compton, president of M.I.T.; Under Secretary of Navy Ralph A. Bard; Assistant Secretary of State William L. Clayton; and James F. Byrnes, representing the president. As late as March 3, 1945, Byrnes, then serving as head of the Office of War Management, had written a critical memo to the president expressing strong reservations about the continued large expenditures of money for the Manhattan Project.[3]

Roosevelt sent the memo to Stimson saying, "I think you should read this. Please speak to me about it at your convenience."[4] After reading the memo, Stimson recorded in his diary that it was "rather a jittery and nervous memorandum and rather silly. . . ." Both men, of course, knew more about the project than Byrnes did, and in Stimson's meeting with Roosevelt on March 15, Stimson outlined the current status of the work. He went over the two schools of thought existing in respect to control of atomic energy after the war if the project succeeded. He explained that one group favored "the secret close-in attempted control of the project by those who control it now, and the other being the international control based upon freedom both of science and of access."

Stimson said he believed that a decision must be reached "before the first projectile is used and that he must be ready with a statement to come out to the people on it just as soon as that is done." Roosevelt agreed, and so with success by then close at hand, they ignored Byrnes' suggestions. On a whole, the secretary concluded that the discussion "was successful."[5] But before he and Roosevelt could draw up any proposal, the president died. As a result, Stimson had had to educate the new president on the project, and Jimmy Byrnes soon found himself sitting in judgment on the use of the bomb as President Truman's personal representative.

The interim committee first met informally on May 9, in Stimson's office. The secretary explained that in addition to providing recommendations on the use of the bomb, the committee was to study and report on the question of publicity about the bomb if it should be used, postwar atomic research and development, and postwar control of the

3. Manhattan Project Records, Office of the Commanding General Files, RG 77, Box 8, File 20. James F. Byrnes to the President, March 3, 1945.
4. Ibid., memo attached to Byrnes's letter.
5. Stimson diary, April 25, 1945.

bomb. In the committee's second meeting, on May 14, it appointed, on Conant's recommendation, a scientific panel consisting of Compton, Lawrence, Fermi, and Oppenheimer to provide advice from the perspective of the men working on the project. They were also closer to those scientists who were disturbed about the lack of or possible nature of the top policy decisions.

Meanwhile, on May 7, 1945, the German High Command surrendered unconditionally to the Allied Armies under General Eisenhower. This strenghtened our hope that the use of the atomic bomb would hasten the surrender of Japan. To this end, the interim committee met on May 31 and June 1. During the morning session, which Groves, Marshall, and the four members of the scientific committee attended, Stimson led a discussion of the future of the Manhattan Project, continued research on atomic energy, and the issue of international control.

In the afternoon session, with Marshall absent, the committee devoted the entire time considering the use of the bomb against Japan. There was discussion concerning the advisability of giving the Japanese some form of harmless demonstration to convince them of the power of this new weapon. The many arguments against this also were considered. Oppenheimer could think of no demonstration sufficiently spectacular to convince the Japanese to surrender. In addition, the demonstration might be a dud, and probably most important, any harmless demonstration would result in losing the important shock effect of surprise. Ultimately, Stimson expressed the conclusion that the weapons should be used without warning in a manner that would "make a profound psychological impression on as many inhabitants as possible." Conant suggested and Stimson agreed that "the most desirable target would be a vital war plant employing a large number of workers and closely surrounded by workers' houses."[6]

After this decision, Groves reported that the project had been plagued since its inception by the presence of certain scientists "of doubtful discretion and uncertain loyalty." However, the committee agreed that nothing should be done until after the test of the weapon or its use in combat, and then the questionable people would be removed. The committee also decided that the four scientists could tell their personnel about the committee's work in dealing with the problems of

6. Minutes of Interim Committee, May 31, June 1, 1945, Harrison-Bundy Papers, RG 77, File 106.

future control, project organization, future legislation, and publicity about the bomb effort once it had been used. While the scientists were not to identify the committee's members, they were to give their colleagues the impression that the government was taking a most active role in all major decisions.[7]

With all aspects of the project seemingly on track toward a July test of the plutonium weapon and August use, a more personal crisis suddenly emerged. On June 5, Groves asked that I come to Washington the next day. When I arrived at his office, we met alone. He started off rather enigmatically: "I have tied it up." That startled me, since I had never before heard him admit any error. He then went on to explain that on the fourth, Somervell had called him and asked if I could be transferred to a very important job that would call for an immediate promotion to brigadier general.

Groves rather ruefully told me that he promptly responded, "You can't have him." At that Somervell replied, "Who do you think you are to tell me I can't have any engineer officer I want? Orders will be issued today." And he hung up. Groves then produced a copy of the orders dated June 4, directing my relief as district engineer of the MED and assigning me to be chief control officer for Army Service Forces in General Somervell's office.

Groves again acknowledged he had made a mistake, saying, "I never should have told General Somervell that he could not have you. After Somervell cools down, I think I can get the orders revoked. In the meantime, go over to Robinson's office and report to him and find out what your new assignment is all about."

For some reason, I was not particularly disturbed by Groves' unexpected news, at least not sufficiently to bother to alert my wife. I guess I just did not think it would happen. Of course, I would have liked a promotion. But sometime earlier, Groves had told me that Stimson had assured him that if we were successful, I would be promoted to brigadier general. Moreover, I certainly wanted to see the atomic bomb project through to its conclusion, good or bad, and I was confident that we would succeed. Perhaps I also had confidence that Groves would be able to reverse my orders.

In any case, I immediately went to the Pentagon to see Robbie (Major General Clinton F. Robinson) and find out what the new assignment was all about. Robbie was a very good friend. We had

7. Ibid.

served together in Nicaragua. The summer I came to Cornell, he was there, and Mary Robinson had arranged my first date with Jackie.

Robbie thanked me for reporting to his office so promptly, saying that he had asked Somervell to transfer him to the Pacific Theater so he could get some field experience. Somervell had agreed, provided Robbie could find a suitable replacement. Robbie had heard that considerable doubt existed in the minds of some very important people about the chances of success for the Manhattan Project. As a result, he felt he was doing me a favor by getting me out of the project.

He knew I had signed most of the contracts and was responsible not only for administration but also for many of the technical decisions and thus would be likely to get much of the blame for a failure. He said, "Let 'Goo Goo' Groves take the blame all by himself." He concluded, "I don't know how you feel about your chances of success, and I am not asking you. Also, I am not asking you whether you prefer your present job, which is like a big frog in a little puddle, or would like my job, which is a little frog in a great big puddle. I understand that it is all settled and you have your orders." I confirmed that Groves had shown me a copy but that I had not yet received the original orders. "Fine. Report for work in about two weeks," he said.

I didn't have long to wait to hear about my future. Groves first asked General Marshall to revoke the orders or have Somervell do so. Marshall refused to intervene, telling Groves that he never should have made such a reply to Somervell. However, Groves had better luck with Stimson. The secretary agreed to ask Somervell to postpone my reporting date until September 1, 1945, because it was critical that I remain district engineer until then. As a result, on June 12 my orders were revised to state that I should report to Somervell's office on September 1, 1945.

On June 14, in preparation for the Potsdam Conference, the Joint Chiefs of Staff again agreed that a November 1 invasion of the southernmost main Japanese island, Kyushu, should be the main initial effort against Japan. On the eighteenth, President Truman received a briefing on the invasion plans. The atomic bomb also was considered, but at that time there still was no certainty that the bomb would succeed.

On June 16, the scientific panel met at Los Alamos to prepare recommendations for the interim committee on the future use of atomic weapons, on the immediate military use, and on future research after the war. The scientists were in favor of using the bomb when ready "to promote a satisfactory adjustment of our international relations. At the same time, we recognize our obligation to our nation to use the weapons to help save American lives in the Japanese war."[8]

The panel recommended that all the Allies be advised of the planned use of the weapon and then noted that diverse opinions now existed on the use of the weapon, ranging from a technical demonstration to direct military action. While some were recommending that the United States not use the bomb under any circumstances, Oppenheimer, writing for the panel, noted that others "emphasize the opportunity of saving American lives by immediate military use, and believe that such use will improve the international prospects, in that they are more concerned with the prevention of war than with the elimination of this specific weapon. We find ourselves closer to these latter views; we can propose no technical demonstration likely to bring an end to the war; we see no acceptable alternatives to direct military use." The scientists concluded that despite working on the project they had "no claim to special competence in solving the political, social, and military problems which are presented by the advent of atomic power."[9]

On June 21, the interim committee considered the scientific panel's comments concerning the use of the bomb as well as the information that a group of scientists at Chicago feared that dropping the bomb might impair the chances for international control. This group believed that the United States should limit itself to a purely technical demonstration. Since the scientific panel had offered no acceptable alternative to direct military use, the interim committee reaffirmed its position that the weapon should be used at the earliest opportunity, without warning, and against a war plant surrounded by homes or other buildings most susceptible to damage. The committee also accepted the recommendation that Truman inform Stalin about our progress on the bomb and its proposed use against Japan.

On July 4, in accordance with the Quebec agreement, Great Britain and Canada were informed at a meeting of the Combined Policy Committee that the United States intended to use the bomb, and the committee approved the decision. All through these discussions, the cost in American lives of an invasion of Japan was the principal argument in favor of using the atomic bombs directly against Japan. Other factors,

8. Reports of the Scientific Panel to the Interim Committee, June 16, 1945, RG 77, Harrison-Bundy Papers, File 76.
9. Ibid.

such as the effect on future international control of atomic weapons, warning Japan, war aims against Japan, informing all our Allies prior to use, Russian participation against Japan, and future Russian relations were considered.

At the Potsdam Conference, plans for action against Japan moved rapidly forward. On July 18, 1945, Groves informed Stimson that the Alamogordo test was successful. In turn, the secretary informed Truman and Churchill. On July 24, the Combined Chiefs of Staff met with Churchill and Truman, and they approved the November 1, 1945, invasion of Kyushu. The Allies informed Stalin of these plans, and in a conversation alone with Stalin, Truman advised him about a new weapon of unusual destructive force, and Stalin concurred in use of such a weapon against Japan. This vague reference to a new weapon of unusual destructive force was not quite as precise as to what the scientific panel had in mind. However, in view of our later discoveries about Russian espionage at Los Alamos, I feel certain that Stalin understood exactly what Truman was describing.

With the approval of Churchill and Truman, on July 25, Stimson and Marshall approved the operational orders for the first use of the atomic bomb as soon after August 3, as weather permitted, on one of the four approved targets: Hiroshima, Kohura, Niigata, or Nagasaki. (Stimson had eliminated Kyoto on the grounds that it was the center of Japanese culture and religion.) The Potsdam Proclamation, which the Big Three issued, called upon the Japanese government to proclaim unconditional surrender of all Japanese armed forces and was more generous than the original call for just "unconditional surrender" and warned that the alternative was "prompt and utter destruction." In response, on July 29, Japan started broadcasting that the government would ignore the proclamation and continue the war.

During the step-by-step process by which Truman reached his decision to authorize the first use of the atomic bomb, I had little part to play. Groves did keep me informed of developments, however, and I was more directly involved in responding to the unrest of the scientists at the met lab. I had several discussions with Groves and Conant about the issues the scientists were raising, and in particular about moral aspects of the radiation effects of the bomb.

From the standpoint of morality, the three of us could see little difference in death by radiation as compared to death by heat or blast. All three can be lethal and would vary depending on the magnitude of the bomb and the height of the burst. A low height of burst would increase radiation effects by increasing the amount and the duration of the residual radiation. In contrast, a greater and optimum height could be calculated for maximum blast effect and a corresponding maximum area of destruction. For optimum height, it was estimated that radiation would be lethal for a radius of about two thirds of a mile, while the blast effect would cover a larger area. However, even with the best of calculations, we could not be certain of the actual energy release that would be achieved by either of the two types of bombs.

We were anxious to obtain the maximum shock effect from the first use of the bomb. If the impact were sufficiently devastating and if the surprise should catch the Japanese completely off balance, chances were that Japan would find it futile to continue their resistance. Further, to achieve even greater shock effect, it was felt necessary to deliver the second bomb as soon as possible after the first and to follow up with additional bombs until the Japanese government surrendered.

To the military, the objective was to destroy the enemy's will to fight in order to gain a quick capitulation. Destroying cities and killing civilians, even if there is a military or industrial target involved, is not a pleasant assignment. But the military man is more conditioned for such events. He has to be. He has been trained to carry out his assignments. Fire-bombing and high-explosive bombing had resulted in more casualties than we anticipated or that actually did occur at Hiroshima and Nagasaki. An important difference was that in these conventional bombings, such as of Leipzig and Tokyo, five hundred to six hundred planes had been required instead of the single one we intended to use.

Despite this military reality, during June and July many scientists continued to be agitated about the anticipated use of atomic weapons. On June 12, Compton forwarded to Stimson's assistant George Harrison a memorandum for the secretary of war from the met lab scientists titled "Political and Social Problems." In his cover letter, Compton said the scientists wanted a technical rather than a military demonstration "preparing the way for a recommendation by the United States that the military use of atomic explosives be outlawed by firm international agreement. It is contended that its military use by us now will prejudice the world against accepting any future recommendations by us that its use be not permitted." Compton noted, however, that the failure to make a military demonstration may make the war longer and more expensive of human lives and concluded that "without a military demonstration, it may be impossible to impress the world with the need for national sacrifices in order to gain lasting security."[10]

During a visit to Chicago in June, Kay Tracy, Compton's secretary, told me that many scientists favored using the bomb as soon as possible. By her rough check, they were individuals who like herself had a close relative or a close friend in the Pacific. She was anxious for peace and hoped that counterpetitions would be signed.

However, Leo Szilard at the met lab drew up a petition on July 3, 1945, addressed to the president, requesting that the bomb not be used. In his cover letter of July 4, requesting scientists to sign the petition, Szilard wrote, "The fact that the people of the United States are unaware of the choice which faces us increases our responsibility in this matter since those who have worked in 'atomic power' represent a sample of the population and they alone are in a position to form an opinion and declare their stand."[11]

Criticism of the original petition caused Szilard to redraft it, and the final version, dated July 17, 1945, contained seventy signatures. It stated: "The war has to be brought speedily to a successful conclusion and attacks by atomic bombs may very well be an effective method of warfare. We feel, however, that such attacks on Japan could not be justified, at least not unless the terms which will be imposed after the war on Japan were made public in detail and Japan were given an opportunity to surrender." The signers concluded that the United States has "the obligation of restraint and if we were to violate this obligation our moral position would be weakened in the eyes of the world and in our own eyes. It would then be more difficult for us to live up to our responsibility of bringing the unloosened forces of destruction under control."[12]

Ironically, the petition had the opposite effect from what Szilard expected. It stimulated a great deal of discussion and reaction. Farrington Daniels, the new director of the met lab, did a survey for Compton on opinion there, giving the responders five choices: (1) The use of the weapon in the most militarily effective manner. This received twenty-three votes, or 15 percent of the sampling. (2) Giving a military demonstration against one of the Japanese cities, to be followed by a renewed opportunity for surrender before full use of the weapon was employed. This received sixty-nine votes, or 46 percent of

10. Harrison-Bundy, File 76.
11. Ibid.
12. Ibid.

the sampling. (3) Giving an experimental demonstration in this country with representatives of Japan present followed by a new opportunity for surrender before full use of the weapons is employed. This received thirty-nine votes, or 26 percent of the sampling. (4) Withholding military use of the weapons but making public experimental demonstrations of their effectiveness. This received sixteen votes, or 11 percent of the sampling. (5) Maintaining as secret as possible all developments of the new weapons and refraining from using them in this war. This received three votes, or 2 percent of the sampling. The bottom line was that 61 percent of the scientists responding to the questions favored using the bomb against Japan in a military setting.

The petition and polling of the scientists produced some individual responses on both sides. On July 14, one individual working at the Clinton lab wrote to Whitaker, director of the lab under the University of Chicago contract. Responding to the argument that the bomb should not be dropped on a technologically inferior nation, the writer noted, "Some of us can remember the days in the dismal thirties when Japan rode ruthlessly over the peaceful and defenseless Chinese." Since the goal of the war remained unconditional surrender, he concluded, "Therefore, it behooves us to support our brothers and buddies overseas with the best and most potent weapons it is in our power to devise, to end this damnable confusion and strife, and to allow the world to return to peaceful pursuits as soon as possible."

He saw the atomic bomb as "essentially nothing more than extrapolation of existing devices (although that extrapolation may be almost LOGARITHMIC!)." But its very power was to him the crux of the matter: "It is hard to imagine anything more conclusive than the devastation of all the eastern coastal cities of Japan by fire bombs; a more fiendish hell than the inferno of blazing Tokyo is beyond the pale of conception. Then why do we attempt to draw the line of morality here, when it is a question of degree, not a question of kind?" As a result, he argued that we should use the bomb "in whatever manner will produce optimum results in the way of shortening the war and saving American lives."[13]

Commenting on the debate, another of the Clinton lab scientists wrote directly to Compton on the sixteenth: "I believe that the issues at stake including any brought to light by recent discussions here, demand none other than the greatest strategic use of the weapon for the

13. Harrison-Bundy, File 76.

smallest loss of American life and the most conclusive victory over Japan.'' He said that he had formed his opinion after evaluating four major points: (1) Any other unrestricted and formidable use would constitute a serious breach of faith with those still fighting ''the fanatical methods of Japanese warfare.'' (2) An impressive victory over Japan with an impressive weapon should inspire American diplomacy and world opinion to tame effectively the growing Russian intransigence. (3) It was only remotely possible to have a desirable effect on world opinion by simply telling the world about the bomb. And (4) since no political or moral issue ''should take preference over winning the war,'' he concluded that the [Army] General Staff should have all discretion to use the weapon to bring about the Japanese surrender.''[14]

During its discussions the interim committee had considered future ramifications of using a new weapon, and those of us in the military were not oblivious to the postwar situation. Both Groves and I always had believed we would face a serious threat from the Soviet Union once it had begun to recover from the destruction Germany had inflicted. Nevertheless, our immediate concern remained ending the current war at the earliest possible time. We knew full well the magnitude of what conventional and incendiary bombs had done to Dresden in February and to Tokyo in March and felt that atomic weapons would be no more horrible and offered the probability of ending the killing for good.

Despite our conviction that the bomb should be used, we were, of course, aware of the differing opinions among the scientists, and I had kept Groves abreast of the progress of Szilard's petition. On the twenty-third, Groves called me from Washington, wanting to know ''at once'' the results of Daniels' poll. Although Compton had received both the petitions and the poll, he had not yet turned them over to me. Consequently I went to his CEW office, where he provided me a written summary as ''objectively'' as he could.[15]

After I read it I called Groves to advise him of the figures. He then asked me to confirm where Compton personally stood on the use of the bomb. Returning to his office, I told him, ''Washington wants to know what you think.'' Instead of answering, he asked me, ''Where do you stand, Nick?'' I believe he asked the question to give himself more time to think because I am sure he already knew my position. I had

14. Ibid.
15. Compton, *Atomic Quest*, p. 246.

reached my own conclusion long before and told him, "I believe we should use the atomic bomb in the most effective manner to bring prompt Japanese surrender and save American lives." Apart from any desire to have my contribution to the war effort successfully utilized or my professional belief that a nation must use all means at its disposal to end a war, I simply had too many friends in the Pacific Theater and already had lost too many others to have much compassion for our enemy.

As I told him that, I could see that Compton wanted more time to give his answer, and I did not press him. I waited patiently. He recalled in *Atomic Quest,* "What a question to answer! Having been in the very midst of these discussions, it seemed to me that a firm negative stand on my part might still prevent an atomic attack on Japan. Thoughts of my pacifist Mennonite ancestors flashed through my mind. I knew all too well the destruction and human agony the bombs would cause. I knew the danger they held in the hands of some future tyrant. These facts I had been living with for four years. But I wanted the war to end. I wanted life to become normal again. I saw a chance for an enduring peace that would be demanded by the very destructiveness of these weapons. I hoped that by use of the bombs many fine young men I knew might be released at once from the demands of war and thus be given a chance to live and not to die."[16]

The next day at Potsdam, Truman and Churchill made the final decision to use the bomb and also to inform Stalin thereof. I did not know whether Groves would have forwarded a negative opinion to Stimson or whether such a stand would have made a difference. Given Compton's stature and involvement in the bomb project from its inception, however, reluctance on his part to advocate use of the bomb would undoubtedly have had an impact on the secretary and might have led to further discussions and perhaps a delay in using the weapon. But Compton's vote was not negative; he finally told me, "My vote is with the majority. It seems to me that as the war stands the bomb should be used, but no more drastically than needed to bring surrender."[17]

The next day, Compton gave me the signed petitions and the poll results, noting in his cover letter that "the strongly favored procedure is to give a military demonstration in Japan, to be followed by a

16. Ibid, p. 247.
17. Ibid.

renewed opportunity for surrender before full use of the weapons is employed. This coincides with my own preference, and is, as nearly as I can judge, the procedure that has found most favor in all informed groups where the subject has been discussed.[18]

Thereupon I sent to Groves on July 25 Compton's letter, the survey, Szilard's petition, and the two letters quoted previously. In my cover letter I wrote, "It is recommended that these papers be forwarded to the President of the United States with the proper comments. It is believed that by such action and example, it will be more nearly possible to control the individual activities of the various scientists who have ideas regarding the political and social implications concerning use of the weapon and to confine their activities to proper channels where security for the project will not be jeopardized." I also noted that contrary to Szilard's hopes, the collected papers "generally support the present plans for use of the weapons."[19]

On August 1, Groves delivered the petitions and Compton's summary of the poll to Stimson's office. As I understand the sequence of events, Stimson filed the papers in the S-1 files and never sent them to the president, who did not return from Potsdam until after August 6, when the first bomb had been dropped on Hiroshima. I do not think it is logical to have expected Stimson to take any other action on the petitions. The decision to drop the two available atomic bombs on the approved targets had been made at the highest level after careful and great consideration of all factors. The scientific panel had expressed its opinion. Truman and Churchill had made their decision to accept the recommendations of the interim committee just a week before. Morever, the majority opinion of the met lab scientists was not materially different from the decision made and the manner in which it was executed. However, as usual in public issues of this type, a vociferous minority makes itself heard to the extent that the public gets the impression that most if not all scientists opposed the use of the bomb to end the war. Many did, but many more approved use of the weapon they had worked so long to perfect.

18. Harrison-Bundy, File 76.
19. Ibid.

11

THREE WEEKS
ONE SUMMER, 1945

THE DECISION OF TRUMAN and Churchill to drop the atomic bomb came after the success of the Trinity test on July 16, 1945, near Alamogordo, New Mexico. Many writers have described the preparations for the test, and some have very well captured the tensions that built up in the last few hours before the scientists detonated the device sitting on the one-hundred-foot steel tower. I have very little to add to the narrative, since I did not attend the test.

Many people probably believe I must have been greatly disappointed that I was not present at this dramatic scene. Except for Groves and Fermi, who was at Los Alamos at the time, there was no one there who had any major responsibility for producing the plutonium used in the Alamogordo bomb. Franklin Matthias, responsible for the plutonium operation at Hanford, has expressed understandable disappointment about not being there, while acknowledging that "there were lots of things I wanted that I never got."

Frank had not known the specific plans or date for the test. However, he had been making shipments of plutonium twice a week to Los Alamos since February 1945, and his contacts with the laboratory probably would have provided him enough information to know that the test would be taking place during July. In addition, John Lansdale, Groves' security man, had asked Frank to help him prepare a cover story about an explosion in the New Mexico desert. Consequently, when Groves, Bush, and Conant showed up at Hanford in mid-July with Los Alamos as their next destination, Frank knew the reason for the trip and

191

asked Groves, "Hey, how about letting me go along to see the test."

Groves responded, "What test?" Matthias told him, "Well, it must be about time. I have spent some time writing up a phony news report about an explosion of an ammunition dump out in the desert." Frank recalls that Groves "kind of snickered" and told him, "You can't go. I decided I am not going to let you or Nichols go because you are running production. And if we lost you, it is a lot worse for the program than if I get killed." Looking back, Frank says, "It was just that simple. I didn't argue with him."[1]

Groves never told me I could not go. In going over the list with me in early July of who shall be included, he began to worry about the number of individuals who were planning to go "fishing." He told me he feared that if Matthias and I went to Trinity, it might alert too many people at the CEW and the HEW about what was taking place. The leak might endanger security, not to mention creating even more requests from people wanting to attend. I agreed not to go, thereby eliminating the prospects of many key men.

In hindsight, should I regret having missed Trinity? The answer is no. However, I was disappointed when I missed the first chain reaction in Chicago. To me, that marked a greater milestone in the development of atomic power and atomic weapons.

Compton also chose not to go in order to limit the number and preserve secrecy. As a matter of course, he had received an invitation from Oppenheimer "to accompany him on a fishing trip." Although his met lab had done all the research and development for Hanford, Compton also believed that it was "highly important that news of the test should not get around" the lab and so concluded, "I could not absent myself at that time without giving rise to questions."[2]

Compton and I did not have to wait long for the satisfaction of learning that our efforts had met with success. Once he returned to Los Alamos, Oppenheimer called Compton with the news: "You'll be interested to know that we caught a very big fish." I also made an effort to find out about the fish. When I agreed not to go to Trinity, Groves suggested that I not make any arrangements to get information about the results. He said he would inform me personally when he returned to Washington.

1. Lawrence Suid interview with Franklin Matthias, April 4, 1983, on file at Office of History, OCE.
2. Compton, *Atomic Quest*, pp. 214–15.

To that, I made no response because I wanted to know the results as quickly as possible. Therefore, I asked both Staff Warren and Hymer Friedell, who went to Trinity to monitor the radiation caused by the explosion, to contact me as soon as possible to tell me the results.

To advance the deception of the people at Oak Ridge, Jackie, Virginia Olsson, and I went to Atlanta on July 15 to have a night on the town. It also gave the ladies some time for shopping in the big city, and I diverted my mind from events in Alamogordo by visiting Fourth Army headquarters in the morning. We returned to Oak Ridge in early afternoon so I could receive calls from Warren and Friedell.

Dr. Friedell was not able to phone me because an intelligence officer held him incommunicado in his hotel in Albuquerque. But Dr. Warren called me with the good news that ended a tense period of wondering whether my optimism really was justified.[3]

Groves called me on July 17, and he followed up the phone account by sending a courier with a letter marked "Secret—destroy after reading." He did not want it to be entered into the files at the CEW.

In his letter, he advised me that the test was "a far greater success than anyone had anticipated . . . everything we had hoped for was proven insofar as the test at New Mexico permitted." He said he was trying to keep the information as secret as possible. He explained that he had instructed Lawrence not to talk to anyone who was not at Trinity about the test, saying, "I think what he would tell you would be of no value to you and would merely encourage him in future discussions with other people. For that reason, please do not talk to him." He then directed me: "After you have personally destroyed this note I would like to have a letter informing me of that fact and of the fact that no one else has been made acquainted with the contents." Groves was, of course, not insensitive to my role in the project and closed, "I hope that it will not be too long before I can properly handle the question of recognition of certain people who are so responsible for our success, particularly yourself."[4] I did obey his instructions, destroying the letter and sending him a note to that effect. However, later, when I replaced Groves as chief of armed forces special weapons, I found a copy of the letter in the files.

Although I was not present at Trinity, I was able to gain an appreciation of the tremendous power of the device by screening the film as

3. Lawrence Suid interview with Dr. Hymer Friedell, August 23, 1983, on file at Office of History, OCE.
4. Groves to Nichols, July 17, 1945, KDN Files, Office of History, OCE.

soon as it arrived in Washington. I also read the eyewitness accounts of several people who had been there, particularly those of Groves and Farrell.

In his memorandum to Secretary Stimson on July 18, Groves did not provide "a concise, formal military report, but an attempt to recite what I would have told you if you had been here on my return from New Mexico." In an exuberant mood, Groves exclaimed, "What an explosion." He continued, "The test was successful beyond the most optimistic expectations of anyone. Based on the data which it has been possible to work up to date, I estimate the energy generated to be in excess of the equivalent of 15,000 to 20,000 tons of TNT, and this is a conservative estimate. Data based on measurements which we have not yet been able to reconcile would make the energy release several times the conservative figure. There were tremendous blast effects. For a brief period there was a lighting effect within a radius of 20 miles equal to several suns in midday; a huge ball of fire was formed which lasted for several seconds. This ball mushroomed and rose to a height of over 10,000 feet before it dimmed."[5]

After this uncharacteristic burst of enthusiasm, Groves provided Stimson with details of the test. Of particular interest was his description of ground zero: "A crater from which all vegetation had vanished, with a diameter of 1,200 feet and a slight slope toward the center, was formed. In the center was a shallow bowl 130 feet in diameter and 6 feet in depth. The material within the crater was deeply pulverized dirt. The material within the outer circle is greenish and can be distinctly seen from as much as 5 miles away. The steel from the tower was evaporated. Fifteen hundred feet away there was a 4-inch iron pipe 16 feet high set in concrete and strongly guyed. It disappeared completely."

In preparing for the test, Oppenheimer had finally decided not to use Jumbo, a massive steel cylinder weighing 220 tons, which had been built to hold the device and contain the plutonium in case the explosion fizzled. Instead, the scientists anchored the container within a strong steel tower seventy feet high and embedded in concrete about one-half mile from ground zero. Groves explained that the blast "tore the tower from its foundations, twisted it, ripped it apart and left it flat on the ground. . . . None of us had expected it to be damaged."

5. Manhattan Project Records, Office of the Commanding General Files, RG 77, Box 2, Folder 4. Memorandum for the Secretary of War, July 18, 1945. Subject: The Test.

Jumbo itself was not damaged. Nevertheless, Groves observed that the effects on Jumbo's tower "indicate that, at that distance, unshielded permanent steel and masonry buildings would have been destroyed" and so concluded, "I no longer consider the Pentagon a safe shelter from such a bomb."[6]

Groves attached to his memorandum General Farrell's impressions of the event. To me, they best capture the period immediately before and after the test: "The scene inside the shelter was dramatic beyond words. In and around the shelter were some twenty-odd people concerned with last-minute arrangements prior to firing the shot. . . . For some hectic two hours preceding the blast, General Groves stayed with the Director, walking with him and steadying his tense excitement. Every time the Director would be about to explode because of some untoward happening, General Groves would take him off and walk with him in the rain, counseling with him and reassuring him that everything would be all right."

Farrell reported that twenty minutes before zero hour, Groves left for the base camp site because it provided better observation and "because of our rule that he and I must not be together in situations where there is an element of danger, which existed at both points. . . . As the time intervals grew smaller and changed from minutes to seconds, the tension increased by leaps and bounds. Everyone in that room knew the awful potentialities of the thing that they thought was about to happen. The scientists felt that their figuring must be right and that the bomb had to go off but there was in everyone's mind a strong measure of doubt. The feeling of many could be expressed by 'Lord, I believe; help Thou mine unbelief.' We were reaching into the unknown and we did not know what might come of it. It can be safely said that most of those present—Christian, Jew and atheist—were praying and praying harder than they had ever prayed before. If the shot were successful, it would be a justification of the several years of intensive effort of tens of thousands of people—statemen, scientists, engineers, manufacturers, soldiers and many others in every walk of life.

"In that brief instant in the remote New Mexico desert, the tremendous effort of the brains and brawn of all these people came suddenly and startlingly to the fullest fruition. Dr. Oppenheimer, on whom had rested a very heavy burden, grew tenser as the last seconds

6. Ibid.

ticked off. He scarcely breathed. He held on to a post to steady him-self. For the last few seconds, he stared directly ahead and then when the announcer shouted, 'Now!' and there came this tremendous burst of light followed shortly thereafter by the deep growling roar of the explosion, his face relaxed into an expression of tremendous relief. Several of the observers standing back of the shelter to watch the lighting effect were knocked flat by the blast.''

According to Farrell, the blast relieved the tension in the room and set off a wave of celebration: ''No matter what might happen now, all knew that the impossible scientific job had been done. Atomic fission would no longer be hidden in the cloisters of the theoretical physicists' dreams. It was almost full grown at birth. It was a new force to be used for good or for evil. There was a feeling in that shelter that those concerned with its nativity should dedicate their lives to the mission that it would always be used for good and never for evil.'' More important for the moment, Farrell noted that ''there was a feeling that no matter what else might happen, we now had the means to ensure it's [the war's] speedy conclusion and save thousands of American lives.''

Like Groves had done in describing the blast to Stimson, Farrell captured his impressions in unmilitarylike terms: ''The effects could well be called unprecedented, magnificent, beautiful, stupendous and terrifying. No man-made phenomenon of such tremendous power had ever occurred before. The lighting effects beggared description. The whole country was lighted by a searing light with the intensity many times that of the midday sun. It was golden, purple, violet, gray and blue. It lighted every peak, crevice and ridge of the nearby mountain range with a clarity and beauty that cannot be described but must be seen to be imagined. It was that beauty the great poets dream about but describe most poorly and inadequately. Thirty seconds after the explosion came, first, the air blast pressing hard against the people and things, to be followed almost immediately by the strong, sustained, awesome roar which warned of doomsday and made us feel that we puny things were blasphemous to dare tamper with the forces hereto-fore reserved to The Almighty. Words are inadequate tools for the job of acquainting those not present with the physical, mental and psy-chological effects. It had to be witnessed to be realized.''[7]

Bad weather in the hours before the Trinity test prevented the two B-29 observation airplanes from being over the area when the device

7. Ibid.

was detonated. As a result, as Groves explained to Stimson in his memorandum, "Certain desired observations could not be made and while the people in the airplanes saw the explosion from a distance, they were not as close as they will be in action. We still have no reason to anticipate the loss of our plane in the actual operation, although we cannot guarantee safety."[8]

In regard to the matter of radioactive fallout from the test, Groves reported that the resulting cloud "deposited its dust and radioactive materials over a wide area. It was followed and monitored by medical doctors and scientists with instruments to check its radioactive effects. Where here and there the activity on the ground was fairly high, at no place did it reach a concentration which required evacuation of the population. Radioactive material in small quantities was located as much as 120 miles away. The measurements are being continued in order to have adequate data with which to protect the government's interest in case of future claims. For a few hours, I was none too comfortable about the situation." Groves explained that he had stationed observers as far as two hundred miles away from Trinity to check on blast effects, property damage, radioactivity, and reactions of the population. In fact, once reports began to come in, he found that no people had received injuries and no property had been damaged.[9]

Groves had brought in Drs. Warren and Friedell to monitor the radiation fallout and determine if and when it became a danger. Friedell recalls that there "was considerable discussion about this. As a matter of fact, both Stafford Warren and I were there and we both were at the test and reviewed the possibilities of what might happen if the winds were in the wrong direction and so on. And as a matter of fact, Stafford Warren, who eventually made the decision, insists, and I have no reason to doubt him, that he really postponed the test for a while because he felt the winds were in the wrong direction, and might result in radiation over some considerable population."[10]

According to Friedell, there were already people stationed around the various areas, mostly under the direction of Louis H. Hempleman (of the University of California) and his group, who were already in the process of preparing for possible fallout problems by the time he and Warren arrived at the test site. As a result, he said that they were at a minor disadvantage "because we didn't have a chance to really go

8. Groves memo, July 18, 1945.
9. Ibid.
10. Friedell interview, on file at Office of History, OCE.

through all the information and see whether we really had any objections—whether we agreed with it or challenged it or wished to modify
it. What happened is we were notified very late that we were to be there
and then we went through the data with them to see whether we
thought there was some rational approach to the way they were looking
at it."[11]

Ultimately, Friedell says that he and Warren "set a pretty high
level of radiation that was acceptable before we would evacuate for the
reason that we didn't want to reveal the fact that there was an atomic
explosion. Consequently, we set the levels fairly high. However, by
July 21, Warren had completed his initial survey to the radioactive
fallout and advised Groves that he had found no dangerous intensity in
populated areas. He noted that monitors "took considerable risks
knowingly and may have received exposures of considerable amounts,
i.e., 8 R total. This is safe within a considerable margin." Nevertheless, he advised that they should not be exposed to additional radiation
for the next month.[12]

As a result of the test results, Groves and the Air Force planning
group decided that the atomic bombs should be detonated at two thousand feet over the Japanese cities to minimize residual radiation on the
ground. But, as I have already noted, few of the military and scientific
leaders directly involved with building the bomb and also responsible
for the decision to use it questioned the ethics or morality of dropping
the weapon just because it created radiation. The purpose of the bomb
was to destroy cities, to kill Japanese, and to destroy the Japanese will
to continue the war. So long as mass killing was considered necessary,
it should not make any material difference whether people died from
the blast, the heat, and the fires created, or the radiation. War itself is
horrible. We wanted to end the war as quickly as possible and minimize the overall casualties, particularly for Americans; at that time we
all remembered Pearl Harbor.

In any case, from the Trinity test onward, events moved rapidly.
Later on July 16, at San Francisco, the uranium bomb minus the last
necessary portion of U-235 was put aboard the cruiser U.S.S. *Indianapolis,* which set sail for Tinian, arriving without incident on July 26.
On July 23, the plutonium hemispheres for the plutonium weapon

11. Ibid.
12. Ibid; RG 77, Box 2, Folder 4.

destined for Nagasaki were completed. On July 27, two days ahead of schedule, the last of the enriched uranium for "Little Boy" was completed and began its journey by plane to Tinian. The last of the enriched uranium necessary arrived on Tinian the night of July 29, and final preparations began for delivery of the first atomic bomb.

In May, I had attended a Military Policy Committee meeting at which it was decided that our production schedule would permit shipping most of the U-235 weapon by ship, which we thought was safer than using air transport. Tragically, as it proceeded on its journey after delivering the weapon, the *Indianapolis* was sunk by a Japanese submarine on July 30. It was the last major U.S. ship to be lost during the war and the greatest single disaster in the history of the Navy.

Meanwhile, in Potsdam and Washington, the decision to use the bomb had been finalized and directives written. Groves' report on the Trinity test reached Potsdam on July 21. In his diary, Stimson recorded: "It was an immensely powerful document, clearly and well written and with supporting documents of the highest importance. It gave a pretty full and eloquent report of the tremendous success of the test and revealed far greater destructive power than we expected in S-1." According to Stimson, who read the report to the president and James Byrnes, who then was secretary of state, they "were immensely pleased. The President was tremendously pepped up by it and spoke to me of it again and again when I saw him. He said it gave him an entirely new feeling of confidence and thanked me for having come to the Conference and being present to help him in this way."[13]

When Groves received word back from Potsdam of the decisions made there, the momentum of preparations for use of the bomb accelerated. On July 23, Groves prepared the final written directive governing the bomb operation on Tinian. Over the signature of General T. T. Handy, the acting chief of staff during General Marshall's absence in Potsdam, the orders stated to General Carl Spaatz, CG, USASTAF, that the 509 Composite Group, Twentieth Air Force was to "deliver its first special bomb as soon as weather will permit visual bombing after about 3 August 1945, on one of the targets: Hiroshima, Kokura, Niigata and Nagasaki. . . . Additional bombs will be delivered on the above targets as soon as made ready by the project staff."

13. Stimson diary, July 21, 1945.

Only the secretary of war and the president were to disseminate information about the use of the weapons against Japan. Spaatz was asked to "personally deliver one copy of this directive to General MacArthur and one copy to Admiral Nimitz for their information."[14]

That evening Harrison had sent Stimson a telegram giving the dates the bombs would be ready for delivery. The next day, Groves sent General Marshall at Potsdam a memorandum to obtain his final approval of the plan of operation. The memo gave the probable date of the first bombing as between the first and tenth of August, as soon as "Little Boy" could be assembled and the weather permitted. Groves also indicated that there would be a gap of at least three days between successive bombs without explaining that we would need the time to assemble the second bomb without being rushed. When Truman received the news that morning, he told Stimson that the information was what he wanted, that he was "highly delighted and that it gave him his cue" for warning Japan to surrender or face complete destruction."[15]

When Stimson returned from Potsdam on the twenty-eighth, he found that "everything seems to be going well." In fact, with the bombs being assembled, all that remained to do in Washington was the completion of the several announcements about the development of the bomb and its future control, which would be released after the first weapon was dropped. The success of the Trinity test caused Stimson to make some changes in the president's announcement "induced by the difference of psychology which now exists. . . . We put some more pep into the paper and made it a little more dramatic."[16]

In the meantime, I continued to work on increasing the rate of production of the fissionable materials. I hoped that dropping one or two bombs would end the war, but no one could be sure. The Japanese showed no signs of quitting, and while we had launched no major campaigns since taking Iwo Jima and Okinawa, fighting continued unabated in China and Southeast Asia. In addition, fighting still was going on in the Philippines. In summarizing events there, the August 6 communiqués from the Southwest Pacific areas, number 1217, stated:

14. T. T. Handy to General Carl Spaatz, CG, USASTAF, July 23, 1945; Groves, *Now it Can Be Told* (New York: Harper & Brothers, 1962).
15. Stimson diary, July 24, 1945; ibid., p. 310.
16. Stimson diary, July 30, 1945.

"Since last reported a week ago, an additional 4,740 Japanese dead have been counted and 444 prisoners taken in mopping-up operations in the mountains of north Luzon and Mindanao. Our own casualties were 27 killed and 61 wounded."[17]

Planning for the invasion of the main Japanese home islands had reached its final stages, and if the landings actually took place, we might supply about fifteen atomic bombs to support the troops. So I was busy in my office early on August 6 while awaiting the news about the Hiroshima mission. As with the Trinity test, my optimism was tempered with anxiety. The Trinity test had proved that one implosion-type plutonium device could be set off under controlled circumstances on a tower. Moreover, we did not test the gun-type weapon, since the production of U-235 took so long, and we had made every possible component test and were sure everything would work and we knew we could bring the U-235 portions of the "Little Boy" (Hiroshima) bomb together in such a way that if the theories of atomic energy were correct, the bomb should go off. But without a test and under combat conditions, I knew that anything could go wrong.

After waiting through an anxious Sunday evening due to communications problems from the Pacific, Groves received a coded message from Deke Parsons at 11:30 P.M.: "Results clear-cut, successful in all respects. Visible effects greater than New Mexico tests. Conditions normal in airplane following delivery." At the same time, a message that had been sent from the *Enola Gay* (the B-29 that dropped the Hiroshima bomb) to Tinian also reached Groves and reported: "target at Hiroshima attacked visually. One-tenth cloud at 0523152 [6:15 P.M. Washington time]. No fighters and no flak."

At 4:15 A.M, Groves received a detailed cable from General Farrell containing information from the debriefing of Tibbets' crew. The message reported: "Sound—None appreciable observed. Flash—Not so blinding as New Mexico test because of bright sunlight. First there was a ball of fire changing in a few seconds to purple clouds and flames boiling and swirling upward. Flash observed just after airplane rolled out of turn. All agreed light was intensely bright and white cloud rose faster than New Mexico test, reaching thirty thousand feet in minutes it was one-third greater diameter." Farrell concluded: "Parsons and

17. Communiqués of August 1 and September 9, 1945, Box 1783, RG 165, War Department and General Staff Operations Division, 1942–45.

other observers felt this strike was tremendous and awesome even in comparison with New Mexico test. Its effects may be attributed by the Japanese to a huge meteor.''

Groves delivered his report to General Marshall when he arrived in the Pentagon at 7:30 A.M. Marshall then informed Stimson, who was at his home on Long Island, New York. Groves then remained in the Pentagon, using Stimson's office to deal with revising the presidential release of the story before the Japanese might do so. While there, he called me to report he had received the first word from Farrell after inexplicable delays in the communication lines from Tinian. With my own doubts removed and without taking much time to think about the significance of the event or even my own feelings, I set in motion a plan I had made to fulfill a promise I had made to my wife sometime earlier when she had said to me, "Look, when this secret comes out, please tell me. I want you to tell me. I mean, I want to know right away. Whatever this is, I hope I'll be one of the first to know."

On the morning of August 6, Staff Warren's wife was visiting Jackie when a messenger knocked on the door with a large envelope. Handing it to Jackie, he said, "Colonel Nichols has sent you this envelope and he wants you to read its contents." Jackie recalls, "Well, since Mrs. Warren was there, I didn't open the envelope. We chatted for a while longer and then, just as she was leaving, the phone rang and it was my sister. She said, 'Well, what do you think?' I said, 'About what?' She asked, 'Haven't you heard the radio?' 'No.' ''

Only then did Jackie learn about the bombing of Hiroshima, which the White House announced to the world at 11:00 A.M. As soon as I heard from Groves, I sent Jackie our whole set of press releases, which gave the history of the Manhattan Project, information about the production facilities, and biographies of the key people, information we were planning to release over a three-day period. But because Vi Warren was there, she had delayed opening the envelope.[18]

Once Jackie did start to read the material, the secrecy and my absences and all the rest fell into place for her. As she recalls, she was excited and happy that the war probably would be ending quickly. She

18. William L. Laurence, a science reporter for *The New York Times,* had worked with us for several months prior to Hiroshima. He was fully indoctrinated with the need for secrecy, and then he reviewed our work and visited our installations. He was at Alamogordo and Tinian. He prepared the news releases and statements to be made in Washington, Oak Ridge, Hanford, and various other locations. He did a superior job, and I have never heard any implications that he violated secrecy. It was a fine example of military and press cooperation.

was "terribly disappointed that the bomb had been dropped on civilians," but she accepted the necessity to use it: "Some of my close friends had suffered terribly at the hands of the Japanese, husbands had been on the death march or were missing in action. I had shared the suffering with them. I was so relieved it was finally over, and glad that my husband had had such a big part in it. So I can't say that I wasn't overjoyed; it had been a success and all this effort had worked, and worked in time. I am perfectly reconciled to the fact that this needed to be done and I'm glad they dropped it. Certainly it was horrible but— no more horrible than war itself."

"Little Boy" had not ended the war, and preparations continued on Tinian for the second mission. One of our major decisions had been whether to use each bomb as soon as we had sufficient material and could assemble the components, or to store up several bombs and then employ them in a massive show of power. Military men generally will go to any lengths to avoid piecemealing away their resources. In this case, however, we decided that the very strength of the bomb and the long-standing goal to use it at the earliest possible time justified using each weapon as soon as it became available. Moreover, we knew that we would have three bombs ready for use during August and certainly hoped they would do the trick, assuming they all worked.

Many nonmilitary people have argued that we should have delayed dropping the second bomb to ascertain whether Japan was willing to surrender after feeling the power of the first weapon. In fact, we had made the military decision to expend our weapons as quickly as possible until Japan capitulated, the president had approved the plan, and since the first bomb did not do the trick, we had no reason not to continue the operation. If we had not dropped the second bomb when we did, the Japanese might well have concluded that we had only one bomb and would have continued their fanatical resistance. In using "Fat Man" (the Nagasaki bomb) so quickly, we demonstrated our commitment to bringing total destruction to the Japanese homeland.

The delivery of the second bomb did not go easily, however, and serves as a reminder that while we did succeed in completing our mission in time to help end the war, we were never that far from failure. Unlike Tibbets' mission, which went with textbook precision, the flight of Bock's Car was plagued by problems even before it left the ground when it was discovered that a fuel transfer valve was inoperative, meaning that six hundred gallons of fuel could not be used.

Moreover, weather reports from the primary target, Kokura, and the only alternate, Nagasaki, were not promising.

Navy Commander Frederick Ashworth, the weaponeer on the mission, recalls that while August 9 proved to be one of the more significant dates in history, during the mission such thoughts "were furthest from our minds as we approached Nagasaki on that fateful morning. We had a job to do and the realities of the moment were paramount." Apart from the fuel problems, the crew found first Kokura and then Nagasaki under a thick cloud cover. As with the first mission, *Bock's Car* was "under specific instructions from the highest levels of authority in the War Department . . . that under no circumstances must the bomb be dropped by other than visual means."

According to Ashworth, by the time the plane turned to Nagasaki, it was the target of sporadic antiaircraft fire, and Japanese fighters were being vectored toward it: "The bomb had again, for the fifth time during the flight, been electronically placed in a fully armed condition requiring only the proper signal from the fuzing equipment to cause it to fire in a full-scale nuclear detonation. Obviously we had little motivation for philosophical thinking about our mission." Ashworth had received the responsibility from Groves to ensure that the bomb was dropped successfully on the assigned target under the rules and restrictions placed on the mission. While he would consult with the plane's commander, Major Charles Sweeney, on any decision that had to be made, Ashworth explains that "the final determination was to be mine. Fortunately Sweeney and I agreed on all decisions required that day."

Ultimately, as the fuel reached a dangerous level, Ashworth and Sweeney agreed that they had to drop the bomb on the assigned target regardless of their own prospects of returning to friendly territory. This meant that they had no choice but to make the bombing approach by radar and release the bomb on the signal from the radar-activated electronic bomb director. Because Nagasaki was as cloud-covered as Kokura had been, Sweeney and Ashworth had briefly considered attacking Tokyo, but Ashworth explains, "We discarded that possibility as one far beyond my responsibility to assume. Further, our third-priority assigned target, Niiagata, surely was too far away to reach with the limited fuel that we had aboard. The radar approach on Nagasaki was continued."

Ashworth continued to check the radar to satisfy himself that the bomber actually was approaching Nagasaki. At the same time, the radar operator continued to update the bombardier so that the Norden bombsight would be kept up to date with the electronic approach. Despite this procedure, Ashworth acknowledges that "there were in my mind the nagging thoughts—we were violating our precise instructions in making the bombing approach and possible drop on radar, would the radar bombardier perform his mission with the same accuracy that was expected of the visual bombardier, what about our fuel supply—could we reach a friendly air field or would ditching be required, had our numerous tests of the bomb arming and firing systems resulted in any deterioration of its reliability. And yet each of these worries was the result of completely appropriate actions on our part." Ashworth recalled a motto he had learned from a Navy friend: "Fortunate Aids the Bold."

At the last moment, after the bomb-bay doors had opened, the Norden bombardier called over the intercom: "I have the target. I am taking control." Ashworth remembers what happened next: "Twenty seconds or so later we felt the jolt as the bomb was released from its shackle—for better or worse it was on its way. The forty-eight seconds time of fall seemed like an eternity. All hands aboard had donned their eye-protection goggles and we waited. Then a brilliant flash of light, brilliant even through our welders-type protective goggles, and we knew at least that the bomb had detonated in an air burst as it was supposed to do and technically probably as anticipated."

To avoid the effect of the anticipated shock wave, Sweeney put the B-29 into a sixty-degree bank and headed away: "The shock wave arrived, then a second and a third—no big deal—more noise than anything—but why three? Had the Japanese antiaircraft batteries picked us up? Later we rationalized the three—one the direct shock wave from the exploding bomb, the second reflected from the ground under the detonation, and the third reflected from the side of the hill that made up one side of the Urakami River Valley. But wait! Nagasaki City, the assigned aiming point, was not in the Urakami Valley—that was a mile and a half or two beyond the city along the line of our bombing approach. Had we missed the target?"

Given the fuel situation, *Bock's Car* made one circle of the mushroom cloud that had already reached thirty thousand feet and then headed to Okinawa, landing with the fuel gauges on empty. As the plane taxied to the ramp, two engines stopped. But the mission was over and Ashworth set out to send back his report. When the communications officer refused to clear a line, Ashworth sought out General Jimmy Doolittle, the commander of the Eighth Air Force, and recounted the saga of the mission. After explaining that they may have missed the target aiming point, Doolittle stood up, put his arm around Ashworth's shoulder, and told him, "Don't worry, son, about the miss of the target. I am sure that General Spaatz will be very happy that the city of Nagasaki was spared and the destruction mostly restricted to the military targets in the river valley."[19]

"Fat Man" had been delivered. Would the Japanese surrender?

19. Admiral Frederick Ashworth, unpublished memoir.

K. D. Nichols as a cadet
at West Point

After the earthquake in Managua in 1931, we had to carry out extensive
demolition to save portions of the city. Colonel Sultan is on left; Lieutenant
Talley is on right.

The S-1 executive committee meeting at the Bohemian Grove, September 13, 1942. Left to right: T. T. Crenshaw, J. R. Oppenheimer, H. C. Urey, E. O. Lawrence, J. B. Conant, L. J. Briggs, E. V. Murphee, A. H. Compton, R. L. Thornton, K. D. Nichols

Left to right, Sir James Chadwick of Great Britain, Major General Leslie R. Groves, officer in charge of Manhattan Engineer District, and Dr. Richard C. Tolman, dean of the Graduate School of California Institute of Technology, meet after the atomic bomb was first dropped on Japan.

U.S. ARMY PHOTO

Secretary of War
Henry L. Stimson

U.S. ARMY PHOTO

Colonel Wilhelm D. Styer

U.S. ARMY PHOTO

Colonel E. H. Marsden

COURTESY OF THE AUTHOR

Colonel Stafford L. Warren

COURTESY OF THE AUTHOR

Lieutenant Colonel Charles
Vanden Bulck

COURTESY OF THE AUTHOR

Lieutenant Colonel Charles E. Rea

Rafferty of the gas diffusion plant

Martin D. Whitaker of the semiworks for plutonium

Lieutenant Colonel Harry S. Traynor of the heavy water plant

Colonel Franklin Matthias of the separation unit

Lieutenant Colonel J. S. Hodgson, head of Y-12 construction at peak period

James C. White, president of Tennessee Eastman

The chapel at Oak Ridge after a Sunday service

The Mitsubishi steel works in Hiroshima after the first atomic explosion

The underwater test at Bikini Atoll

PHOTO BY ABBIE ROWE—COURTESY NATIONAL PARK SERVICE

President Dwight D. Eisenhower signs the Atomic Act of 1954

COURTESY OF THE AUTHOR

Secretary of War Frank Pace, Jr., awards the Distinguished Service Medal to K. D. Nichols. Major General Leslie R. Groves is at left.

Watching the test of
the NIKE missile.
K. D. Nichols is third
from left.

Left to right: T. E. Shea, vice president and general manager of Sandia
Corporation; Joseph Campbell, commissioner, U.S. Atomic Energy
Commission; K. D. Nichols, general manager of U.S. Atomic Energy
Commission; James W. McRae, president, Sandia Corporation; and
Brigadier General Kenneth E. Fields, director of the Division of Military
Application, U.S. Atomic Energy Commission

12

TRANSITION: WAR TO PEACE, 1945–46

EVEN AFTER WE HAD DROPPED the plutonium bomb on Nagasaki, we received no immediate indication that Japan would accept our peace terms. G-2 recommended on August 10 that pending acceptance by the Japanese of the Allied demands, military and other pressure to the maximum extent possible be everywhere applied to the Japanese enemy. Groves sent the following memo to General Marshall on August 10: "The next bomb of the implosion type had been scheduled to be ready for delivery on the target on the first good weather after 24 August 1945. We have gained 4 days in manufacture and expect to ship from New Mexico on 12 or 13 August the final components. Providing there are no unforeseen difficulties in manufacture, in transportation to the theatre or after arrival in the theatre, the bomb should be ready for delivery on the first suitable weather after 17 or 18 August."[1]

Anticipating Japanese acceptance of peace terms, Marshall wrote on the memo: "It is not to be released over Japan without express authority from the President." On the thirteenth, Groves sent a memo to General Handy that he would remind him on Wednesday, the fifteenth, "that I am ready to start shipment. I will arrange for planes so that they can depart from New Mexico on Thursday if the decision is to send the materials. This will change date in theatre from (subject to first good weather) the 20th or 21st." Meanwhile, Los Alamos was preparing for their next shipment. On the fourteenth, with a car waiting

---

1. Manhattan Project Records Office of the Commanding General Files, RG 77, File 290.

outside, Dr. Robert Bacher was in the ice house completing the packaging of the core for the third bomb when friends burst into the room with the news that the war had ended.[2]

Early on the fourteenth, Groves asked me to come to Washington. I had already approved the text of this news release to the workers of the CEW when Japan surrendered: "We have accomplished our initial objective—VICTORY. Our next objective is to continue operations at the Clinton Engineer Works after V-J Day as long as need be to make secure the defense of our country. While future official policy in this regard has not yet been announced, no thinking American can imagine that our Government will neglect to exploit to the fullest extent the most powerful and revolutionary discovery in all history. We cannot afford to cease operation at our plants for a single moment to celebrate."

Groves wanted me in Washington to discuss the various actions that would be required upon the termination of hostilities. Anticipating the president's announcement, I invited myself to dinner with friends, and a group of us were together when at 7:00 P.M. on August 14, President Truman announced that he had received a message from the Japanese government that he deemed unconditional surrender, a full acceptance of the Potsdam Proclamation. We were all elated by the news. With the official Smyth Report released, now there was plenty of leeway for discussion and answering questions about the A-bomb. All my friends had been wondering what I had been doing. But the big news was that the war was over. Wives no longer had to worry about their husbands transferring to the Pacific Theater. We joined the exuberant crowd in the streets of Washington. It was a great evening that lasted into the early-morning hours.

I stayed in Washington until the eighteenth. We knew it would require considerable time for Congress to establish long-range policy. However, plans had to be made for the immediate future. We no longer had the justification of war to warrant an all-out urgent effort; emphasis had to be put on making economies.

For the CEW I recommended shutting down the alpha stages at Y-12 and the S-50 liquid thermal diffusion plant. Both of these production units were less efficient than the K-25 gaseous diffusion plant. I also recommended completing the authorized beta plants, the K-25 and K-27 plants, and the housing, central facilities, and paving of

2. Suid interview with Dr. Robert Bacher, September 21, 1982, Office of History, OCE.

roads under way. Groves was apprehensive about shutting down any plants at Oak Ridge while continuing construction. However, he agreed that the alpha plants and S-50 should be shut down in September. All heavy-water plants except Trail were shut down. Hanford continued to operate, but modifications were made to meet stricter specifications desired by Los Alamos. In December, one reactor was shut down for maintenance, but with the diagnosis of the Wigner disease (the swelling of graphite due to radiation), it was decided to keep one reactor in a standby status and operate only two reactors. We did this to ensure that we would have at least one reactor available for irradiating material for bomb initiators. As a result of these decisions, the production of U-235 actually increased and that of plutonium decreased during 1945 and 1946.

Of more immediate concern was determination of the effects of the two atomic bombs at Hiroshima and Nagasaki. Various survey groups were sent to Japan to determine the extent of destruction and the effects of radiation. The radiation survey teams under Staff Warren were organized and left San Francisco on August 14. General Farrell, Warren, and others reached Hiroshima on September 8 and Nagasaki on October 8, and returned to the United States on October 15. Warren reported:

Japanese physicians and scientists were extremely helpful during the survey, and the general population, including those injured by the bomb, acted as patients act everywhere. There were no incidents, even though the survey party were in the bomb-shattered cities and elsewhere in Japan before the country was occupied by U.S. troops.

It was the consensus of the U.S. survey team and of all the Japanese with whom its members came into contact that a coastal assault on Japan could not have been made without tremendous losses of ships and men, including U.S. casualties of perhaps 500,000, two to four times as many Japanese casualties, and complete destruction of Japan. The use of the atomic bomb, some observers held, gave the Japanese Government the opportunity to surrender without loss of face or need to commit hara-kiri. Fewer were killed by the bomb than had died in the Tokyo-Yokohama raids with conventional bombs. The ethics of the use of the atomic bomb had been raised by U.S. newspapermen in Tokyo, but many Japanese told the survey team they could not understand why the question should have been raised at all: Their own forces would have used it without the slightest qualm if they had had it themselves.

It is realized that this information was not the concern of the survey team, but the discussion came up when fixed coastal gun installations and Kamikaze stations were visited in the downwind area, and it is included for record.

Radioactivity and Blast Damage

In spite of primitive transportation conditions and almost continuous rain, the instruments brought with them by the members of the survey team functioned satisfactorily and lasted well enough to permit an extensive survey of the detonation area in Nagasaki and a somewhat less complete survey of the Hiroshima area, where team activities were hampered by lack of roads in the down (northwest) wind areas.

In all the areas examined, ground contamination with radioactive materials was found to be below the hazardous limit; when the readings were extrapolated back to zero hour, the levels were not considered to be of great significance. The explanation was that the detonation occurred at altitude 1,800 ft., and the fireball therefore did not actually touch the ground. Vaporized materials arose from the ground in the updraft and mixed with fissioned materials, but at that, the amount of radioactive contamination was lower than had been expected.

In Nagasaki, where the affected area was examined more thoroughly than in Hiroshima, the approximate center of the detonation was indicated by a uniform charring of the top and sides of a single fencepost. Other posts in the same area were more charred on one side than on others. Trees, walls, and other standing objects leaned outward spokewise from this central point. The effect was particularly notable to the northeast, up the low hill on which the prison was located.

Induced radioactivity from neutron bombardment could be demonstrated in sulfur insulators, copper wires, and brass objects, in human and animal bones, and in the silver amalgam in human teeth, for a distance of about a thousand meters from the assumed mid point of the destroyed area. The neutron effects ceased rather sharply. It is unfortunate that it was not possible to make precise enough measurements to determine the full extent of the affected area.

The wind at the time of the Nagasaki detonation carried the debris to the east. It could be traced along the roads to the ocean for 90 to 100 miles in a path at least 40 miles wide at the seashore.

Nagasaki lies in the Urakami Valley, which is generally narrow but is about 2,500 meters wide at the detonation point. The eastern wall of the valley rises almost 2,000 feet above the valley floor. The hills were covered by terraces and rocks, and there were almost no dwellings to be damaged. In the next valley, however, the fallout path crossed the north end of the reservoir that supplied the city and also the houses to the north of it. The remainder of the town south of the northern reservoir and some part of the lower Urakami Valley and the harbor area were virtually unaffected by contamination.

Blast effects were well marked for 2,000 meters north and south of the central detonation point in the Urakami Valley. Many peculiar concentration and skip effects were clearly evident, especially in a long series of steel frame buildings of the steel and torpedo works that ran north and south toward the harbor from the central area. Other stronger concrete and steel buildings had suffered obvious structural damage. Most concrete buildings had lost their steel window frames, which, it was evident, could become dangerous missiles inside the buildings.

In the Nagasaki Medical School, bodies were found entangled in the twisted window frames of the laboratory wing, which faced the blast. The contents of many rooms consisted of the wainscoting, the window frames, the ceiling, equipment, linens, and papers, which were all distributed over the floor in a somewhat circular pattern. Many fuses had apparently been replaced with metal coins, and the fixtures hanging from the ceiling had therefore been violently twisted by the blast. The resulting short circuits had apparently lasted long enough to set the ceiling afire in many rooms, with the further result that the contents of the rooms burned along with the bodies of the staff. The prevailing wind carried the fire to the northeast part of the building, along the maple flooring and even up the maple treads of the staircase. The maple-floored ward areas on the upper floors were also burned but only downwind from the staircase.

Many fires apparently occurred from similar short circuits.

Overturned stoves caused many others. In both Hiroshima and Nagasaki there was considerable testimony to the effect that the fires started in multiple places at once but did not burn vigorously until about half an hour after the detonation.

The blast wave apparently put out the flames produced by infrared radiation in ripe brown wheat and smoldering wooden and dark surfaces before the fires from this source grew to any size.

Clinical Considerations

When the survey team arrived in both Hiroshima and Nagasaki, it found feeble evidence of first aid efforts. Injured casualties lay wherever any sort of roof offered shelter from the elements. Mats were laid on the floor, and the Prefectorial Government in charge of the country delivered rice and tea to the patients. Helmets had apparently been used for carrying water to them. Later, some of the supposed patients were obviously malingerers, who had come to the aid stations for foods.

From their own observations and from testimony of Japanese, members of the survey team divided the morbidity and mortality of the atomic bombs that were dropped on Japan into the following phases:

1. Very large numbers of persons were crushed in their homes and in the building in which they were working. Their skeletons could be seen in the debris and ashes for almost 1,500 meters from the center of the blast, particularly in the downwind directions. The remains of large numbers of bodies were seen in poorly constructed trench shelters along the main roads. An occasional fresh body, with evidences of purpura, was found in ruined buildings. Collections of shoes (geta) were seen outside many of the first aid stations, where piles of human ashes were left from the extensive cremations carried out in the first few weeks after the bombings; all bodies were cremated, at first, for military reasons, to conceal the number of dead, and later to clean up the area for sanitary reasons. Parties from the Japanese Army and the Prefectorial Government were still searching for bodies as late as 25 September.

2. Large numbers of the population walked for considerable distances after the detonation before they collapsed and died. Many who crowded on the trains that left both cities several hours after

the blast died promptly, and their bodies were taken off at the first and second stops.

3. Large numbers developed vomiting and bloody and watery diarrhea (vomitus and bloody feces were found on the floor in many of the aid stations), associated with extreme weakness. They died in the first and second weeks after the bombs were dropped. These manifestations gave rise to fear of typhoid and dysentery, neither of which developed.

4. During this same period deaths from internal injuries and from burns were common. Either the heat from the fires or infrared radiation from the detonations caused many burns, particularly on bare skin or under dark clothing.

5. After a lull without peak mortality from any special causes, death began to occur from purpura, which was often associated with epilation, anemia, and a yellowish coloration of the skin. The so-called bone marrow syndrome, manifested by a low white blood cell count and almost complete absence of the platelets necessary to prevent bleeding, was probably at its maximum between the fourth and sixth weeks after the bombs were dropped (that is, between 10 and 20 September). Most patients with purpura died within a few days after it appeared, but some of them were observed in the aid stations by the survey team. Deaths from purpura occurred a few days earlier and stopped a few days earlier at Hiroshima than at Nagasaki, but otherwise the reactions in both cities were much the same in respect to both clinical manifestations and timing.

Epilation, anemia, and purpura were only occasionally seen in general surviving population, the assumption being that radiation sufficient to cause these pathologic changes was likely to be lethal. Nonetheless, an occasional patient with purpura, particularly if it developed late in September, seemed to have some power of recovery.

6. The death rate after 20 September was much lower than in the preceding weeks, though many casualties continued to die from protracted anemia, secondary infection, burns, and other complications.

As soon as patients with bone marrow and other injuries died in the aid stations, the spaces they had occupied were filled with patients with severe burns who had survived and who were brought in by farmers and others of local population, chiefly to take advantage of the rice and tea available there, as well as of occasional visits by physicians.

7. No count could be made of those who died outside of the devastated area, in public schools or other buildings to which they had been taken for care.

8. Occasional survivors (misses) in the devastated area showed little or no effects of radiation. Some of them had been in deep shelters or inside large buildings, but some escapes could not be explained.

9. The real mortality of the atomic bombs that were dropped on Japan will never be known. The Japanese had no accurate census at the time of the bombing. Afterward, no census was possible. Bodies were hastily cremated, as already mentioned. The destruction and overwhelming chaos made orderly counting impossible. It is not unlikely that the estimates of killed and wounded in Hiroshima (150,000) and Nagasaki (75,000) are overestimated.

Therapy

Occasional attempts to treat casualties showing bone marrow injuries with transfusions, plasma infusions, and penicillin were soon discontinued, chiefly because any needle puncture resulted in serosanguinous oozing that continued to death. Even pricks to obtain blood for blood counts caused oozing that could not be checked. It was thought that if the platelets were not too greatly reduced, because some functioning bone marrow was left, supportive treatment might be useful in carefully selected patients. If laboratory tests showed that the bone marrow was completely destroyed, the treatment available at the time the bombs were dropped on Japan was of no value at all.[3]

3. Stafford L. Warren, M.D., "The Role of Radiology in the Development of the Atomic Bomb." A reprint of a chapter in *Radiology in World War II*, edited by Kenneth D. A. Allen. Issued by The Surgeon-General's Office in 1966 as a volume in the series *Medical Department of the U.S. Army in World War II*.

When Warren returned to Oak Ridge, he presented me a very fine Japanese saber. He told me that when he first visited the Japanese hospital in Hiroshima, the commanding medical officer surrendered his saber to him. Warren was reluctant to accept it, but when the Japanese insisted he must surrender to someone and Warren was the first American officer to reach the hospital, Warren finally accepted the saber on behalf of his commanding officer and would present it to him at Oak Ridge, Tennessee. It was a fine saber, my only war trophy. Unfortunately, it along with my officer's saber, cadet sword, and my Nicaragua machete were stolen from my country home in 1960.

Although the largest number of casualties at Hiroshima and Nagasaki were caused by blast and heat from the explosion and the resulting fires, the effects of radiation have been emphasized as the most horrible and the most feared aspect of atomic war. We were fairly well informed about radiation, but there was plenty that we did not know about it at that time. Our knowledge about the amount and nature of the radiation that would be generated by the bomb was not precise. The Japanese did not know how best to treat those suffering from much larger doses of radiation than man had ever encountered before. We had done extensive research to improve our knowledge, but research data available could not possibly answer all the questions that would be created by the first use of the atomic bomb on a city. We did know that the bomb would generate various types of radiation and neutrons. We had a good idea about what would be a lethal dose, and we knew that there would be many deaths and injuries caused by radiation as well as by heat and blast. We thought correctly that the blast effect of the bomb would be the most effective in causing casualties and destruction, so we tried to maximize that by setting the height of burst at approximately two thousand feet. With this height, we knew there would be radiation and heat casualties. How many we did not know. We also believed that at this height the residual radiation on the ground would be minimized. We wanted little or no residual radiation because we hoped that our troops would soon occupy the city, and we did not want to endanger them.

The number of casualties to be caused by radiation was considered but was not a major factor of the decision to use or not to use the atomic bomb. However, the number of radiation casualties, the general effects of radiation, the action of radiation on living tissues, the genetic effects, and the number of cases of cancer among the survivors have been matters of controversy for over forty years. To get some of the answers, both Warren and Stone advocated more research. The Atomic Bomb Casualty Commission and its successor, the Radiation Effects Research Foundation, have conducted studies and have found an excess of about 250 cancer fatalities attributable to radiation exposure over the years 1950–78. This excess can be compared with about forty-eight hundred normal cancer deaths, and 23,500 deaths from all causes for the A-bomb survivors under study for the same period. This excess number may increase with time, and the study continues. But the number is relatively small compared to some of the exaggerated claims. The studies also have failed to find any statistically significant increase of genetic effects. Likewise, there have been no irreversible environmental effects, and as anticipated, there was little dangerous residual radiation in Hiroshima or Nagasaki.

Recently I heard of new findings concerning the relative hazards of neutrons as compared to ionizing radiation. However, my informant hastened to add that this new finding does not have any material effect on the overall casualties. We are still learning. Such controversies and new findings will continue, but they do not materially change the horrible effect of nuclear or thermonuclear war.

The most important military effect was that it required only two atomic bombs to end the war. The planned invasion of Japan was not necessary. The two bombs probably saved many tens of thousands of lives, and ending the war certainly justified the decision to use them. It also made the world aware of atomic weapons. The initial public reaction undoubtedly exaggerated the effect of an atomic bomb. The shock effect and the fact that Japan was already defeated destroyed their will to fight. I have always believed that if the bomb had not been used, any demonstration or test probably would have been underrated and, as a result, atomic weapons might not have been as effective as deterrents for avoiding war between the United States and the Soviet Union. Moreover, not using the A-bomb would not have avoided the race for nuclear supremacy between the United States and the USSR Although we did not know it for sure at the time, the USSR was already working on their own A-bomb before the war ended.

There has been considerable discussion about how many bombs would have been available for use against Japan if she had not surrendered and had the Kyushu invasion been executed as planned on November 1, 1945. We expected three to be produced in September, with the rate per month possibly accelerating to seven per month by December. This acceleration would result from the increasing rate of production of fissionable materials and design improvements made possible by the Alamogordo test and utilizing the implosion method instead of the gun for U-235.

The period between the end of the war and December 31, 1946, was marked by the transition from wartime to peace, planning for a change in organization from the Manhattan Project to the U.S. Atomic Energy Commission, adoption of a national security policy of relying on collective security, and an effort to achieve international control of atomic energy.

The biggest problem Groves and I faced as the war ended was the partial demobilization of the Manhattan District and, at the same time, carrying on essential operations. Scientists were eager to go back to their universities, industrial contractors were eager to pursue more profitable peacetime objectives, and many of our reserve officers were eager to revert to civilian status. Without the wartime sense of urgency, development and production of weapons practically stopped. Los Alamos had lost many key people, the Bikini tests required a major effort, and there was no clearly stated national objective. It was recognized, however, that a need existed for development of a more efficient implosion bomb for plutonium, an even greater need for an implosion bomb for U-235, and that we should continue to produce U-235 and plutonium.

One of the first major steps after the war was reduction of the armed forces: From June 1945 to June 1946 they were reduced from twelve million to three million, and by June 1947 they were further reduced, to one and a half million. In 1945 the United Nations was organized, and in 1946 the United States, through the U.N. Atomic Energy Commission, advocated the Baruch Plan for international control of atomic weapons. Our national security depended upon the strength of our economy and collective security. The conventional armed forces were too weakened to have a significant role in supporting our foreign policy in case of an emergency. Congress passed and the president signed the Atomic Energy Act of 1946, which established civilian control over atomic development and separated the development, production, and control of atomic weapons from the military. This was advocated by the scientists and was in accord with the political concept of eventually having international control.

My activities during this transition period changed materially. I was no longer coordinating an outstanding team of individuals and organizations. Instead, a real effort was necessary to keep everything from falling apart. To keep the Manhattan District functioning, Groves secured authority from the new secretary of war, Robert P. Patterson, and over the objections of many on the Army General Staff, to select by name about fifty young, outstanding regular officers, mostly West Point graduates, to fill key positions in the MED vacated by the departure of reserve officers who wished to resume civilian status as soon as possible. Groves and I independently prepared a list of individuals in order of priority and then compared our results. Colonel K. E. Fields headed both lists, Groves' because he ranked first in his class and was a football player and an all-around athlete, mine because I knew him well, having served with him in the Department of Engineering at West Point. The chief of engineers cooperated with us, and orders were issued promptly. These officers were assigned to key positions where they would acquire experience in the design, manufacture, and handling of atomic weapons. That we selected wisely is evidenced by the fact that more than half of these men later became general officers. Colonel Frederick J. Clarke, who replaced Matthias of HEW, became chief of engineers in 1969.

In the autumn of 1945, Groves and I were busy making "E" awards to the MED, contractors who had excelled in performing an important task. For the "E" awards at Oak Ridge, Secretary of War Patterson was the main speaker and presented the awards. This was followed by a banquet in Knoxville for all the key government and contractor personnel. To my delight and great surprise, I was awarded the Army Distinguished Service Medal by the secretary of war. I was also surprised to see that Jackie had joined the party for this special event and was presented a gorgeous bouquet of red roses. She deserved a medal, too.

More than twenty Medal for Merit awards were made to key civilian scientists and contractor personnel, and the Legion of Merit was awarded to many deserving military personnel. The military, civil service, and contractor personnel were awarded silver or bronze A-bomb pins and a Certificate of Service signed by the secretary of war. Shortly after V-J Day and upon my return to Oak Ridge, I was besieged by a delegation from our WAC detachment. They had designed a shoulder patch and made an ardent appeal that I get it approved and produced in a hurry so they could have some way of showing that they had spent the war on the atomic bomb project. As one expressed it, "We are tired of hiding our light under a bushel basket and not being able to tell our friends and relatives what we are doing. We want the world to know that we helped end the war." At peak strength, we had more than 275 in our WAC detachment under Captain Arlene G. Scheidenhelm. They had performed exceptionally well in various assignments throughout the district. I knew that Captain Scheidenhelm was on our recommended list to be awarded the Legion of Merit, and I was only too pleased to take on the assignment to get something for her command that also would be appreciated by all the men in uniform throughout the district.

Groves supported the idea and immediately called the Quartermaster Corps for the man in charge of heraldry to come over and check our design. We had little trouble with him because we had incorporated the Army Service Force blue star encircled by white and red that already met heraldry requirements, and there wasn't too much military tradition about how to portray splitting the atom. The heraldry man drew up our design in a form suitable for production. I then took it to General Somervell, who was pleased that we had incorporated the ASF design. I asked him if his approval was adequate to have it produced, and he said, "No, for a new shoulder patch you need the approval of the secretary of war." So I hastily went to the secretary's office. I had no appointment so I requested his aide to take the letter in and ask the secretary if I could see him about it. He immediately returned with a smile on his face. "The secretary says you folks in the Manhattan District can have anything you want. He approved it."

However, the heraldry man protested, "Oh! You didn't get the adjutant general's approval. You should have done that first." Groves laughed and told me to go see the adjutant general. He also commented something about "normal channels" being a waste of time. The aide in the adjutant general's office told me it would take a month to make the necessary study before the adjutant general could recommend approval. At that point "normal channels" was no longer a laughing matter to me. I asked him if he questioned the secretary of war's authority to approve it without the adjutant general's recommendation and told him I had to leave Washington in one hour and suggested that the adjutant general stamp it either "recommend approval," "recommend disapproval," or just "noted." If he refused to do any of these, I would take it back to the secretary of war. I recommended that he use the "noted" stamp. He took it to the adjutant general and came back with it stamped "noted" and initialed. The quartermaster then did a rush job to get the shoulder patches produced. My mission was accomplished in an elapsed time of about six hours in Washington, but it was a good warning that "normal channels" were going to take longer for many of our actions in the future.

When Farrell returned to his civilian status, I was designated as Groves' deputy but continued as district engineer, with Kirkpatrick resuming his position at Oak Ridge as deputy district engineer. I found myself spending more and more time in Washington. Turnover of the MED to a permanent agency took much longer than anticipated. Considerable controversy arose in Congress concerning the provisions of the Atomic Energy Act, mainly involving patents, information, and civilian versus military control. After the Atomic Energy Act was passed on August 1, 1946, it took until October 28 to appoint the five commissioners, and they refused to take any responsibility until January 1, 1947. As a result, Groves had to continue operating from August 1945 to December 31, 1946, without adequate authority to make long-range decisions. However, during this period many decisions were necessary.

At Los Alamos the situation was critical. Practically all of the key personnel returned to a more normal life in universities. Their mission had been accomplished. After much discussion, Norris E. Bradbury was first selected on an interim basis and, finally, was designated as director of Los Alamos. He did an amazingly good job of reorganizing for the development of weapons.

Groves asked me to spend more time on production facilities and storage of weapons. Accordingly, action was taken to select industrial contractors and construct facilities for the manufacture of weapons. A new underground plant at Dayton, Ohio, to produce initiators was designed and constructed. I recommended that Sandia Air Base near Albuquerque, N.M., be transferred from the Air Corps to the Manhattan District to provide a base for training military assembly teams as well as to be a site for industrial participation in the engineering design and production of weapons. A military unit was established at Sandia to take over assembly of weapons, surveillance, and storage of weapons. Construction of underground storage depots at three locations was begun.

The decision was made not to relocate the facilities at Los Alamos, and improvement of the living conditions and facilities was initiated. Los Alamos continued to have the full responsibility for development of atomic weapons.

The stockpile of complete but unassembled weapons remained small throughout the turnover period; figures frequently quoted are seven to nine. Many question why this figure was so low. During the turnover period, the only weapon stockpiled was the Mark III plutonium weapon. If the war had continued, more efficient implosion-type weapons for both plutonium and U-235 would have been developed and used. The lack of urgency and reluctance to stockpile obsolete weapons delayed production and stockpiling of weapons until after the Sandstone tests in the spring of 1948. However, fissionable material continued to be produced and stockpiled. U-235 and plutonium were the vital materials that took time to produce. In an emergency, weapons of the best available design could be produced in a relatively short time. Besides, in 1946 and 1947, we thought we had plenty of time, since the war was over. In briefing Army Chief of Staff General Dwight D. Eisenhower, which seemed to fall to my lot, in addition to informing him about the weapons for which we had the component parts, I also informed him about the potential weapons that could be made from the fissionable material available if components were manufactured.

On December 17, 1945, while at Los Alamos, I received a request from Groves to return to Washington the next morning and bring Bradbury and six of the other key scientists to Washington to plan for an effects test in the Pacific against naval targets. I alerted Pete Young, the pilot of my B-25, that I wanted to take off from Sante Fe for a nonstop flight to Washington's National Airport. (We had an extra gas tank in the bomb bay, so we could cross the continent nonstop in the B-25). Pete called back that a snowstorm was sweeping down from Canada and would cover practically all of the United States, but if we drove to Albuquerque, he had arranged for a B-29 from the 509 unit to fly us to Washington. We would have a hundred-mile-an-hour tail wind and would reach Washington two hours before the storm. (The B-25 could not make it fast enough.) So we took off from Albuquerque in a B-29 fully loaded with fuel, with the destination, Washington, alternate airport, return to Albuquerque. I had never flown in a B-29, so I chose to ride up front with the two pilots, the engineer, the navigator, and a major in the 509th, who wanted a ride to Washington. Bradbury and the scientists were in the rear. It was a beautiful night flying high above the storm that completely blanketed the Midwest. We were making fantastic speed with the tail wind. Over Kentucky, our left outboard engine literally exploded and a piece of metal hit a cylinder of the left inboard engine. We lost speed and the pilot told me it would be impossible to beat the storm to Washington. We were the only plane in the area that night, so we had plenty of attention from ground control. It was impossible to return to Albuquerque with only two and a half engines against the strong wind. Bermuda would be risky. Ground control thought the weather was clearing in Canada and we should head for Dayton and get instructions there. The pilot asked the navigator for the course. The navigator asked me, "Where is Dayton? I have only a strip map, and Dayton is not on it!" I looked at his strip map, and he pointed to our location. I put my finger on the location north of the map area where I thought Dayton might be, so following that course, we finally established radio directional contact with Wright Field at Dayton. When we approached Dayton, they recommended that we try to land at Wright Field rather than nearby Patterson, the visibility seemed good enough to warrant the try. As the young captain, who was obviously nervous, started down, he switched the wheel over to the copilot, who promptly returned it to the pilot. That was repeated twice. The major sitting behind me leaned over and asked, "Did you see that?" I said, "Yes." He then asked, "I am just a passenger on this flight. Will you back me if I take over the plane? I don't trust these guys." I readily replied, "Go ahead." All it took was a tap on the shoulder and the pilot gratefully surrendered his position.

It was quite a spectacular landing, on two and a half engines. Both Patterson and Wright had turned on their landing lights. We spotted a field but weren't sure which one. The pilot was not about to lose sight of the landing lights; he flipped the B-29 around like it was a trainer plane and came down on a runway covered with about twelve inches of snow. The tower told us to stay on the runway. They sent a truck out to pick us up. Upon our arrival in the office, the officer there offered to arrange another flight to Washington the next day. Bradbury's response was, "We'll take a train. Help us arrange for reservations."

I stayed over in Dayton the next day and visited the underground plant for producing initiators, then left by train for Washington that evening.

It had been a close call and the chief of the Air Corps heard about it; he also learned that I had been hauling key scientists around in my B-25. As a result, our status and ground service improved a great deal, which Pete and I greatly appreciated.

In January 1946 I planned to take a short vacation, my first vacation in four and a half years, but Groves at the last moment intervened. Berkeley had a problem, and either he or I had to go. He could not go and suggested that I take Jackie with me and spend a few extra days on the West Coast. So Jackie donned a suit of coveralls and we took off for a most pleasant trip. We visited friends in El Paso on our way west, then flew up the Pacific Coast at cliff-height elevation to San Francisco. Jackie and I were in the nose of the plane in the bombadier's compartment. The view was splendid. The weather was perfect and permitted visual flying both ways. Ernest Lawrence and Molly made our stay in the Berkeley-San Francisco area one to remember. It was Jackie's first trip to California. We saw the big trees, stayed at the del Monte Lodge at Monterey, and finally dropped Ernest off at Palm

Springs before heading east. After passing over Las Vegas, we flew down into the Grand Canyon for an unusual inside view of it. We stopped at Los Alamos, and Jackie observed how different it was from Oak Ridge. The trip provided a relaxing break plus a bird's-eye view of a good portion of the United States. We couldn't have planned a more exciting vacation. And Ernest and I did solve the problem.

Upon my return I learned that I had been promoted to brigadier general, an event that was properly celebrated with friends and family. It was the last promotion list for World War II. Shortly thereafter I was notified that with the reduction in the size of the Army, reduction in grade of all officers, in reverse order to the promotion, should be anticipated. In April I moved the family to Washington. Rentals were practically nonexistent, so we bought a house.

Early in 1946, du Pont insisted that we find someone to take over the Hanford operations. The du Pont contract specified that they be relieved of contract obligations six months after cessation of hostilities. It would have been better if our successor could select a contractor, but that would take too long. General Electric and Westinghouse appeared to be the most appropriate choices. These two companies had a basic interest in commercial atomic power development that I felt was essential. After the war most industrial companies were more interested in getting back into commercial production than in getting a new government contract. The best indication of this is that no one made any effort to be awarded the Hanford contract. General Electric was selected over Westinghouse primarily because I believed General Electric had more chemical capabilities than did Westinghouse. Groves was only lukewarm to my choice because he thought General Electric tended to be too expensive. Also, General Electric was demanding that we support a nuclear laboratory adjacent to their research center at Schenectady, N.Y. However, he said, "Go ahead." In June it was officially announced to the public that General Electric would take over the operation of the HEW from du Pont on about September 1, 1946, with no major change in policies or personnel.

The Clinton laboratories had been transferred from the University of Chicago to Monsanto on July 1, 1945. Union Carbide was willing to continue to operate K-25 and K-27. And Tennessee Eastman continued to operate the beta units of Y-12.

Although Arthur Compton had become chancellor of Washington University in St. Louis, he was willing to remain as director of our projects at the University of Chicago. Farrington Daniels became director of the met lab, and Walter Zinn stayed at the Argonne lab, where he was busy developing a fast breeder reactor. When Compton was inaugurated as chancellor, he invited Groves, Bush, Conant, Fermi, Thomas, Peagram, Greenewalt, Briggs, and myself to the banquet and, as a preliminary, he paid off his champagne bet he had lost because the first kilogram of plutonium was produced six weeks later than his prediction. We all enjoyed the event.

At Berkeley we had no problem. Lawrence had a thriving organization when the atomic bomb development started, and he was ready to continue indefinitely. He welcomed our continued support.

Thomas and Compton had urged me to support the concept of regional laboratories for nuclear development. The Argonne was to continue as the first of this concept by bringing in universities in the Chicago area as participants. Likewise, at the Clinton labs we organized regional participants. We sold Groves on the idea in such a subtle manner that he later claimed the idea was his own. Be that as it may, he did support the plan, and I assigned Colonel George W. Beeler and Peterson to find a group of universities and a site in the Northeast for another regional laboratory. The Brookhaven lab was the result. They worked on the formation of a similar laboratory in Southern California but did not succeed in getting a project started.

Early in 1946, I proposed that Groves appoint an Advisory Committee on Research and Development. The committee consisted of Bacher, Compton, Lewis, Ruhoff, Thomas, Tolman, and John A. Wheeler, all participants in our war effort.

Our first meeting was in March 1946, and the committee made broad policy recommendations concerning the type of research we should support and where the work should be conducted. They supported the national or regional laboratory concept and in particular supported a lab for the Northeast. They made specific proposals for support of various projects, and they endorsed the distribution of radioisotopes at cost. These recommendations were helpful in getting our fiscal year 1947 appropriation approved by Congress. With this appropriation we were again able legitimately to support desirable research programs.

Concurrent with our gaining authorization for general research and development in the nuclear field, Admiral Earle Mills, chief of the Bureau of Ships, sought our support to organize a training program at Oak Ridge for naval officers, with the primary objective of the Navy ultimately developing atomic power for submarines. Hyman

Rickover may be called the father of the atomic submarine, but Abelson, Mills, and others certainly participated in the conception at an earlier date.

Groves was receptive to Mills' proposal and authorized Mills and me to work out the details and proceed. Mills outlined in greater detail just what we should do. Initially we would concentrate on training and educating the officers in practical atomic engineering at the Clinton labs. His idea was to develop a group of naval engineers who would become advocates of atomic power for naval propulsion. I agreed fully with his concept but felt that I did not have the time to wet-nurse thirty or forty naval officers. I preferred to deal with only one officer who could direct the group and who had experience with industrial organization. Further, I wanted to have the right to veto the selection of this officer. Mills told me that he had a Captain Rickover who could head the group, and he soon had six industrial references that were known to me. I checked all six by telephone and found that three of them thought Rickover was an outstanding, imaginative, and capable engineer but at times difficult to get along with, whereas the other three admitted he might have some capabilities but claimed it was impossible to deal with the man.

I told Mills that the references he had given me were very frank and conflicting but that I would like to meet Rickover. Mills said, "He will be right over." I told Mills, "That won't do. I have already set up my plane at National Airport to be ready to leave in twenty minutes." Mills replied, "Make it twenty-five minutes and he will be there and ride with you."

Rickover joined me and we talked all the way to Knoxville and Oak Ridge about what he had been doing for the Navy and problems he encountered in the Bureau of Ships and with the industrial contractors.

After I reached Oak Ridge, I called Mills and told him, "Rickover will do, so start the ball rolling. It will be interesting to see what happens."

Early in June, Groves told me I should go to Bikini for the tests. I asked, "Why should I go? The task force is completely organized. I don't want to join the multitude of military observers." He replied, "You will represent me and be on Admiral Blandy's command ship, the *Mount McKinley,* and if the bomb does not go off, YOU take over." I didn't ask him how a brigadier general takes over from a vice admiral on his own command ship. However, the bomb probably would go off, so why worry about it?

The president had approved the atomic bomb tests, Operation Crossroads, on January 10, 1946. As a result, Los Alamos had a new priority activity. The tests would do much to determine weapons effects on naval vessels but little to further the actual development of more efficient atomic weapons. However, the tests would develop means for collection of data that were essential for confirming results and for future development. Also, a greater number of military personnel would become acquainted with atomic weapon logistics and effects. Admiral William H. P. Blandy was appointed task force commander; Parsons, now a rear admiral, was appointed commander for technical direction; Major General A. C. McAuliffe was named ground force adviser; and Major General W. E. Kepner was designated deputy task force commander for aviation. In addition, there was a full staff to coordinate the activities of more than forty-two thousand men, two hundred ships, and one hundred and fifty aircraft. Bikini Atoll in the Pacific was the site for the tests, and the curtain was to rise on July 1, 1946, for the first test of this spectacular drama, an air burst over an array of Japanese and American warships. An underwater shot was to follow.

I had paid very little attention to the preparations for the tests. Groves was handling it, and I was busy with other things. However, on June 23 I left on a leisurely trip to the Pacific on a C-54 plane with the Joint Chiefs of Staff evaluation board consisting of Lieutenant General Lewis W. Brereton, commanding general, First Air Force; Dr. Karl T. Compton, president, Massachusetts Institute of Technology; Rear Admiral Ralph A. Ofstie, senior naval member, bombing survey, Naval Analysis Division; Vice Admiral John H. Hoover, assistant chief of naval operations for materiel; Major General Thomas F. Farrell, U.S.A. (Ret.); General Joseph W. Stilwell, commanding general, Sixth Army; and Mr. Bradley Dewey. I wrote Jackie from San Francisco: "Our trip was fine while in the air, but the AAF tied it up as usual on the ground. Somebody forgot to tell somebody that our plane was to continue immediately beyond Topeka, so it took three hours to get that straightened out. General Farrell and I are sharing the same room. He is provided with envelopes already addressed to his wife. You should have thought of that, too. . . . We are having lunch with E.O.L., taking a tour through the lab, then E.O.L. is taking us to Trader Vic's tonight." I continued on June 25: "At Trader Vic's I ate six spareribs, three of them for you. Also drank two Tongas, a Scorpion, and a few Atomic Bombs. . . . We still hope to stay two days in Hawaii. Nobody on this trip seems to be in a hurry to get to Kwajalein.

I'm in no hurry, either." We landed at Pearl Harbor and spent a few days there. Bradley Dewey was on the board of directors of the big Hawaiian Pineapple Company, so he arranged for a tour of the island of Oahu. My guide was the pineapple research director, and I learned all about the sex life of a pineapple and the progress being made to control ripening in order to extend the canning season and how to make the pineapple more the shape of the can in order to avoid waste in canning pineapple slices. We had a delicious lunch out on the plantation, and I tasted for the first time delicious sun-ripened hunks of pineapple. After a late afternoon spent surf riding in an outrigger canoe, we enjoyed drinks and dinner at the home of the Hawaiian manager, high on the slopes of Diamond Head. We then adjourned to our quarters at Schofield Barracks. Except for finishing our flight to Hawaii on three engines, it had been a perfect way to start my new assignment, or perhaps I should call it a vacation.

When we left Hickam Field 2:30 A.M. on the twenty-eighth, everyone shed their ties; apparently Hickam Field is the dividing line between ties and no ties in the Pacific. We stopped at Johnston Island for breakfast, repairs, and gas, again on three engines. We reached Kwajalein, having crossed the international date line on the twenty-ninth. After lunch we flew over Bikini to see from the air the tremendous array of target ships, and in the evening I took the evaluation board to see the A-bomb being assembled. It was just at an early stage, so they could see most of the components. The next day we flew back to Bikini in a PBY so we could land in the water and inspect the ships.

I wrote home, "First on the schedule was a press conference on the *Appalachian*, then after lunch we inspected the target ship, battleship *Nevada*, the aircraft carrier *Independence*, and a Jap battleship, *Nagato*. The *Nevada* is painted orange and white so you can see it better. We inspected just as they were evacuating the ships. They sure have put everything on board except the kitchen sink: airplanes, bombs, tanks, teletypes, rats, goats, a pig, armor plate, clothing, etc., etc.

"Saw W. L. Lawrence at the press conference, also Kelly plus a couple of other MED's. I didn't get a chance to get over to where Dr. Warren or the rest of our outfit hang out but will do so tomorrow.

"Everybody in God's Christendom is here. 350 VIP's arrived today."

We returned to Kwajalein for a briefing on the test operations. Everything was all set for tomorrow, weather perfect, bomb checked out, everyone briefed for his part in the big event. We were to see the shot from the air at a distance of about twenty miles.

I spent that night on the *Albemarle,* Admiral Parsons' flagship for test operations. We took off immediately after the bomb plane at 6:00 A.M. and flew around and over Bikini until time for the drop. We put on our dark glasses, and the test came off right on schedule and as soon as we saw the flash we took off the glasses to observe the fantastic sight of an A-bomb explosion, the shock wave, and the mushroom cloud. Actually, I was disappointed. We were twenty miles away, and I must say that the movies I had seen of the Alamogordo shot were much more impressive. Again, when we looked down at the impressive array of target ships, the *Nevada* was floating majestically in the tropical sunlight, apparently unharmed. The only visible results from the air were two ships sunk and one on its side, plus four more ships burning. Vinegar Joe Stilwell, sitting in front of me, turned and commented, "The damned Air Corps has missed the target again." Everyone had the feeling that something had gone wrong.

We returned to Kwajalein, where I left the Evaluation Group to go out over the reef to board a destroyer for my trip back to Bikini. The destroyer carried the radioactive filters that had been collected by airplanes flying through the edge of the radioactive cloud. These filters were rushed overnight to Bikini to be analyzed to determine the chemistry and magnitude of the explosion. The captain of the ship turned over his small cabin to me. He said that running full speed in the moderate rough water would require that he be on the bridge. I enjoyed the ride and had a fairly good night's rest, but I never before realized that a ship could bounce, turn, rock, roll, pitch, and jump in so many violent ways and still stay afloat and remain in one piece. I finally joined the captain on the bridge to see the wave action from the vantage point.

After we arrived at Bikini at 9:00 A.M., Deke Parsons met me and took me to the *Mount McKinley,* where I was assigned my quarters, a very nice air-conditioned cabin plus a living room I shared with General Kepner. I wrote home, "It is the corresponding suite to Admiral Blandy's, who has presently loaned his to Sec of Navy Forrestal. Nice service I am getting here. I should have joined the Navy." In the command room the big question was, how big the bombing error was and the magnitude of the explosion. Bill Penney, a member of the British group at Los Alamos, came up with the first and also the correct answer about where the center of the explosion had occurred. He had placed empty five-gallon gasoline cans on the decks of the target ships. All he had to do was fill the cans with water and measure the volume of water, which varied due to the cans being crushed by the explosive pressure in proportion to the distance from the center. He plotted the pressures, drew lines connecting equal pressures, and had nice concentric circles that very clearly showed the center of the explosion. A few days later, analysis of more precise instrumentation verified that his crude method was correct. The miss distance was fifteen hundred to two thousand feet west of the assigned target. A controversy immediately arose: Was the bomb aerodynamically defective, or had the Air Corps bombardier missed the target? The result was that only the battleships *Nevada* and *Arkansas* and the heavy cruiser *Pensacola* were within one-half mile of the bomb explosion point. Little damage had been done to the main turrets of these ships but their superstructures had been badly wrecked. There was relatively little damage to ships more than three-fourths mile away.

When I went on an inspection trip with Admiral T. A. Solberg to see the damage, we found the carrier *Independence* still smoldering. It was badly wrecked, gutted by fire, and further damaged by explosions of its own ammunition, including torpedoes. The carrier still was too "hot" from radiation for us to board, although it had been boarded earlier for a brief interval. Generally speaking, however, there was relatively little residual radioactivity remaining from the air burst. The Japanese cruiser *Sakawa* suffered major damage to the hull and superstructure. She had already sunk to the bottom when I visited her.

The submarine *Skate* was close to the explosion point of the bomb. It had been on the surface, and the blast and heat had turned the superstructure into a mass of scrap. I was invited aboard, and I enjoyed a cup of coffee with the skipper. It was my first opportunity to go on board a submarine. Although it still was slightly radioactive, the crew put it back in operation. It could not submerge, however, due to damage to the superstructure.

The overall result of the Able shot was to bring more realism into the military and media appraisal of the effects of atomic weapons that had been so dramatically overestimated by the publicity given to the Hiroshima and Nagasaki bombs. But the pendulum probably had swung too far back.

On the fourth I went over to the *Appalachian* for the concluding conference for the press before they went on a cruise while we prepared for the underwater test, shot Baker. I wrote home asking Jackie to save the newspapers so I could see how the press covered the tests as compared to the facts and answers to questions we had given them. After a visit to the *Albemarle* for a briefing on the test preparations, I went for a swim. I wrote home, "Our whole outfit was at the beach this afternoon. Saw Staunton Brown, Flaherty, Hunter Kennedy for the first time. After swimming, went over to dinner and to see the movies of the shot with Admiral Parsons on the *Albemarle*.

"I think I've been in the wrong service. By comparison duty at Oak Ridge is—well there is no comparison. The Navy really has a very enjoyable caste system. I don't even have to put my shoes outside the door at night, they come in and find them after I'm asleep. I haven't yet learned all the official etiquette of how to be piped on board, but I'm gradually learning. Maybe I need a copy of *The Navy Wife*."

Life at Bikini turned out to be one of the most interesting and educational periods in my life. In addition, it was relaxing and enjoyable. I always had a yen for the Navy, and this was a splendid opportunity to see how they function. Concerning the Navy etiquette, I soon learned that you board a small boat in reverse order of rank and as soon as the top rank gets on board, the boat starts to move, but when you pull alongside a larger ship you board it in order of rank. In fact, even when walking the beach "shelling," the ranking admiral walks ahead. He gets first choice of the shells. Admiral Blandy welcomed me aboard, and Groves' prediction proved to be correct. I participated in all conferences on the daily course of action and also was invited on all the recreational and fishing trips organized by Admiral Blandy each weekend. The Army has nothing to compare with the service provided in the flag section of a "command" ship. Every day we went to the beach to swim and enjoy a drink at the officers' club there.

On July 6, Admiral Blandy, Parsons, and I went to inspect the *Independence*. I wrote home, "Had to climb a Jacob's ladder to get aboard her. She was pulled away from the rest of the target, so was no longer in the lee of Bikini island, so there was quite a swell and catching a Jacob's ladder on the fly is some sport. We wandered all over the ship, she is a mess, everything burnable burned, and the blast effects of the A-bomb and the internal explosions twisted things out of shape. Getting back into the boat we all made it but Admiral Blandy: following Navy etiquette he was last, and was left hanging on the ladder when our line broke and it took about five minutes to get back in close enough to pick him up.

"After lunch I went over to the *Haven* to see Kelley and Doc

Warren. He showed me his set-up on the *Haven* which is a hospital ship. Old Doc is working as hard as ever. I tried to get him to go swimming, but he is too absorbed in all the samples and data that he has acquired.''

When we heard that the Bikini natives were discontented with Rongerik Atoll, Admiral Blandy organized an expedition to meet with their king. Senator Hatch accompanied us. We landed the seaplane in the lagoon; used an inflated raft to go ashore; four in a two-man raft, and passed through a double line of men, women, and children to an area where the king welcomed us. My first brush with royalty! The Navy had built a new village for all the Bikini natives. Everything looked new and shiny but we learned that the chief complaint was that the coconuts on Rongerik were smaller than the Bikini coconuts. After about an hour of discussion and an exchange of gifts, Admiral Blandy invited the king to visit the task force and see for himself the tremendous size of the operation that necessitated the evacuation from their Bikini homes.

Before returning to the seaplane, we enjoyed a ride in an outrigger canoe to another island to swim and have lunch. I had never been in such clear water before and thoroughly enjoyed the outing. On the way back to Bikini, someone asked Admiral Blandy where the king would sleep when he came to visit. Admiral Blandy quickly responded, ''Out on the forward deck.''

The Baker underwater shot required extensive preparations and good weather. I frequently accompanied Parsons as he checked on the preparations. Dr. Ralph A. Sawyer was technical director. Dr. Marshall G. Holloway was the head of the Los Alamos Laboratory Group at Bikini. R. S. Warner, Jr., was responsible for preparing the bomb and suspending it in a caisson beneath the zero-point ship *LSM-60*.

My only official act occurred when Parsons and I went out to sea on the *Cumberland* to watch a dry run of Los Alamos procedures, the signals for activating all instrumentations, coordinating all test activities for photography, and finally throwing the electronic switch to detonate the bomb. Parsons expected to find a written countdown procedure, but some of the scientists thought they didn't need one, explaining that each one knew his role and would do his part on time. It was amazing how casual they were about such a highly important procedure. After three failures to carry out a successful trial—someone always tied it up—Parsons gave instructions to prepare a written procedure for the test and have a formal countdown, with Dr. Ernest W.

Titterton announcing the time for each step. We returned to the lagoon for preparation of the countdown procedure. When it was completed, we went out to sea again. Everything went perfectly on the first trial. Parsons then signed the procedures as deputy task force commander for technical direction, and I signed it as contracting officer for the University of California contract for the Los Alamos operations.

I wrote more about this situation in a letter to Jackie on July 20: "I still have the sniffles. Tony McAuliffe advised bourbon as the best cure so although we were too late to swim, Admiral Parsons, Captain Rivero, and Captain Ashworth were glad to join me on a trip to the club to get a highball. The beach was empty, due to the time and the weather, but the club was overcrowded. The bar, as a result, was short of glasses but we finally got a double Scotch each, just about the same time double bourbon each was delivered, a lot more than we bargained for. The extra drink helped break down the shield of reserve that surrounds Warner so I got to know him a little better. He is the guy that puts the bomb together. He is a rare one, not a scientist but a temperamental engineer. He usually is very aloof but he expounded on a good many subjects tonight. He has a fear that one in ten bombs may go off when assembled so he wants to call it quits with four. He has assembled three so far and he starts the fourth tomorrow morning." I continued later on in the letter: "We are having quite a problem with a few of the scientists. Never before have I wished that I were a bit more knowledgeable about some things so that I could decide if it were safe to give one of them his walking papers. Warner advised me today that we couldn't get along without this particular fellow but he agreed that my natural reaction to fire him was the correct one. When I heard this particular scientist's reaction to the countdown procedure worked out by fellow scientists, I wondered why his boss hadn't fired him.

"One advantage of staying on here is that I am learning this end of the game so I can overcome some of the built-up mystery that L.R.G. has encouraged or at least tolerated. This fellow was trying to spread a little hysteria about the dangers of certain operations. Most of these scientists are jittery enough without trying to get them more upset. That, plus the fact that the *Nevada* was still afloat is the main reason I stayed on out here, but I am glad I did for I got a lot of ideas about how the weapon end of the game should be run. There undoubtedly is plenty of room for improvement."

About the twenty-second, all the observers were flooding back to

Bikini. The JCS evaluation board had been on an excursion to Australia and New Zealand. The press returned, and news conferences were resumed. As B-Day, July 25, approached, one of our chief concerns was the weather. Each morning we had a weather briefing by Colonel B. G. Holtzman. Rigid requirements had been established for determining safe and favorable weather conditions, but the weather refused to cooperate.

On the twenty-third, I wrote: "Parsons and I reviewed the film we propose to release for newsreels. It shows the crew leaving the bomb barge, the countdown, pushing the button, etc. It was all taken today because we didn't want the cameraman to say 'just one more now' on B-Day. Parsons was saying that he hopes to remember to wear the same hat on B-Day in case they get his picture again. We had Hanson Baldwin plus two other newspapermen over to dinner tonight. This is the last chance for a while because the *Appalachian* (the press ship) leaves the lagoon tomorrow at 6:30 A.M. We start putting the bomb and bomb barge in position as soon as they leave.

"Earlier in the day I went out to inspect the target with the President's and JCS Boards. We watched the Navy get ready to submerge a sub, then on to a press conference. In the early P.M. we let them look at the bomb barge and also the firing links. Later, I took General Kepner over to look at the completed A-bomb. He had seen lots of dummies and was glad to see a real one. We both autographed it for luck. The weather forecast this morning for the twenty-fifth was good so I hope it holds.

"Today, B-Day, July 25, has been a long day but a great one. Last night the weatherman still predicted good weather but it was doubtful. Admiral Blandy made the right decision. He decided to keep the test on schedule. I gave Admiral Parsons your letter, so it slept on the *LSM-60* with 'Helen of Bikini,' in the A.M. he gave it to R. Warner to mail on the *Albemarle* and is now en route to Kwajalein so it should get in the air sometime tomorrow. Rather a circuitous route.

"This A.M. at 5:30 the weather looked OK so everything went ahead on schedule except for a 10 minute delay in clearing the *LSM-60*. We followed the procession out to sea, had breakfast at 7:15, then watched the target from 8:00 to 8:35. This time I knew just exactly where the bomb was going to go off because we had put it there. It sure did go off too, we were 8.9 miles away and for the first few seconds I didn't think we were far enough. If the movies make Baker-bomb as much more impressive as compared to the actual as they did for Able-bomb it sure will be some show. Except for no visible fireball, the underwater shot had everything the air burst had plus a lot of special features, some of which had not been predicted. First there was an explosion shooting up water and black smoke shaped like a cauliflower, then the visible shock wave and then the condensation cloud and vapor blocked out the view of everything for a few moments. As it cleared a Niagara Falls in reverse shot up over an area fully 2200 feet in diameter, millions of tons of water rose about 5,000 feet and finally vapor and steam came out of the top. As the tons of water came tumbling back into the lagoon, what appeared like a tremendous breaking wave broke out of the mass of water and advanced toward the next circle of target ships. Momentarily I thought, 'My God we have miscalculated the height of the wave, the alarmist may be right,' but as the 'wave' hit the first ships, the masts did not roll with the force of the 'wave.' It was not solid water, merely steam and spray. What a relief.''

Some alarmists claimed that a wave might endanger the Pacific Coast of California. Before the test, Parsons had asked me to review the model predictions that the initial wave would be a maximum of eighty to a hundred feet high and would not be dangerous outside the lagoon.

"About 38 seconds after the explosion, we heard the rumbling boom. Then we could taste it; Doc said that was ozone. A few minutes later the waves hit the beach and were just about what we had expected.

"After about eight minutes, visibility improved and we could see the target again. I was particularly interested in two ships: the *Arkansas,* which was gone completely, and the *Saratoga,* which had a slight list. All day long we watched this Lady die with dignity. It seemed that just about everyone in the Navy had been aboard her at one time or another. There were volunteers to go in the lagoon and tow her nearer the beach so she would not sink, but Staff Warren and Admiral Parsons vetoed that idea because of the expected radioactivity. The *Saratoga* gradually settled and at 3:45 P.M. everyone was called on deck to see her go down. She settled slowly by the stern. At 4:00 P.M., her flight deck was awash and settled steadily until her stern rested on the bottom and her bow was in the air. At 4:10 P.M., everything had disappeared except the very top of her mast. She remained a Lady until the last, not once did she turn her bottom up for the world to gaze upon. I won one buck on the *Sarah* going down and the *Nagato* staying up. So my score for Tests A and B is $7 won and $0 lost. We can buy the kids new shoes [with that].'' The *Nagato* sank four and half days later.

"All in all, it was a pretty good test and lived up to my expectations. Even the press should be a little bit more impressed again. The tests place the A-bomb just about where it should be.''

Overall, the results of the underwater shot were far more spectacular than the air burst, chiefly because the air burst missed the center of the target array.

The unexpected result was the base surge of spray and steam that momentarily had alarmed me so much. It spread over the target area, drenching everything with its dangerous radioactive spray. This hazard was far greater than expected. The radioactivity in the lagoon and on the atoll persisted for weeks.

On the next day, July 26, I wrote Jackie, "After lunch Staff Warren, Admiral Blandy, Parsons, a few of the press, and I went out to look at the target in a small boat. We were lucky and found a safe path all the way across the target array. We used up one day's tolerance in one hour.'' Staff Warren watched his Geiger counter and his watch and after passing over the center of the target where the *Arkansas* was somewhere below, said to me, "We had better get out of here as soon as we can.'' We did. "As a result of our trip, Admiral Solberg sent his crews in and towed the *Hughes* to the beach so they can study the damage to her hull,'' I noted. The large area affected by radiation gave the radiological section a wonderful opportunity to learn many lessons.

I also wrote, "I saw some films of B-Day. The colored shots are beauties. Spent about 15 minutes talking Capt. Quackenbush into making some arrangements with Eastman to produce extra copies of a good series. I am looking forward to seeing more movies of B-Day. It will make A-Day look like a pop gun.''

Now that I had seen the overall results of the tests, I was getting anxious to go home. Admiral Hussy, chief of the Bureau of Ships, offered me a ride home in his flagplane leaving on July 28. Because of the radioactivity in the lagoon, we still were out in the open sea and had to make a rocket-assisted takeoff in what I considered a rather rough sea to reach Kwajalein. The takeoff was a thrilling climax to a wonderful month with the test task force.

In the interval between the Able and Baker tests, I had received a letter from Jackie stating that my name was on the list of postwar demotions published in the *Washington Star*. I told Admiral Blandy. He asked, "Have you received the order demoting you?'' I replied,

"No, but my wife saw a list in the newspaper." "Well, don't take off your stars until you receive official orders. It is just too complicated to rearrange your station on the *Mount McKinley*. I doubt if any such message will be delivered to you while you're on board my ship." None was. I replaced my stars with eagles when I reached Hawaii. The trip back on Admiral Hussy's plane was very pleasant. I again had a comfortable berth. Once again we had engine trouble on every leg of the trip. On the last overwater leg en route to San Francisco, we turned back to Hawaii to get a replacement plane. Out of the six legs on the round trip to Kwajalein, on only one did the plane finish the trip on four engines. Was this the effect of rapid demobilization on air maintenance, or was I a jinx?

I was happy to arrive home on July 30. I learned from Jackie that I was to be awarded the "Most Excellent Order of the British Empire, degree of Commander," and that I should arrange a date with the British embassy for the award.

On October 28, 1946, President Truman appointed the five Atomic Energy Commission (AEC) members: David E. Lilienthal, chairman; Sumner T. Pike; William W. Waymack; Robert F. Bacher; and Lewis L. Strauss. Groves designated me as liaison with the commission to turn over the Manhattan District to the AEC. I had prepared a plan for the turnover, including a list of the items involved. The military position was that all property under the custody of the MED, with the exception of Sandia and the storage bases, should be transferred immediately to the AEC. It was considered essential to our national defense that the military retain custody of all weapons stockpiles, including all components and fabricated fissionable materials. The commission took a strong position against the exceptions and insisted that everything be transferred to them and that the AEC assume custody of all weapons. The Army, the Air Corps, and the Navy would not agree to this. I represented their case to the best of my ability but failed to convince Lilienthal. It was up to the president to decide the issue.

The Atomic Energy Act provided for a military officer to be appointed as director of military application. Groves had recommended to both the secretary of war and the secretary of the Navy that I be the sole nominee for this position. Also, General Maxwell Taylor, superintendent at West Point, had offered me a position as a permanent professor there. Jackie and I were weighing the merits of each of these prospective positions. It would have been a difficult choice, but I never had to make it. Lilienthal and I did not see eye to eye on very many aspects of atomic energy. We tangled on the division of responsibility between the AEC and the military, particularly on the issue of military custody of weapons. I considered it entirely impractical and dangerous to national security for the AEC rather than the military to have custody of weapons. At Berkeley, while on our trip to all the facilities, Lewis Strauss tried to convince me that I should yield on this issue, and as he put it, "work from within" to rectify the situation eventually. My belief was and still is that AEC custody was not only too dangerous to national security to tolerate for even a short interval but also was an absurd concept, and I would never indicate in any way that I would accept it as a correct solution.

There also were many other issues that apparently annoyed Lilienthal in the turnover. It was at times a difficult situation. I was still working for Groves, and also representing the military in negotiating the transfer. However, even some events that were not related to the transfer upset Lilienthal. One such involved the U-235 plants at Oak Ridge. Right after I returned from Bikini, Felbeck and Clark Center, his deputy, told me that they had good reason to believe that K-25 could produce enriched U-235 of high enough enrichment to be used for weapons. They proposed an experiment. Supply to the Y-12 beta plant would be cut off and the K-25 and K-27 plants would continue to operate without any product being withdrawn, to see if the enrichment would rise to a high enough percentage of U-235 to be satisfactory for weapons. The beta plant would continue to operate on the excess feed material that had been accumulated. This feed supply would run out about the middle of December, and if the experiment was successful, the electromagnetic plant would be shut down and the gaseous diffusion plant would supply U-235 for weapons. If the experiment was not successful, the flow of material from K-25 to Y-12 would be resumed. No production would be lost by the experiment. It certainly was worth trying. I hoped it would be successful.

When the commission visited Oak Ridge in November, I explained this to them, but like many other details of operation, it apparently made little or no impression on the individual members. They were more interested in other aspects, such as development of atomic power.

Just before Christmas, Peterson and Clark Center told me the good news that the experiment was a tremendous success. The electromagnetic plant should start shutting down immediately and be completely shut down early in 1947. The saving in cost of U-235 was extremely large. I prepared a letter for Groves' signature notifying the AEC that unless they ordered Groves in writing not to do so, he would order the shutdown and begin the layoff of personnel immediately.

The AEC, particularly Lilienthal, was surprised and annoyed to be confronted with this situation. He asked why they had not been informed, and I had little trouble convincing them that it had been included in their briefing at Oak Ridge and that they had been notified then that there should be no delay in making the decision; otherwise there would be a tremendous waste of government funds. More than ten thousand employees were involved. I told them that we were trying to make it easy for them (the AEC). No action on their part was required. We were taking the responsibility, as well as the credit for eliminating most of the large Tennessee Eastman payroll. Lilienthal seemed to take the attitude that Groves and I had arranged this just to embarrass the AEC. As a matter of fact, Groves had paid little attention to it. All during the last year of the war he relied on me, assisted by Peterson, to operate the combination of plants at the CEW to get maximum production. The experiment we had run was just a continuation of this practice. Perhaps the mistake I had made was in assuming the AEC would be very pleased with any action that cut the cost of producing U-235. Groves was capable of realizing the significance of it, whereas the AEC, with little interest in the details of plant operations, failed to grasp the fact that success would mean the layoff of so many employees at Oak Ridge.

The discussion ended when Lilienthal told me they would not take any action in regard to our letter. However, he did ask if I would delay the announcement until a few days after Christmas. I readily agreed. I was pleased that they did not interfere with what I thought was quite an accomplishment.

Strauss arranged for another meeting for me with the AEC concerning custody of weapons. At the end of my presentation, Strauss commented, "I have never heard a more lucid enumeration of a man's position." There were practically no other comments.

After the meeting he came to my office and told me, "I hope you realize that you have eliminated any chance of being appointed director of military application." I agreed with him completely. I understood the situation.

However, that was not the end of the matter. I went to see General Handy to tell him that Lilienthal did not want me as director of military application; therefore I wanted the approval of General Eisenhower and Secretary of War Patterson to accept appointment as a professor of mechanics at West Point. I met with them; they both thought that Lilienthal would change his mind, but I didn't. However, both Eisenhower and Patterson promised me that if I didn't receive the appointment to the AEC post I would be permitted to go to West Point. Max Taylor agreed to hold the position open.

Lilienthal soon made his intentions clear and requested additional nominees from the secretary of war and the secretary of the Navy. He selected a good friend of mine, Colonel James McCormack, to be director of military application. Unfortunately, at about the same time, Groves requested leave and announced he was planning to retire. Lilienthal's opponents in Congress raised objections to both Groves and Nichols leaving Washington. They claimed that the two men best informed about atomic matters would no longer be available. I heard about this when General Handy sent for me. He made it clear that both General Eisenhower and Secretary Patterson remembered their promise to me, but would I consider staying in Washington in order to alleviate the political situation. I came up with a countersuggestion: Why couldn't I go to West Point and still be available for part-time consultation with Congress, the military, and the AEC? Handy immediately liked the idea. He pointed to the telephone and asked me to call Max Taylor. I did, and he readily agreed. Also, General Groves finally agreed to come back to Washington to become chief of the Armed Forces Special Weapons Project. That ended that political crisis but not the political opposition to Lilienthal.

The president ruled that all property be transferred to the AEC. Accordingly, the MED organization, contracts, facilities, all property including weapons, and appropriated funds were transferred on December 31, 1946. I was to remain on duty with the AEC as deputy general manager for the month of January 1947, and other military personnel were to be available until July 1, 1947, to facilitate the transfer.

After completing my one-month transition period with the AEC, I departed for Florida with the family on a much-needed vacation. I had orders to report to West Point on May 1, 1947.

13

INTERLUDE, 1947

THE YEAR 1947 PRESENTED the prospects for a complete change of pace in my life compared to the past five and a half years of exciting activity and great responsibility. I thought that wars were over, I looked forward to the academic life at the U.S. Military Academy. I liked to teach, and I enjoyed in particular the prospect of educating the future leaders of our Army. In addition, West Point offered many advantages as a location to raise our two young children, Jan and David. I expected to have plenty of time to enjoy the pleasures of family life. My appointment as professor of mechanics was by law a permanent assignment, until retirement age of sixty-five. I was looking forward to almost twenty-six years of service at West Point, with a lovely, comfortable set of quarters overlooking the Hudson. I hoped that the condition put on my appointment by Secretary Patterson and General Eisenhower, that I be available for consultation for the initial year, might set a precedent and might be broadened and accepted as a general policy for all professors at the Academy. Max Taylor seemed receptive to this idea. Such a policy would enhance the position of professor and would offer me the potential for continued participation in the atomic energy field.

I was lucky; my Florida vacation was interrupted only twice, once when my services were requested by the U.S. representative to the U.N. Atomic Energy Commission, who was responsible for negotiating an international agreement for control of atomic weapons. I was ordered to New York to meet with this man, General Frederick Osborn. He was a reserve officer who had risen to the rank of brigadier general during the war and was considered an excellent negotiator. Bernard

Baruch had resigned from his position at the U.N. when he became convinced that the Soviet Union would never agree to any method of international control of atomic weapons that involved any effective verification to ensure observance of such a treaty. Osborn had taken Baruch's place to continue the effort to find some basis for agreement with the USSR, or failing that, to establish the terms of an agreement that would be acceptable to the other nations represented on the U.N. Atomic Energy Commission. Osborn was an impressive, distinguished-looking figure, six feet, seven inches tall. He had the nickname "high pockets." I took to him immediately; he was extremely friendly and had a sense of humor that appealed to me. Osborn wanted me to give up my leave and report at once, but we compromised on my reporting to West Point on schedule on May 1, and then commute to New York City for duty at the United Nations.

The second interruption was a request to deliver a lecture at the Air University, Maxwell Field, Alabama, on Atomic Weapons. It was scheduled for the first hour in the morning, March 27, 1947, with a question period, but the following lecturers were grounded by the weather so I answered questions for the entire morning. I enjoyed the session thoroughly, and judging by the questions, the student officers seemed to like it, too.

In private conversation with General Orvil A. Anderson, an outstanding World War II, Pacific Theater bombing commander, I heard his reasons for advocating what could be called "preventive war," but I disagreed with him. He never gained much support for that idea and was critized for it by many.

Reporting to Lieutenant General Maxwell Taylor, Superintendent of the U.S. Military Academy, I learned that in addition to three to five days a week at the U.N., I was expected to attend the joint Military Liaison Committee meetings with the Atomic Energy Commission in Washington, be a consultant to the Army Research and Development Division, the War Department General Staff, the Joint Strategic Survey Committee, and when requested to do so, to meet with Senator Bourke B. Hickenlooper, who was chairman of the Joint Congressional Committee on Atomic Energy. I also was to meet with others such as Allen Dulles, director of the Central Intelligence Agency, who already had sent me an invitation to lunch in New York for the following week.

Colonel O. J. Gatchell, Head of the Department of Mechanics at West Point, agreed that until my consulting load decreased I need do little except offer whatever advice on the current courses I thought might be pertinent. I was the junior professor in the department. Ten new professorships had been established by Congress at the recommendation of Max Taylor. Initially we all were called junior professor, one in each department, to distinguish us from the senior professor or head of the department. Only the head of each department was on the academic board, which established academic policy at West Point. Colonel Gatchell had no difficulty continuing as he had in the past without any assistance from me.

Life at West Point was as pleasant and comfortable as I had anticipated. It was great to enjoy the spectator sports and the active social life with many friends new and old. Commuting to New York City was no hardship. A small boat provided ferry service across the Hudson to reach the New York Central Railroad, and on my return Jackie could see my arrival and had plenty of time to drive down the hill to meet me at the ferry landing. General Osborn lived on the hills on the eastern side of the Hudson, so occasionally we commuted together. Generally he spent most of the week at his New York house. He invited me to use a small apartment on an upper floor whenever I desired to stay in town overnight, and he encouraged Jackie to join me whenever she cared to visit New York City.

Osborn's technique for gaining support for our proposed plan for control of atomic energy was to invite a delegation to breakfast to discuss a current issue. Usually he would ask me to present the issue and the reason why such a provision in the treaty was necessary. This technique worked very well, and frequently the French, British, or Canadian delegate would take on the task of supporting the provision at the meeting of the whole Atomic Energy Commission. I always was impressed by François de Rose, the French delegate, when he decided to support a critical issue. Then I would hear many of my arguments coupled with his own presented in a most logical way that only the French can do. It would be far more convincing than if anyone in the U.S. delegation had made the presentation. However, we had little luck in gaining the support of the Soviet delegation. I soon learned that Baruch was right: They were not interested in any agreement to ban the bomb that was verifiable. If they would agree to anything, it might have been just a simple statement agreeing not to use atomic weapons. But what would that mean? Who could rely upon such an agreement? The terms of the agreement that we were developing with the free nations were generous and equitable, and it was difficult to understand why the Soviet Union was not interested if they really desired a world at peace, as they claimed. Of course, a Communist world completely dominated by the Soviet Union would be considered by them as a world at "peace"—but at what a price in human spirit and freedom. We finally succeeded in receiving a ten to two vote in the Security Council for approval of the second report of the U.N. Atomic Energy Commission. This vote showed the plan to be reasonable, but in the absence of Soviet agreement it was not an advance toward control.

I never considered my time in trying to get a U.N. agreement on control of atomic weapons wasted. It was an effort that had to be made, a great step forward if agreement had been reached. Subsequent efforts by our State Department to obtain agreement with the USSR on the numbers and kinds of weapons to be produced by the Soviet Union and the United States neglect the main issue, which was and is to eliminate completely atomic weapons from the arsenals of all nations. Until all atomic weapons are eliminated and all nations have a better understanding of war and peace, we will continue to live under the threat of nuclear and thermonuclear war. Ban-the-bomb statements, freezes, promises not to be the first to use nuclear or thermonuclear weapons, and bans on testing are just window dressing and only allow the general public to have greater peace of mind because they are led to believe that something is being accomplished. Only steps toward reduction of weapons, definitely leading toward complete elimination when combined with an adequate defense and conventional forces, can achieve any meaningful effect. The USSR-U.S. agreements to the present date in 1986, in my opinion, are agreements only for the sake of agreement. They have not reduced numbers but simply have limited the growth of weapons to a figure that generally was in accord with what the Soviet Union planned to do. The result has been that we have not only lost our position of superiority but also have permitted the USSR to achieve superiority in the nuclear weapons field. Moreover, the agreed limitation on defensive weapons has been based on the faulty thinking that if we possess defensive weapons, that would incite the Soviet Union to believe that we are planning a first strike. The result has been that we have installed no defensive weapons against ballistic missiles, whereas the Soviet Union has installed a ring of defensive weapons around Moscow and is establishing a national network of radar that exceeds the limitation in the ABM treaty. True to form, they in turn are criticizing President Reagan's Strategic Defense Initiative (misnamed by the media as "Star Wars,") as a violation of the treaty. The SDI concept is the only proposal since the Baruch plan of 1946 that offers any hope of reducing nuclear weapons to zero or at least to a more reasonable level than exists at the present time. However, we should keep in mind that a reduction to zero would require an agreement on verification and reducing the Soviets' conventional forces.

In addition to U.N. duty, I periodically went to Washington for Military Liaison Committee and MLC-AEC meetings. Thus I was able to keep abreast of developments in the atomic energy field.

After the AEC took over responsibility for atomic energy on January 1, 1947, it was faced with recruiting new personnel, organizing, and making many critical decisions we had to defer during the transition period. They also were faced with the difficult problem of how to cooperate with and gain cooperation of the military in the fields of research, development, production, stockpiling and storage, and use of atomic weapons. However, before they could devote full time to any of these important matters, they were confronted with long and controversial confirmation hearings. The five commissioners finally were confirmed by the Senate on March 10, 1947.

Early in 1947, the Armed Forces Special Weapons Project was established by the secretary of war and the secretary of the Navy. Groves organized the unit and was the first chief of the AFSWP. Admiral William S. Parsons and Colonel Roscoe C. Wilson, Army Air Corps, were assigned to the AFSWP as deputies. The AFSWP was a joint Army, Navy, and Air Corps unit. It was to be the main contact with the AEC on the operational level and be responsible for many aspects of the armed forces' participation in the military uses of atomic weapons. It served in a staff and also a command capacity. The chief of the AFSWP reported directly to the chief of staff of the Army, the chief of the Army Air Corps, and the chief of Naval Operations. In addition, the Military Joint Research and Development Board established a Committee on Atomic Energy consisting of Conant, Oppenheimer, Crawford H. Greenewalt of du Pont, and the six military members of the MLC.

In the weapons development field, the AEC made progress during 1947. Development proceeded on improved implosion weapons for both plutonium and U-235. Studies on the thermonuclear bomb continued at a slow pace. Los Alamos plans and preparation for the Sandstone tests were in hand, the site had been selected, the task force commander had been selected, and work at the site had been initiated. However, weapons component production and the Sandia operations were far from being satisfactory.

Production of U-235 at Oak Ridge continued to exceed expectations. Production of plutonium at Hanford, however, still had problems, and construction of additional reactors was being planned and initiated.

Considerable progress was made in recruiting personnel and organizing for the tasks confronting the commission. In developing procedures for good relations, cooperation, and delineation of lines of responsibility between the AEC and the military, there was little progress. However, there were exceptions, such as setting up the task force for the Sandstone tests. The issue of custody still existed. Friction existed at the policy and operational levels. Civilian versus military control still was the fundamental cause for much of this dissension. The military, including civilian leaders, as well as the commission were unhappy with the situation. Late in the year, with the international situation becoming more critical, it was recognized that to achieve greater operational strength in atomic weapons, greater effort and better cooperation between the military and the AEC were absolutely necessary.

David Lilienthal, chairman of the Atomic Energy Commission, was particularly disturbed by the criticism and controversies he blamed on Groves. Lilienthal made high-level approaches to have Groves removed as chief of the AFSWP and as a member of the MLC. I knew that Groves wanted to retire, but he hoped to be promoted to lieutenant general before doing so. Even the Senate got into the act. Many senators felt that Groves should be promoted and took steps to ensure that he was. Senator Hickenlooper sought my advice and assistance on how it could be accomplished, and I was happy to cooperate with him. I felt that Groves deserved promotion.

In the autumn of 1947, Groves visited West Point to attend a football game and stopped by our home. When he noticed that we were installing a stair carpet at our own expense, he chided me, "You don't expect to stay here long enough to get much use out of that carpet, do you?" A little later, Captain Tom Hill, an assistant in the office of the chief of naval operations, asked me to get a ticket for him for a football game. After the game, he told me that he came to West Point at the request of Admiral Denfeld, the chief of Naval Operations, to ask me if I objected to the Navy advocating that I be promoted to a major general and be designated as chief of the AFSWP; I told Tom that I didn't object but that I didn't think the Navy had a prayer of accomplishing it. He said that he believed the Air Force also would support my selection to be chief of the AFSWP because the Navy and the Air Force each were vetoing the other's candidate to succeed Groves. (Jackie now tells me that I never informed her about all this. Maybe I didn't; she had not been well and was from time to time hospitalized at West Point and early in 1948 at Walter Reed in Washington.)

I warned Max Taylor that efforts were being made in Washington to have me replace Groves. I had no desire to leave West Point but thought that he should know about these rumors. He was quite firm in his comment: "You made a definite choice to become a permanent professor. You made your bed, and I expect you to remain in it." I replied, "That's what I expect to do." My various consulting assignments were requiring less of my time. I was prepared to enjoy devoting more time to being a professor. But that was not to be.

14

THE ERA OF ATOMIC SCARCITY, 1948–53

IN MID-JANUARY 1948, I received a long-distance call from Max Taylor. "Where are you?" he asked. I replied, "Walter Reed. Jackie has been here for a month. We are returning to West Point today." Taylor then said, "The rumors you heard are correct. Report to Joe Collins tomorrow morning. They want to promote you to major general. Disregard the previous conversation I had with you. You are a free agent to make whatever kind of deal you like with the chief of staff. Let me know what you decide."

Jackie was not at all pleased with the news. She definitely did not want to leave West Point. It was such a wonderful place to raise Jan and David. We talked about it long into the night. Finally I decided that I should accept the new assignment, with the understanding that I would return to West Point after two years.

The next day I proposed this to General Collins. He could not understand why I would want to go back to West Point. However, when the judge advocate general confirmed that I could be promoted a temporary major general while retaining my position as a professor and also explained that shifting me from professor back to the regular Army would take an act of Congress, Collins agreed to defer for a year the decision about my return to West Point.

As I had been informed by Captain Hill, the Air Force and Navy had agreed that I be designated chief of the AFSWP; the Army had concurred. Groves had written the criteria for the selection of an Army officer to be promoted to fill the position. A special promotion board had been established, and I was selected. The secretary of the Army forwarded the recommendation to President Truman.

In addition to being assigned as chief of the AFSWP, I was to be assigned as deputy director of plans and operations for Atomic Energy, General Staff, U.S. Army, senior Army representative on the MLC, and senior Army representative on the Atomic Energy Committee of the Research and Development Board. My Navy and Air Force deputies in the AFSWP were to be given similar assignments within their services. Pending the promotion and the retirement of Groves, I was to become informed on the current situation. I would rank quarters at Fort Myer, Va., but General Bradley suggested that I not ask for them. I would be the youngest major general in the Army, and there had been criticism from some of the older colonels. He recommended that I buy a house in Washington and said that I would be in the Pentagon a long time.

The MLC was to be given a new charter, and Donald F. Carpenter had been selected to be the first civilian chairman (William Webster was scheduled to replace him in the autumn of 1948). Carpenter, a vice president of Remington Arms, was noted for his ability to achieve harmony and cooperation.

Looking back on this period, I always am amazed at how rapidly the Cold War developed. At West Point I had been looking forward to a long, relaxing career as a professor during a long period of peace. Suddenly I found myself involved in a rapidly changing international situation that required that the United States prepare for a possible conflict with the Soviet Union. In regard to all this I will not give a complete historical account, nor will I relate a complete story of all my activities. I plan to concentrate on the key events in which I personally was involved, to illustrate how we achieved greater readiness in the atomic field.

In February, the crisis in Czechoslovakia increased tension. Groves retired on February 29. My promotion had been held up for reasons that I suspect originated in the AEC. On March 1, 1948, General Omar N. Bradley, chief of staff of the Army, asked me if, pending my promotion, I would have any difficulty taking over the AFSWP without written orders. In view of the fact that my Navy and Air Force deputies, Admiral Parsons and Colonel Roscoe Wilson, had participated in my selection, I did not anticipate any difficulty and did not have any. The international situation developed too rapidly to afford any question about the appropriateness of Bradley's verbal orders. Colonel Sherman

V. Hasbrouck, many years my senior and my chief of staff, signed all correspondence in the usual form, "For the Chief, AFSWP."

On March 5, Kenneth C. Royall, secretary of the Army, invited James V. Forrestal, secretary of defense; John L. Sullivan, secretary of the Navy; Stuart Symington, secretary of the Air Force; Donald F. Carpenter, future deputy secretary of defense for atomic energy and chairman of the Military Liaison Committee; David Lilienthal, chairman of the AEC; and me to dinner, arranged for the purpose of promoting better cooperation. However, the better motivation for generating cooperation turned out to be a cable received that day from the European commander, General Lucius D. Clay, which read as follows: "For many months based on logical analysis, I have felt and held that war was unlikely for at least ten years. Within the last few weeks I have felt a subtle change in Soviet attitude which I cannot define but which now gives me a feeling that it may come with dramatic suddenness. . . ." The cable was the main subject of conversation that evening.

At 10:00 A.M. on March 11, I received a call to come immediately to Secretary Royall's office. When I arrived, Royall, without any comment, led me to an elevator and we descended, then entered his car. As we left the Pentagon, he asked, "Why are we going?" I replied, "Mr. Secretary, I don't even know where we are going, so I certainly can't tell you why." He then said, "I had a call to bring you over to the White House, so I assumed that you would know why." When we arrived, Lilienthal met us in the anteroom and we proceeded into the president's office. Truman indicated that Lilienthal and I should take seats on two chairs that were directly in front of his desk and facing him. After we sat down, as I recall it, the president said to both of us, "I know you two hate each other's guts." Then, looking me squarely in the eyes, he said, "Nichols, if I instruct Mr. Lilienthal that the primary objective of the AEC is to develop and produce atomic weapons, do you see any reason why you cannot cooperate fully with Mr. Lilienthal?" I replied: "There is no problem if that is the primary objective." Then the president turned to Lilienthal and said: "Dave, I am signing the letter appointing Nichols a major general and he is to be chief of the AFSWP and a member of the MLC. You will have to forgo your desire to place a bottle of milk on every doorstep and get down to the business of producing atomic weapons." Then, looking at both of us, he said, "I expect you two to cooperate." The president then excused Royall and me and continued a conversation with Lilienthal.

On March 12, General A. C. Wedemeyer phoned me to say that I had been approved as a member of the MLC, that I had been approved as his deputy in plans and operations, and that he would like me to start joining their staff meeting every Tuesday and Friday at 8:30 A.M. That meeting plus other contacts were a basic source of information concerning current military operations and Army views concerning papers of the Joint Chiefs of Staff. Colonels David Parker and Jay Dawley were my assistants for performing my duties in that office.

On March 16 I met with General A. M. Gruenther, director of the Joint Staff, to discuss my ideas concerning AFSWP activities for the future. I told him I thought we should be doing more specific planning for use of atomic weapons, that details should be worked out concerning AEC transfer of weapons to AFSWP in case of emergency, and that we should have more maneuvers with the AEC, the AFSWP, and the Air Force participating. He surprised me when he said that I was violating a presidential order not to plan on the use of atomic weapons. I asked him if he was telling me to stop working toward these objectives. He replied: "No, I am not telling you to stop, I just want you to know that you are not in accord with present presidential policy."

On March 19 I had lunch with Secretary of the Army Royall, Secretary of the Air Force Symington, Secretary of the Navy Sullivan, Secretary of Defense Forrestal, Ernest Lawrence, and Dr. David B. Langmuir to discuss the prospects for developing radioactive warfare. Lawrence recommended that we give it greater consideration.

The Berlin crisis continued to intensify. On March 31 I was hastily summoned to Forrestal's office for lunch. Forrestal, Royall, Symington, the three JCS members, Under Secretary of State Robert A. Lovett, and others were there when I arrived. Eisenhower, who was President of Columbia University then, came in near the end of the luncheon, when more serious discussions began. I was greatly impressed with how easily Eisenhower assumed leadership of the entire group as he entered into the discussion. Clay had reported further restrictions on our land transportation, and he threatened to confront the Russians if they stopped and boarded any train, to shoot if necessary. I was at the meeting to supply information about whether, if the crisis grew worse, we were in a position to deliver any atomic weapons. We were not. I told them that the only assembly teams, military and civilian, were at Eniwetok for the Sandstone test and that the military teams were not yet qualified to assemble atomic weapons. I was told in very definite terms by Eisenhower to accelerate training and improve the situation at once.

As a result of this new information, the next morning I met with General James McCormack, director of military application at the AEC and also, at a joint AEC-MLC meeting, briefed the commission on this latest crisis. Thus, both at an operational level and a policy level, action was initiated to perfect plans for transfer of atomic weapons to the military in case of emergency and to expedite training and equipping the military assembly teams. When Forrestal called at twelve-thirty for me to see him, I was able to report on the actions already taken. My assignment in the AFSWP was taking on a real sense of urgency.

Throughout this period, I found McCormack most cooperative. I had known him as a cadet and had spent a most pleasant slow-boat return from Europe with him in 1935. He was returning from a Rhodes Scholarship assignment, and I from a year in Germany. With the sense of urgency developed by the international situation, cooperation and agreement on objectives were much easier to attain at both the operational and the policy levels. Due to my several different assignments in the atomic field, I had access to policy levels in the Pentagon and the AEC and to operational levels at the AFSWP field command at Sandia and Los Alamos in New Mexico. Although I reported directly to each of the three chiefs—Army, Navy, and Air Force—Secretary Forrestal set up a direct red line to my office and sent for me on many occasions to discuss atomic matters. I enjoyed my contacts with him. He was friendly, warm, and sincere. As my promotion still was being delayed, he asked me if I knew the source of the opposition. I thought it was due to the fact that Lilienthal and I had such a difference in concept concerning utilizing atomic weapons for maximum security of our country. Forrestal suggested that I call two friends of his who knew me and who had good access to President Truman, which I did.

Forrestal took his overall responsibilities very seriously, perhaps too seriously for his own mental health. He believed that atomic weapons must play an important role in our defense plans. He also desired to achieve unification of the armed forces and spent considerable time impressing me with the importance of unification. In the reorganization of the MLC and increased reliance on the AFSWP, Forrestal was setting an example for greater unification of the Army, Navy, and Air Force. Certainly, unification was working in the AFSWP. The multiple assignments I had in addition to being chief of the AFSWP, and that my Navy and Air Force deputies had, facilitated communication of ideas and provided a means for resolving differences. Frequently the question came up about how I could report to the three chiefs as individuals and not have problems carrying out their individual requests. My routine answer was, "Show me where I have not been able to comply because of any conflict." For our first maneuver I sent a Navy weapons assembly team to the Strategic Air Force. General Curtis LeMay, the SAC commander, asked me why, and I told him that I sent the Navy team because it was the best team available. His only comment was, "Expedite training the Air Force teams so they are equally or more competent." In my experience with three joint Army, Navy, and Air Force (earlier Air Corps) organizations, from 1942 until my retirement in 1953, I found, like many others, that in an operations organization having a specific objective there is no difficulty in getting good teamwork, particularly if the commander or chief writes the efficiency reports for all the Army, Navy, and Air Force members of his command. Staff organizations are something else. There the officers tend to represent the views of their own services.

On April 8 Donald Carpenter arrived in Washington to be chairman of the MLC and deputy secretary of defense for atomic energy. I had awaited his arrival, expecting that we would go to the first Sandstone test together but, due to the Berlin crisis, Forrestal canceled our trip. Although Carpenter's tenure of office was short—approximately six months—his leadership ability and personality assisted greatly in achieving better cooperation between the military and the AEC.

Late on the afternoon of April 22, I received the good news that the Senate Armed Services Committee had finally approved my promotion to major general. This enabled me to get orders for a change of station and move to Washington.

The Sandstone tests were very successful. The first shot was on April 14, followed by a second on May 1 and a third on May 15. Preliminary reports indicated that a large increase in efficiency could be obtained by improvements in design of the implosion weapons that had been tested.

Late in May, Carpenter arranged for a joint AEC-MLC trip to Sandia and Los Alamos. Lilienthal did not attend. Generally it helps to get policy-level people away from Washington to observe what goes on at the operational level. Sandia was the location of AFSWP field command for training assembly teams and for contact with the AEC units at Sandia and Los Alamos for weapons development. The visit permitted briefings on storage, surveillance, assembly of weapons. weapons development, emergency transfer of weapons procedures, and the prospects for new weapons as a result of the Sandstone tests. The subject of military custody of weapons was thoroughly discussed by both sides. Dr. Norris Bradbury stressed technical considerations against military custody, and, of course, the military emphasized the military considerations. Carpenter thought he had a good answer to Bradbury's technical considerations but was not yet fully aware of how important the commission thought the issue of civilian custody to be. The civilian control or custody issue involved not only military versus civilian but also which civilian organization. The Department of Defense is basically a civilian organization.

On June 18 the USSR imposed a complete blockade on land access to West Berlin. The airlift for coal and other vital supplies turned out to be the only feasible solution. Our existing military resources did not permit the choice of any action that might precipitate war.

On June 23 Forrestal and Lilienthal met for lunch at the Pentagon but could not agree on the issue of custody. On June 30 the full commission met in Forrestal's office with Forrestal, Royall, Bush, Carpenter, and me. Bush was invited at my suggestion because Lilienthal had been raising technical issues concerning the competence of the AFSWP to handle surveillance of atomic weapons. As I expected, Bush took the position that if technical problems were involved, they should be identified, and that military personnel could be trained to handle them competently. This had been my position from the beginning, but Bush's support was most helpful. Lilienthal then admitted that the real issue was civilian control of atomic weapons. He was not influenced by the emergency created by the Berlin blockade or by any of the military arguments for custody, such as the need for the military to become completely familiar with the weapons in peacetime, the risks of the transfer plan in case of emergency, the need to locate weapons for strategic use, etc. Emotion rather than reason was the basis for his position. The two sides still were too far apart on the issue; only the president could decide.

The issue was finally resolved when Forrestal, Carpenter, and Lilienthal met with the president on July 21. I rode with Forrestal to the White House. He asked me if I thought the custody issue was so important that he should resign if he received an adverse decision from the president. I replied, "It is certainly important enough, but I hope you will not resign over the issue. Your leadership is vital to national security. I think you should continue as secretary of defense." I did not attend the meeting. The president decided in favor of the AEC. I now realized that patience, persistence, and real threat of war would be required to obtain the right decision.

In all these discussions, I never knew of any top leader in the Defense Department, civilian secretaries, or chiefs of the Army, Navy, or Air Force, ever questioning the necessity and desirability of a presidential decision prior to use of atomic weapons. That is basically civilian control. However, these discussions did not seriously jeopardize the improved cooperation between the AEC and the military. Instead, the decision made it even more important to achieve greater cooperation, particularly as it applied to emergency transfer of weapons.

Carpenter created a Long-Range-Objectives Panel consisting of Oppenheimer as chairman, the three senior MLC members, and several civilians. It reviewed many of the military-related activities of the AEC, including supply of ore, production of fissionable materials, and all atomic energy activities of the military. The purpose of the panel, as its name implies, was to recommend long-range objectives. Oppenheimer demonstrated his leadership and was a master at summarizing the discussions and recommendations of the group. The recommendations of this panel were very helpful both to the AEC and the military. Oppenheimer gave me considerable support for developing tactical use of atomic weapons. I enjoyed working with him.

From June 21 to 23 I accompanied Generals Bradley, Gruenther, and McAuliffe on a visit to Sandia and Los Alamos. The objective was to get the top rank of the Army familiar with some of the special requirements of atomic weapons. I remember in particular Gruenther's remark to me at lunch, after he had seen the Mark III. He asked, "When are you going to show us the real thing?" He added, "Surely this laboratory monstrosity is not the only type of atomic bomb we have in stockpile?" I told him that we soon would have better weapons. After the astonishingly good results of Sandstone were available, stockpiling of improved weapons resumed. My deputies in AFSWP and I encouraged many other such trips to Los Alamos and Sandia by Army, Navy, and Air Force officers. Such visits were vital if the military was to rely on atomic weapons for the security of the United States. The top rank had to be better informed.

Although the increasing tension of the international situation facilitated better cooperation between the AEC and the military, many national policy problems remained unsolved. Our national policy in the Cold War was "containment." However, our budget restriction did not provide for the forces necessary. No decision was yet available on the presidential level to establish policy or establish the condition that would warrant the use of atomic weapons. Forrestal and the JCS were unquestionably moving in the direction of planning for the use of atomic weapons for retaliation in case of all-out war. The Air Force would have the mission of carrying out such an attack. The Navy's part, if any, in the atomic attack was not resolved. This led to a bitter controversy between the Air Force and the Navy. The Navy made every effort to participate in the atomic offensive by securing approval of a supercarrier and also by insisting that certain naval targets be assigned to them. I saw to it that the AFSWP cooperated with both the Air Force and the Navy, training their assembly teams and recognizing that the Navy probably would need smaller weapons for carrier planes. The Air Force generally opposed Navy participation because of the shortage of atomic weapons and their desire to take over full responsibility for delivery of all atomic weapons. However, as a result of Sandstone, the potential supply of atomic weapons increased significantly, and General Carl Spaatz, a retired chief of staff of the Air Force, finally agreed with Vice Admiral John H. Towers, retired naval aviator, that the Navy should have carrier-launched atomic weapons capability for strategic targets. The stockpile on June 30, 1948, was fifty weapons.[1]

However, the JCS was unable to solve the budget problem and could not afford the luxury of providing for the desires of both the Air Force and the Navy. Truman wouldn't yield on his strict budget limitations. Caught in this overall controversy was the future of the AFSWP. To whom should I report, the three chiefs or the Air Force? At the August 1948 Newport conference, attended by the secretary of defense and the Joint Chiefs, the decision was made for the continuation of the AFSWP reporting directly to the three chiefs of the Army, Navy, and Air Force. However, Forrestal approved the assignment of the Air Force as the executive agent for the JCS for strategic delivery of atomic weapons, with the understanding that the Navy would be utilized to assist when it had developed capabilities for delivery. On July 28 Forrestal told the JCS to assign top priority to a war plan

1. Stephen L. Reardon, *History of the Office of the Secretary of Defense: The Formative Years, 1947–1950* (Washington, D.C.: Historical Office, Office of the Secretary of Defense, 1984).

involving use of atomic weapons and low priority to one not involving such use.[2]

On August 25, 1948, in accord with these decisions, I received a directive signed by the three chiefs, General Hoyt S. Vandenberg, Admiral Louis E. Denfeld, and General Omar N. Bradley—directing me "to report to the Chief of Staff of the Air Force without delay for instructions concerning the atomic aspects of the agreed . . . Plan" and also directing that "until further instructions are issued, you will implement the atomic aspects of the . . . Plan in accordance with the instructions received from the Chief of Staff, U.S. Air Force." For other matters, I continued to report to the three chiefs. Immediately upon receipt of the order, I called General W. F. McKee in Vandenberg's office and asked, "When should I report to Vandenberg for instructions?" He called back in less than ten minutes and said, "Vandenberg says there is no necessity for you to report to anyone. Just continue as you have been."

On August 30 I suggested to General Samuel E. Anderson on the Air Force general staff that we have a monthly meeting of our staffs for briefing General Lauris Norstad, Anderson, and me on capabilities of the AFSWP and Air Force for delivering atomic bombs. Anderson called back and informed me that the first meeting should be scheduled for September 15 and that Vandenberg also would attend.

During the remainder of August and early September the interest at higher levels in our atomic activities including stockpile information increased. Interest was focused not only on weapons but also on the amount of fissionable material available. Arrangements were made through McCormack for this type of information to be available monthly.

On September 8, 1948, as a result of a meeting between Forrestal and George C. Marshall (now secretary of state), I received a request from R. Gordon Arneson in Marshall's office. Marshall had asked about installing certain assembly equipment in air bases in the United Kingdom. This raised the question about how involved the State Department should become with arrangements that had already been made informally with the United Kingdom by the Air Force and the MED and, later, AFSWP. I was, of course, happy to see that Marshall was interested in making preparations of the type that had been made informally sometime before. However, I recommended that the ar-

2. Ibid.

rangement for storing assembly equipment at Air Force bases in the United Kingdom continue to remain on an informal basis. If additional steps became desirable, as they ultimately did, more formal arrangements might be necessary.

On September 10, in Forrestal's office, I attended a meeting on the Berlin situation with the three armed services chiefs, the Secretaries of the Army, Navy, and Air Force, Secretary of State Marshall, Generals Gruenther and Norstad, and Carpenter. The situation with the USSR was worsening, and a greater state of readiness to deliver atomic weapons was necessary. This meeting gave me the explanation of why Arneson had discussed the U.K. air bases with me. The situation certainly was getting hotter in the Cold War.

As a result, every effort was made to expedite the training of assembly teams at Sandia and to procure the necessary assembly equipment for the teams. We checked every aspect of the transfer of weapons plan. Some AFSWP personnel were sent to the U.K. air bases. Cooperation among the various units at Sandia, the AEC, the AFSWP, and the Air Force continued to improve. Cooperation was helped by the new sense of urgency and by the judicious selection of officers and civilians for key liaison positions by all these organizations. It is amazing how cooperation can be improved by careful selection of personnel, clear-cut objectives, and a sense of urgency.

Near the end of September, William Webster, an electric utility executive and also an Annapolis graduate, replaced Carpenter as chairman of the Military Liaison Committee and deputy secretary of defense for atomic energy. Carpenter became chairman of the Munitions Board. Webster proved to be most helpful and effective in attaining greater atomic capability. He and I understood each other, became very good friends, and worked as a team to accomplish greater strength and readiness in atomic weapons.

During this September crisis Truman finally recognized that planning for use of atomic weapons was necessary. After being briefed by Forrestal and representatives of the Army and Air Force on the status of our atomic planning, Truman confided in his diary: "Forrestal, Bradley, Symington briefed me on bases, bombs, Moscow, Leningrad, etc. I have a terrible feeling afterward that we are very close to war, I hope not." As Forrestal recalled the scene, Truman prayed that a decision to use the bomb would not be necessary. Yet, significantly, Truman appeared committed to a firm stand and either at this meeting or a later one with Forrestal on September 16, he approved operational planning to proceed.

However, Truman continued to limit defense spending. He knew that Forrestal was relying more and more on atomic weapons and that the JCS considered that atomic weapons were the only means to prevent the USSR from rapidly overrunning Europe and also keep within the manpower and budget limitations of the United States and our allies. Plans for emergency transfer of weapons from the AEC to the AFSWP and to the Air Force had been worked out, and the procedure was tested on December 14. The approved procedure, although complicated, was the best arrangement that could be made under the presidential custody decision. I personally was very uneasy about the plan (so were others), but I felt that with the better personal relationship that had developed between the AFSWP and the AEC personnel at the operational level, we probably could make it work under many but not all emergency situations. There were too many possibilities for a snafu to contemplate making the plan a permanent part of our defense planning. For the present, we had to live with it—I hoped for not too long.

Also on December 14, at Gruenther's request, I briefed Eisenhower on the status of atomic weapons production, training of assembly teams, conversion of aircraft, and future prospects for weapons. He was about to play a more important role in the JCS. I always found him to be a very attentive listener and quick to grasp the significance of new developments. He was most pleasant, but firm.

Overall, considerable progress had been made during 1948 for increasing our atomic capability. The AEC had increased production of fissionable materials. The stockpile of Mark III weapons increased. Mark IV weapons were approaching the production stage. Studies were being made for smaller, lighter weapons. The year 1948 had been an exciting and demanding one. I was glad to be back in a position of responsibility. I enjoyed it.

The most significant events pertaining to atomic weapons in 1949 were the reentry of Eisenhower as chairman of the JCS, a military request for a substantial increase in the number of weapons, the replacement of Forrestal by Louis A. Johnson, the development of defense plans for Europe, the detection of the Russian atomic bomb explosion, and the thermonuclear bomb controversy.

Early in 1949, General Eisenhower acceded to a request from Truman and Forrestal to act as chairman of the Joint Chiefs of Staff.

On Saturday, March 12, I was asked to brief Eisenhower on the proposed JCS approval of the annual Forrestal-Lilienthal letter to the president establishing atomic weapon production requirements. This action is required by law. After explaining the nature of the letter and the difficulty of getting agreement with the AEC, I took the opportunity to tell Eisenhower my personal views on requirements for weapons and the possibility of acquiring a much larger supply at a relatively low cost as compared to more conventional weapons: "More bang per buck." I specifically recommended that a comprehensive study be made to determine the number of weapons to be stockpiled. I recommended that we should be thinking in terms of thousands of weapons rather than hundreds. I hoped that I would gain his future support for an expansion program.

At 9:00 A.M. on March 14, I briefed General Bradley on the same letter and also recommended to him that we needed a comprehensive study to determine the number of atomic weapons to be stockpiled. At 11:30 A.M. that same day, I was summoned by Gruenther to appear at the JCS meeting. I found that I had convinced Eisenhower of the need for an expanded program, only he had moved it from the future to the present. He was quite critical that the draft letter to the president made no mention of any need for expansion. I again explained that the MLC had been negotiating the terms of the letter with the AEC and that we had made the best deal we could for a joint recommendation. He told me to go back to my office and write the letter as it should be written. He stated that he did not want the president to receive any letter that indicated that he, Eisenhower, was satisfied with present rates of production and, if necessary, he would go personally to the president to tell him why he was not in agreement with Lilienthal. I returned to my office and called McCormack, who sent over Commander Edward Hooper (his stockpile man); and Admiral Parsons' assistant, Captain Duke, joined us to represent the Navy. The Air Force had no one immediately available but said to go ahead. The AEC history *Atomic Shield 1947–52* states, "The draft letter to the President evolved from the MLC formal notice that the currently established military requirements for scheduled production should be substantially increased and extended."[3]

The MLC now had better support for a major expansion. We soon

3. Hewlett and Duncan, *Atomic Shield 1947–52: A History of the United States Atomic Energy Commission*, Volume II (University Park and London: Pennsylvania State University Press, 1969), p. 179.

found we had an ally in Senator Brien McMahon, chairman of the Joint Congressional Committee. In due course, a program for expansion was worked out between the AEC and the MLC. Actually, Bill Webster, chairman of the MLC; Walt Williams of the AEC Production Division; Felbeck and Center of Carbide; and I worked out the scope of the fissionable material expansion program at Oak Ridge on May 19. The various staffs then built up the supporting figures. I heard that Lilienthal complained about my having taken this direct action with McCormack and my MED civilian friends now under control of the AEC. This type of cooperation so bothered Lilienthal that he discussed with the other commission members the advisability of replacing military officers on McCormack's staff with civilians. Actually, McCormack had not been involved; we worked only with the production people, all civilians who had participated in the Manhattan Project.

The Advanced Study Group and various other study groups within the military worked on weapons requirements, methods of targeting, and justification for expanded production. On May 26, the MLC sent new requirements to the AEC. On July 26, a special committee consisting of the secretaries of state and defense and the chairman of the AEC was appointed to assist Sidney W. Souers, executive secretary of the National Security Council, in reviewing plans for producing fissionable materials and atomic weapons.

In March 1949, Louis Johnson succeeded James Forrestal as secretary of defense. Johnson took several actions that renewed aggravation between the Navy and the Air Force. He canceled the Navy supercarrier, thus in effect limiting the possibility of any Navy atomic bomb capability. In May he told the JCS that they should plan on an additional $1.4 billion cut in the budget. This precipitated a Navy rebellion in which they attacked the Air Force B-36 and the strategy of relying primarily on atomic weapons.

Meanwhile, real progress was being made with General LeMay and his Strategic Air Command. On June 29, Webster, Major General David M. Schlatter, Air Force MLC member, Colonel Alvin Luedecke, secretary of the MLC, and I visited Curtis LeMay at SAC in Omaha to review their present emergency plans and their planning through 1952. I was particularly interested in arriving at a tentative agreement on the assembly rate requirements that would be consistent with SAC capabilities and the anticipated stockpile of atomic bombs so that the AFSWP could procure equipment and train the assembly teams necessary. The next day I prepared a letter to the JCS outlining the AFSWP assembly capabilities and need for expansion.

On July 6 I met with General Bradley, now the chairman of the JCS, General John E. Hull, Dr. Frederick Hovde, and Dr. Bradbury concerning the Army need for a short-range guided missile with an atomic warhead for tactical use. We also discussed the availability of uranium ore and fissionable material to support such a program. The next day a similar meeting was held with General Collins, chief of staff of the Army.

On July 25 I met with Senator Bourke B. Hickenlooper, member of the Joint Committee on Atomic Energy (JCAE), at his request and learned that he and other Republican senators would oppose the agreement on U.S.-U.K.-Canadian atomic energy policy we had been negotiating. It appeared that a lot of effort I had been expending on the project might go down the drain. He outlined his point of view on the matter, including specific ideas that he considered might be acceptable. However, his ideas probably would not be acceptable to the British. We were trying to get the British to agree to concentrate their fissionable material production program in Canada in exchange for renewal of full cooperation on building a gaseous diffusion plant in Canada and joint weapons development at Los Alamos. The alternative to no agreement was that the United Kingdom would build a gaseous diffusion plant in Britain and set up their own weapons development project.

In addition, I was spending considerable time at the working level of documenting the need for the increase in production of fissionable material that was being considered by the special NSC committee of Lilienthal, Marshall, and Johnson. For the routine operation of the AFSWP I had to rely more and more on my classmate Colonel E. E. Kirkpatrick, who had replaced Hasbrouck as my chief of staff, and Generals Robert M. Montague and Kenner F. Hertford, who were busy with the expanded AFSWP training program at Sandia. Kirk complained that the only time he got to see me was while riding together to and from work. Kirk was not only a close friend but also an outstanding superior officer. I could rely on him 100 percent to make the right decisions whenever necessary. He agreed with me that with the staff available in the AFSWP he could come up with the right decision, but if I were not available, selling the decision to Montague and Hertford, who were both much older, was a more difficult task for him than for me. And he added, "You have two stars to make up for the difference in seniority."

Near the end of August 1949, the USSR exploded their first atomic bomb. The Long-Range-Detection Program paid off. On September 3, the Air Force picked up the first signs of radioactivity. A high-level scientific panel headed by Bush concluded that the Russians had exploded an atomic bomb. On September 23, the president announced the Russian atomic explosion. I am probably responsible for the use of the term "explosion" rather than "bomb." In discussing the proposed news release with James Lay, on the Security Council staff, he asked if it had been dropped as a bomb. I expressed my view that it probably was a ground test similar to the way we tested at Alamogordo. I was being too precise, and as a result the president was misled and later stated the Russians did not actually have a bomb.

I was surprised that the Russians developed their A-bomb as fast as they did, but I was not as surprised as many others. After the war, Groves estimated that it would take twenty years for the Russians to develop an A-bomb. Many of our scientists disagreed. Groves felt that the Russians, although they might have sufficient scientific talent, lacked the industrial know-how and capacity required. I felt the scientists probably were right and my own guess was five years, but I had no way of knowing when they started. Perhaps with all the difficulties we were having with USSR beginning in 1948, we should have anticipated that they were having success with their A-bomb development.

The explosion of their first A-bomb shocked us into more action. We no longer could rely on our monopoly position. The immediate results were initiation of greater effort to develop the H-bomb, approval of increased production of fissionable material, increased readiness for hostilities, and an increase in the military budget.

The move toward developing an H-bomb gained support from many individuals and organizations. On Friday evening I received a call from Senator Brien McMahon, chairman of the JCAE to meet with him at his home on Saturday morning, October 1. When I arrived, there was quite a group seated before a rostrum where McMahon was standing. I took a seat in the rear row. I made a note that General Manager Carroll Wilson, Joseph Volpe, Everett Hollis of the AEC, and Adrian Fisher and Arneson of the State Department, among others, were present. Shortly after McMahon started to talk, I noticed one of his staff members hand him a note. He read it to himself, then announced: "Yes, I know General Nichols is present. I personally invited him. I believe that on this issue we will be on the same side." He then continued with his views on the necessity for developing the H-bomb. After the meeting, McMahon asked me to stay and bring him up to date on the U.S.-U.K.-Canadian negotiations for a new cooperation agreement.

On October 13 I attended a meeting with the MLC and the State Department with the JCAE. Secretary of State Dean G. Acheson brought the JCAE up to date on the status of our negotiations with Canada and Great Britain on the new cooperation agreement on atomic energy. Near the end of the meeting, McMahon changed the discussion to the subject of the H-bomb. He asked me what position the JCS was taking in regard to developing it. In reply, I suggested that since the Joint Chiefs were scheduled to meet with the JCAE the next day, it might be better if he received that information directly from them. He agreed.

I immediately went to Norstad's office and briefed him and Vandenberg on the JCAE meeting and, in particular, stressed that the JCS was to be asked about the H-bomb the next day. After a long discussion concerning the prospects for the H-bomb, Vandenberg asked me to accompany Norstad to the JCS meeting. Vandenberg was unable to attend and instructed Norstad to inform Bradley that he supported proceeding with development of the H-bomb.

I briefed the chiefs on the status of the H-bomb and strongly recommended that the JCS support development. Bradley's first question was, "If you thought development of the H-bomb was so urgent, why haven't you been around to discuss the matter before?" I told him the situation had changed, that before the Russian bomb many scientists were reluctant to work on development of the H-bomb, but that now many more might be willing to participate, and that scientists such as Lawrence, Edward Teller, and Luis Alvarez were urging support for a vigorous program. I thought that there were a sufficient number of scientists who were willing to work on the development to warrant proceeding with it. After further discussion, Bradley stated that he believed the United States would be in an intolerable position if a possible enemy possessed the H-bomb and the United States did not. He said that he personally supported development of the H-bomb; Norstad and the chiefs of the Army and Navy concurred.

After the JCAE meeting I noted in my diary: "Oct. 14, 1949, 10:30

A.M. Attended JCS meeting with JCAE. JCAE indicated changed sentiment and desire to expedite all AEC programs. The meeting was conducted in a friendly atmosphere and JCS indicated satisfaction with U.K. Canadian explorations (exploring talks on new atomic energy agreement) and exhibited proper sense of urgency concerning super (H-bomb). Net result of meeting may be that Congress will expedite appropriating funds for radar fence (50 million dollars) and make it easier for AEC to utilize existing funds to initiate action on expanded production program.''

On October 19, the president, having received the report of the Special Committee of the National Security Council in connection with the proposed acceleration of atomic weapons production, decided that the program was necessary from the standpoint of national security and that it should not be delayed. The conversation I had with General Eisenhower on March 12, 1949, had finally borne fruit.[4]

On October 20, members of the MLC and others met with Bradbury and McCormack and the latter's staff to discuss the program for the H-bomb and other weapons. I came away from the meeting with the feeling that Los Alamos was proposing a good program to meet the new challenge of the Russian bomb. With the JCS and the JCAE supporting development of the H-bomb, everything seemed to be in order for proceeding with it. I did not expect any opposition by the AEC.

From October 24 to November 7, accompanied by Arneson of State and George Weil of the AEC, I visited the United Kingdom in relation to the proposed U.S.-U.K.-Canadian cooperation program. We visited all the U.K. atomic energy facilities, including those for weapons. At the request of Norstad and Vandenberg, I also inspected the buildings and assembly equipment we had at two U.K. bases and received informal permission from Air Marshal Sir Arthur Tedder to send assembly teams to these bases. His only suggestion was that he hoped the personnel would wear Air Force wings instead of engineer castles.

While on the trip, I indicated a desire to visit Oxford. The British complied with my request. I attended a Sunday evening dinner at Christ Church College there. I met with the faculty before and after the dinner and was introduced to ''snuff'' and other interesting customs. I was told that Lord Cherwell would be present after dinner and that it had been arranged that Cherwell and I could withdraw to a secluded

4. Truman to Lilienthal, October 19, 1949.

corner with our brandy to discuss whatever we desired and it would not be overheard by the others. Churchill was about to come back as prime minister, and they hoped I would feel free to discuss the U.S.-U.K.-Canadian proposed atomic energy agreement with him. After a reminiscing session, we got down to the agreement and I found that Cherwell appeared to be in accord with what we were trying to arrange.

After I returned from the United Kingdom, I learned that the AEC's General Advisory Committee had met near the end of October to make recommendations concerning development of the H-bomb. However, no one in the AEC would give me any information about the recommendations. Even McCormack, usually most cooperative, clammed up. He told me he had orders not to discuss the issue. It took considerable effort on the part of all members of the MLC to determine through our various channels that the GAC had recommended against developing the H-bomb and that a majority of the commission was against development. Views varied from Strauss's desire for urgent development, delaying development pending an attempt to get the USSR to agree on no development, opposition on the technical grounds that development was impossible, and questioning the military worth of such a powerful weapon, to opposition on moral grounds that such a horrible weapon never should be developed. Without informing the MLC, Lilienthal took the report of the commission majority to the president.

On November 18, Truman set up another Special Committee to the NSC, consisting of Lilienthal, Acheson, and Johnson, to evaluate the H-bomb in terms of political, military, and technical factors.

On November 21, Senator McMahon wrote a very strong letter to the president, answering many of the objections voiced by members of the GAC and majority members of the commission and advocating speeding up the program to develop the H-bomb. This letter is one of the best letters on supporting development. I believe it greatly influenced Truman's thinking.

On November 23, Bradley sent the JCS position on the H-bomb development to the secretary of defense, who forwarded it directly to the president.

On November 23, Webster and I met with Bush to suggest that Bush visit Truman to advise him to support development of the H-bomb. I remember Bush's comment to me: "Nichols, when are you going to learn to be patient? You don't need my support. The president will know, without any advice from me, that from a national security point of view we must develop the H-bomb if it is possible to do so.'' Also, Bush predicted that all of the present commission would be gone by the following June. He was so right.

During the remainder of November and all of December and January, I attended various working-level meetings to help formulate the secretary of defense's position. Robert LeBaron, who replaced Webster, was in overall charge of this for the secretary. Some meetings involved State Department and AEC representatives. Norstad, or General Frank F. Everett, and Tom Hill, who had replaced Parsons, usually were the Air Force and Navy representatives at these meetings. This took a lot of time, but we had to get the right decision. Smyth and Arneson were involved for the AEC and the State Department, respectively. Bradley took a very active role in meeting with members of the Special Committee and the JCC.

On November 25, 1949, I have a diary note on a personal matter: ''11:30—Gen. Witsell's office. Took oath of office for appointment in Regular Army in grade of Colonel, to rank from 10 June 1948. Assigned to Corps of Engineers.'' I was no longer a professor at the U.S. Military Academy. In January 1949 I agreed with General Collins that I would stay in the Pentagon. The work and situation were too exciting and important to leave, and I had thought up an additional reason, as if I needed one. I did not want to face the cadets and be pointed out as the one who preferred being a professor rather than a major general. The change required a specific act of Congress specifying me by name and exact position on the permanent promotion list. I insisted on returning to the Corps of Engineers, although Personnel wanted it to be ''branch immaterial.'' I retained the temporary rank of major general on the basis of my present assignments.

On January 13, 1950, Bradley sent a memorandum to the secretary of defense, commenting on the views of the GAC and supporting the H-bomb development. The JCS did not recommend a crash program but did recommend an accelerated program.

Secretary of State Acheson proved to be one of the most effective supporters of development, and his views were approved by the president on January 31. Arneson rather than George F. Kennan, State Department expert on the Soviet Union, had more influence concerning Acheson's position. The president made a brief announcement that he had decided to accelerate development of the H-bomb.

Once the presidential decision had been made, a more normal

relationship between the AEC and the military resumed. The resignation of Lilienthal eliminated our biggest obstacle to cooperation. The announcement of Klaus Fuchs's espionage added a sense of urgency. The Fuchs case also dealt a final blow to the U.S.-U.K.-Canadian plan for renewing extensive cooperation in atomic energy that I had spent so much time on.[5] Late in February, the MLC met at Los Alamos, and the H-bomb development program proposed by Bradbury was considered to be satisfactory.

On March 10, Truman requested that the AEC and the Defense Department work out the expansion program required for production of tritium. I was a member of the working group, and little difficulty was encountered in reaching an agreement; a report was submitted to the president on May 25. Again, this event illustrates how a clear objective facilitates cooperation. Early in June the president approved the report, and du Pont was selected as the contractor to build the new Savannah River production plant.

In April, NSC 68 outlined a comprehensive national security policy and strongly advocated a large increase in the military budget. Although Johnson had supported the expanded AEC program, he continued to be obsessed with his economy program for military expenditures. Again, Acheson turned out to be the best support for the JCS and the policies expressed in NSC 68. Except for some serious technical questions concerning the feasibility of the H-bomb development, the program was on its way. All necessary policy decisions had been made; now it was up to Bradbury and Los Alamos to produce results.

In addition to getting the H-bomb program under way, many other projects, particularly for delivery of atomic weapons, defense against atomic weapons, and testing of atomic weapons, were initiated. The various international agreements to set up NATO and to defend Europe led to greater need for tactical use of atomic weapons. The AFSWP added personnel to coordinate development of guided-missile atomic warheads with Los Alamos and Sandia. The MLC met at Los Alamos on February 23 and 24 and, among other things, conferred on an "artillery-fired atomic bomb." On March 1 I attended a meeting in

5. Fuchs's arrest in February 1950 led to the discovery in 1950 of the American spy ring of Harry Gold, David Greenglass, Morton Sobol, and Julius and Ethel Rosenberg, all of whom were convicted. In March 1951 Ethel and Julius Rosenberg were found guilty of conspiracy to commit wartime espionage, sentenced to death, and executed in June 1953. Greenglass, Gold, and Sobol received prison terms.

Gruenther's office on air defense and on guided missile responsibility. All three services competed for position in the delivery of atomic weapons by guided missiles for tactical use, and the Navy and Air Force for strategic use.

Targeting of atomic weapons developed considerable controversy on the Joint Staff and even within the Air Force. Should cities be the main targets, or should targeting be confined strictly to military and industrial targets? One study I saw showed little difference in overall casualties in either case. I was not directly involved in targeting; that was mainly an Air Force and Joint Staff problem. However, I did enter into the controversy over the yield—the size of the bang. Many scientists and some members of the Air Force were advocating lower-yield weapons and greater precision. I resisted that on the basis that higher-yield weapons make up for the inaccuracies occurring in warfare and also provide greater overpressure necessary to destroy hardened targets. This same controversy arose previously in regard to the need for the H-bomb. LeMay supported the need for higher-yield bombs. He seemed to be more practical in regard to what accuracies can be routinely achieved in wartime. The physical size of weapons also became important in order to increase the variety of delivery vehicles. The Navy generally required smaller-size weapons for carrier use, as did the various delivery methods of tactical use. The AFSWP organization coordinated the physical size requirements with the AEC for the various delivery vehicles.

The active feud between the Air Force and the Navy, particularly as represented in the congressional hearings over the B-36, had ended on October 21, 1949, but had not been forgotten. Admiral Louis E. Denfeld was a major casualty in the battle. His replacement as chief of Naval Operations was Admiral Forrest P. Sherman, a naval aviator who was very well chosen to face the task of restoring morale in the Navy. He was a fine officer and had the respect of the Army as well as the Navy. He informed me that he would like to visit Sandia and asked if I would accompany him there. I would, of course, and I appreciated the opportunity to become acquainted with my new Navy commanding officer. We arrived at Sandia on January 21, 1950. After having the usual briefings about AFSWP activities and the readiness of the AFSWP to assemble bombs for the current war plans, Admiral Sherman asked if I would assemble all my Navy officers so he could speak to them. Admiral Hill asked if he wanted me to excuse myself from the meeting. Sherman interrupted rather crisply, "Of course not. Nichols is under my command as well as all the Navy officers.'' When all were assembled, I listened to one of the most inspiring talks on the need for teamwork in the armed forces that I have ever heard. He emphasized the need for pride of service, the need for cooperation on practically all major tasks, and the obligation to give loyal service to your designated commander whether he be Navy, Army, or Air Force.

The need for testing atomic weapons had to be given greater consideration by the AEC, the AFSWP, the MLC, and the JCS. A nucleus of a military test organization and funding provisions for advanced work needed for any task force were initiated by the AFSWP for Pacific testing. This nucleus could be transferred to the designated task force commander and after the tests be transferred back to the AFSWP to assume responsibility for closing out the test activities. Also, action was initiated again to propose a continental test site. Such a test site was first proposed by Parsons in 1947, and the MLC recommended a continental site to the AEC, but the AEC reaction was negative. The MLC again recommended continental testing in 1949, but again it was rejected by the AEC. Lilienthal was adamant, stating, ''It is impossible to conduct a test safely anywhere in the United States.'' I had just stated the opposite. Bacher calmly stated: ''Nichols is right, Mr. Chairman. Remember, we did conduct a test at Alamagordo safely.'' As a result, they qualified their rejection by stating it might be desirable if a national emergency arose. In July 1950, after hostilities in Korea began, Gordon Dean, new chairman of the AEC, wrote to the MLC warning that a national emergency might exist. As a result, new studies were made by the AFSWP, the MLC, Los Alamos, the AEC, and the NSC concerning criteria for site selection, comparison of various sites, reasons for continental testing, etc. The invasion of Korea by the Chinese late in November 1950 increased the state of emergency, and continental testing was approved by Truman in December.

Also during 1950, as a result of studies about destroying air bases, I generated greater interest in a penetrating atomic bomb. Considerable work was done on selection of a site for an underground test. Amchitka Island was proposed, but conservationists' desire to protect the sea beavers complicated and delayed approval of the site.

In the summer and fall of 1950, another expansion program for fissionable materials worked its way through JCS, AEC, Defense Department, and NSC channels for approval by the president on October 9.

Action toward taking additional preparatory steps at advanced bases for use of atomic weapons was taken during 1950. In July, in consideration of the possibility that the Korean War might develop into a world war, the JCS recommended that atomic bomb components, except for the fissionable cores, be stored in Britain, thus decreasing the amount of material that would need to be shipped in case of emergency. The MLC met with the AEC, and the AEC agreed to the proposal. Dean and Johnson met with the president, and the president agreed to the transfer. This was the first time the terminology military need for "availability" of weapons was used instead of military need for "custody." About two weeks later, a second transfer was approved to store weapons at an advance base in the Pacific.

When the Chinese entered the Korean War in November, forcing MacArthur to fall back, I advocated use of atomic weapons and personally discussed the matter with each of the three chiefs, but all of them were lukewarm to the idea. Vandenberg was afraid that the small number available would not be effective enough and might debunk the general impression of their value. Later, consideration was given to their use, but British objection combined with a desire not to widen the war ruled against their use. The crisis did, however, lead to even greater support for expansion of fissionable material production facilities, and, as previously stated, continental testing of atomic weapons was authorized, and the military budget was greatly expanded, but success in the H-bomb development still remained uncertain due to technical difficulties.

During 1950 there was a proliferation of effort by the Army, Navy, and Air Force to develop guided missiles for air defense and for strategic and tactical delivery of atomic weapons. A race to establish the best specifications resulted, and the needs for greater coordination and priority for air defense were obvious.

At the time, the Department of Defense was completing the written records pertaining to how, why, and when the H-bomb development was approved. Secretary of the Army Frank Pace was reviewing the Army record before giving his written concurrence. He sent for me, and among other questions he asked was, "Why shouldn't we be spending as much on defense against atomic weapons as we are going to spend on the hydrogen bomb? Why haven't you recommended developing the Nike air defense missile at a faster rate?" My reply was prompt: "We should be spending more on air defense, particularly the Nike missile. If I were directly responsible for air defense or guided missiles, I would be in here recommending a larger program."

Likewise, Secretary of the Air Force Finletter became interested in guided missiles for both offense and defense, and he asked me, "How do we set up a Manhattan Project to get the job done? What are the authorities that should be given to someone to set up such an organization?" When I started to list the responsibilities and authorities Groves had and the agencies that were bypassed in the normal chain of command, he stopped me time and time again, asking, "Why do you need to do that? It can't be done that way in time of peace." I agreed, saying, "I consider it impossible to set up a Manhattan Project and, in particular, to establish the degree of secrecy that is essential to avoid interference with any such command. You can only do it in time of war." Finletter concluded the conversation by thanking me and added, "You think about how to do it."

The next I heard on the subject was a call from K. T. Keller, chairman of the board of Chrysler Corporation, on October 18, 1950. He invited me to lunch. He told me that president Truman had asked him to make sense of the guided missile program. He told the president he would do it on a part-time basis, but only if he could have me as his assistant. The president pointed out that I already had an important job and besides he didn't believe a president should select any officer to take on a specific assignment. He advised Keller to see if he could persuade me and also get the approval of the secretary of defense. I told Keller I wanted a little time to think it over. I wanted to ascertain if the job were feasible. I said that if I did accept, I would specify two conditions: The first was that he would never see anyone in Washington or elsewhere without my being present. I didn't want to be hearing from others—"Oh, Keller approved that when we had lunch together." As the second condition, I wanted to be certain that it was clearly understood by the president and by the secretary of defense that he, Keller, would resign the first time his recommendations were disapproved by anyone in the Pentagon chain of command. Keller laughed and said, "I am way ahead of you on that one. Truman understands that."

The only one I consulted on this proposed assignment was Larry Norstad, whose support I would need and whose advice I valued as a friend. He commented, "It is a good idea both from the standpoint of defense and for your own personal future." He stated that I probably could remain czar of atomic weapons in the Pentagon for as long as I desired. However, his staff would be pleased to see me leave the AFSWP in hopes that they could get an Air Force officer designated chief. He doubted that would take place without a hassle. He added, "I am certain that Finletter will be pleased if you move to guided missiles. I will, too. I will continue to give you full support."

I told Keller I would accept. On October 24, 1950, George C. Marshall, now secretary of defense, appointed Keller director of guided missiles in the office of the secretary of defense, with the understanding that Keller would retain his position as chairman of the board of Chrysler and serve only part time as director of guided missiles. The main objectives were to select the missiles to be put into production and to give more priority to air defense missiles.

On November 2, 1950, I was appointed deputy director of guided missiles to act as Keller's principal assistant and also to provide assistance as may be designated to the Research and Development Board, the JCS, or other agencies of the Department of Defense. I was not relieved from my atomic assignments until January 23, 1951. As Norstad predicted, there was a hassle between the Air Force and the Navy about my successor. Bradley asked me if I could handle both jobs until the selection of a new chief was decided. I said that I could. After the controversy showed no signs of being resolved by the Navy and Air Force, I suggested to General Bradley that Major General Herbert B. Loper be named. He was already on the MLC, was a very capable engineer, easygoing but firm, and was well liked by his associates. He was accepted by the Air Force and Navy and assigned as chief of the AFSWP.

In establishing Keller's authority, there still was talk about another Manhattan Project, but Keller and I knew that was impossible, so instead we agreed to rely on his prestige and close contact with the president. Robert A. Lovett, deputy secretary of defense, asked me to draft a charter. I worked with Admiral Tom Hill and Major General Frank F. Everest of the Air Force. We were all very good friends. Hill thought it would be difficult to get agreement on a charter. He asked me what I was up to when he read the draft I had prepared. It gave Keller no authority except that my charter required that I carry out his orders. To gain the control we desired, I had concurred with a draft order that Wilfred J. MacNeil, comptroller for the Department of Defense, had independently proposed to the secretary to gain fiscal control over production of missiles. He was surprised that I agreed so readily, but it fit in perfectly with the method of operations Keller and I had planned. The order specified that no funds be expended for production of guided missiles without the express written authority of the secretary of defense. It was just what we needed. Keller did not plan to interfere with research and development. He did not plan to stop any production under way, but by being the sole adviser to the secretary for any new production and knowing that the secretary would approve only what he recommended, we had all the authority we needed to direct what missiles be put into production, with little paperwork involved.

We had a small Army, Navy, and Air Force staff that varied from four to six officers, a master sergeant, and my secretary. Unexpectedly Harry Traynor appeared and reported for duty. He had been called to active duty because of the Korean War and ordered to report to the Pentagon. He assumed that I had requested his assignment. I had not, but I welcomed his availability and had no trouble arranging for proper orders.

Initially we visited test sites to get acquainted with the work under way. Few missiles were ready for production. The first thirteen firings we saw all failed; some of the failures were really spectacular, but as we continued we found a few more advanced in development. In selecting the more advanced missiles for review, the entire staff except for one officer, the master sergeant, and Miss Olsson, who remained behind to mind the shop, accompanied Keller on a visit to the missile contractor's plant. We reviewed the project with the contractor's personnel and the Army, Navy, or Air Force personnel responsible for that particular missile. As a result of discussion and review of the status of research and development, and consideration of the recommendations being made for production, Keller would indicate the extent of the program for production that he would approve, provided the service involved would recommend that program in writing to him. My staff would assist the service involved to see that the written recommendations were in accord with what Keller and I wanted. Upon receipt of such letters for three or four programs, we made a written report, then hand-carried the report to the president and then to Secretary of Defense Marshall or his deputy, Lovett (who later became secretary of defense). I would prepare directives for Marshall's or Lovett's signature authorizing expenditure of funds. This was a simple and fast procedure and came to be known as Kellerizing a missile program. It was effective because there was a definite understanding that Keller would serve only as long as his recommendations were accepted. President Truman seemed to enjoy the procedure. Keller always completed explaining the missile programs in less than ten minutes. The president would thank him for spending time on the program then start discussing other current problems. He appeared to value Keller's opinion on quite a range of subjects. In addition, they always had at least one good story to swap. Keller was a master at having an appropriate story for every occasion. I caught hell for not keeping the meeting short. At the next meeting Connally, the president's appointments secretary, opened the door after the allotted time, but the president waved him away.

At meetings with Lovett, the programs were discussed in greater detail because he was more interested in the progress being made, and he did take the time to read and understand carefully the letters I had prepared for his signature.

Keller's procedure was challenged only once, and that was by the Navy. After reviewing a Navy missile instead of receiving the usual letter from the Navy recommending the production program Keller was willing to approve, we received a letter asking Keller to explain why he was limiting the production program to a much lower figure than the Navy desired. When Keller arrived for his next visit to the Pentagon I showed him the letter, with the comment, "If you answer this letter that will start the usual Pentagon paper mill procedure." Keller replied, "Let me handle it." I assembled our staff, and when we got to the Navy missile Keller commented to our Navy assistant, "It appears that the chief of Naval Operations does not like my program: Tell him that I am willing to discuss the matter further in his office, or he can come here." The assistant replied, "He doesn't want to discuss the matter; he desires a reply in writing giving your reason for your program." Keller smiled and said, "Tell him that I am willing to discuss the matter as long as he desires, but as for answering his letter, I not only won't answer it, I won't even receive it. Here it is, take it back." Shortly thereafter we received the letter in the form desired.

In early 1953, President Eisenhower selected Charles E. Wilson, head of General Motors, as secretary of defense. When I next saw Keller he told me that Wilson had asked him to stay on. Keller said he had agreed to do so, but only for a limited time. He added, "It will be impossible to work with Charlie Wilson the way we have with Lovett. Charlie will want all the details." Prior to inauguration, Keller invited Wilson, Robert Stevens, Dan A. Kimbal, and Harold E. Talbot, appointees to be secretary of the Army, Navy, and Air Force, respectively, to visit Cape Canaveral and then to have a meeting in Nassau where we would brief them on the guided missile program. At the last minute, Wilson found he could not attend because of meetings with Eisenhower. We had a very good demonstration at Canaveral and a very productive meeting at Nassau. We stayed in the Colonial Hotel and, due to the governor's worry about security, we had been assigned the entire top floor. After a late afternoon meeting, Stevens, Kimbal, Talbot, Keller, and I went out for a walk. Suddenly a taxi pulled alongside and Talbot said, "Let's find a good hot spot to have dinner." So we crowded into a cab, thereby unintentionally eluding the governor's security force that was tailing us. The taxi driver took us to a good spot, where we had a jolly evening. At one point where the show was unusually hilarious and we were all laughing, Talbot commented, "If some photographer took a photo of this table with the present scene on the stage, he would have a real scoop. He could label it 'Eisenhower's appointees having a serious discussion on guided missiles at Nassau.' " Just about that time a member of my staff rushed in, saying, "Thank God I finally found you. The governor is in a turmoil. His security men are looking all over town for you."

Working for Wilson turned out about as Keller predicted. Keller did not attempt to establish the same relationship with Eisenhower that we had with Truman, although I understood that Eisenhower did call Keller and asked him to stay on for a reasonable time to avoid the appearance of a conflict between Wilson and Keller. On the occasion of our first program approval, we proceeded as usual except that we went directly to Wilson's office. Keller and I outlined our recommended programs. When we finished, Wilson asked, "What do we do next?" I told him that in the past the secretary had told me to prepare the necessary letters for his signature, which I already had done, and I handed them to him. He read them and then asked Keller, "Do you mind if I take a few days to think this over before I sign them?" Keller replied, "No, that's all right. I am planning to spend the next two weeks at Marathon, Florida, fishing. Let Nichols know when you have signed the letters and he will carry on from there and also call to notify me that everything is in order."

Keller commented as we left the office, "Charlie will have a time with our procedure. Call me a week from today and let me know how he is making out."

A week later I told Wilson's aide that I planned to call Keller that afternoon, and I asked if Wilson had signed the letters. About an hour later he called and said the secretary wanted to see me.

When I first sat down, Wilson said, "I have signed the letters," and handed them to me. He then proceeded to tell me that he had checked with his staff and with the three departments and found that there still were opinions not fully in accord with Keller's recommendations. I told him that the situation was normal, that there always had been differences of opinion, but that it was Keller's function to resolve them or to decide what should be done. Keller was doing what the secretary would have to do if Keller's office didn't exist, or go back to the old system of endless reviews and letter-writing. I called Keller to tell him that the system still worked but should be phased out sometime during 1953 or Wilson would undoubtedly try to revise it.

During the three years Keller served as director, we established production programs for Army tactical missiles such as Corporal, Hermes, Redstone, and Honest John, Air Force and Navy pilotless strategic bombers such as Navaho, Snark, Rascal, Triton, and Rigel, Navy and Air Force pilotless tactical bombers Regulus and Matador, and air defense missiles such as Nike, Sparrow, Falcon, Nike-Hercules, Bomarc, and Talos. Except for Nike, Sparrow, and Falcon all were being designed for atomic as well as conventional warheads. The AFSWP continued to coordinate warhead design with the AEC. During Keller's regime, which ended in September 1953, the first Air Force strategic ballistic missile, the Atlas, was in the study stage, as was the Army antimissile missile. In the air defense field the largest production program was for the Army Nike ground-to-air missile. Extensive installations of this air defense missile were made and maintained until well after intercontinental ballistic missiles became the major threat to the United States. The Sparrow and the Falcon air-to-air missiles were put in production by the Navy and the Air Force. These missiles definitely improved our defense against bombers and other aircraft. My three years with Keller was a most interesting experience. He was a big man in both stature and spirit. He showed me a new and lighter way of accomplishing an objective. He enjoyed working with people and was an excellent judge of men and organizations. His approach to the missile program was not to waste time trying to stop a missile program under way that he did not think would be successful. Instead he relied on our control of production funds and felt that it would eventually fade away for lack of money, whereas blocking tactics would result only in a fight that would distract from our main effort to get critical missiles into the production stage. He did not attempt to resolve the roles and missions of the individual services but was objective and fair in approving any missile ready for production that had promise of fulfilling a military requirement. He encouraged research and development without trying to control it. He was intrigued with the Navy Sidewinder air-to-air missile and encouraged the scientists and engineers working on it to keep it simple. It still is one of our most effective air-to-air missiles. Many of the contractors, such as Don Douglas (of Douglas Aircraft) were his personal friends, and frequently just the three of us would go out to lunch or dinner and Keller would emphasize to Don what he wanted accomplished concerning production methods. He believed in the man-to-man, at-the-site approach rather than relying on paper studies, committee reviews, and detailed programs. His attitude about eliminating paperwork was consistent. On the first day we entered the small office that I had arranged for our use, he sat down at his desk, picked up the pen and pencil set that Virginia Olsson had requisitioned for him, and asked her, "Does everyone get a fancy set like this? I won't need it. I won't be signing many letters. I will use the pen I carry in my pocket. You can take this one back to the supply office or wherever you got it." When Keller left the government, we closed the office. The entire file that contained the programs he had recommended and critical correspondence occupied less than one file drawer.

Frequently contractors or others would ask him why he insisted on having Nichols as his deputy. He always gave the same two incidents. He embellished the story about his contract with the MED and the occasion when he had phoned me to tell me that the costs were $20 million less than had been advanced to Chrysler under our payment schedule and asked what he should do with the money. I immediately told him to send me a check, but final settlement would be subject to our audit. He liked that. The other story was when we had asked him to give us advice at Oak Ridge on maintenance problems for the electromagnetic plant, and we also showed him the gaseous diffusion plant. As Felbeck, Keller, Groves, and I entered K-25 and started to walk down the long, thirty-foot-wide alley between the production units, we were preceded by about twelve stout Negro women who were slowly advancing in a line with push brooms, sweeping the alley. Keller stopped, stared at the broad backsides of the women ahead of him, and asked me, "Nichols, don't you know there is a machine made to sweep a concrete floor like this?" "Sure I do," I replied. "But these gals can do more than one of those machines." I told him that the head of our main construction contractor for K-25 had come to me with a problem concerning the large turnover among his Negro construction labor. They were in a separate construction camp and it was some distance from any outside community, and such communities were mainly white. He said, "On Saturday nights Negra men like to fracas and here there is no place to fracas." He asked my permission to set up a separate black women's camp. He stated that he would fence the camp and not allow any men inside their camp, but the men's camp would not be so guarded. I approved this idea. I told Keller we now had less turnover and apparently the men now had an opportunity to "fracas." I never heard such a hearty laugh in my life. He always told the story with considerably more flourish than I can. Usually he would add at the end, "In addition, Nichols knows how to get things done in the Pentagon. I don't."

In January 1952, Frank Pace, secretary of the Army, asked Keller if he would agree to my being assigned additional duty as chief of research and development of the Army. Keller agreed. Earlier I had made a study of the Army R&D and had recommended a reorganization and moving the office of chief of R&D from G-4 division of the general staff to the office of the chief of staff.

In the past, the Army R&D had been criticized for having concentrated too much on improving existing hardware and paying too little attention to new weapons and the relationships among war plans, tactics, human resources and weapons systems. Lieutenant General Thomas B. Larkin, G-4 General Staff, U.S. Army had concerns about the proposed change, especially that it might downgrade the influence of the technical services. I knew General Larkin from the days in Vicksburg, Miss. and also he had cooperated with my writing the report on R&D. He indicated to the chief of staff that he would be more satisfied with the change if I were selected to be the one to put the new organization into effect. I welcomed the additional assignment. By that time we had practically every missile that was ready programmed for production. We were over the hump, and Keller was spending less time on the missile program. All I needed was a superior man for my chief of staff in R&D and I could handle both assignments. For that task, I selected Colonel A. W. Betts, who had worked with me before; I knew he could do the job.

We concentrated on getting the Army thinking in longer-range terms. Betts developed several briefings, really lectures, on how long it takes from conception of a weapons system to getting the system into the field, the tactics developed, and the men trained to use the system.

We gave greater support to research in the field of radioactive and biological warfare and new chemical agents, even though we knew that probably they never would be used, on the basis that to develop proper defense against these weapons one has to know their offensive capabilities. We also found that the idea of using tactical radioactive materials in war was not very attractive. It took too much to kill or incapacitate. If you wanted radioactivity, some form of the bomb would be a better weapon. I also initiated action to develop an antimissile missile. In Keller's office we were supporting long-range strategic missiles of the pilotless aircraft type, and the intercontinental ballistic missile soon would enter the development stage. My concept of weapons always has been that for every offensive weapon you need a defensive weapon to counter it. I always remember my wrestling instructor at West Point, Tom Jenkins. He used to say: "Show me the holt, and I'll show you the guard agin it. There ain't no holt that can't be broke." I asked Ordnance to start research on an antimissile missile, or as it was later named, an antiballistic missile (ABM). The initial reaction was that you can't hit a bullet with a bullet, so the task is impossible. I asked that they contract the Bell labs for a study of the problem. From the initial study the project gradually evolved into the ABM. My line of reasoning was that in dealing with a Soviet threat the psychological aspects are most important. If there is a threat by them to launch a missile against New York City and all you can tell the New York population is that we have no defense whatsoever except to threaten to blow up Moscow, our decisionmaking process may be paralyzed by public hysteria. If you have an ABM system in the Jersey flats, there may be doubt both in New York and Moscow as to how effective it might be but at least it is not zero. If you can achieve 60 to 80 percent or more effectiveness of the system, the New Yorkers still will have considerable doubt, but so will Moscow.

In March 1952, as chief of research and development, I made a trip to Europe to become more familiar with NATO requirements and to check on our atomic weapon storage sites. I flew over on a military plane, and Jackie traveled on TWA. It was my first opportunity to see and understand the infrastructure and our military organization in Europe. In Paris I was asked by General Gruenther to brief General Eisenhower, then in Command of SHAPE, on atomic weapons and guided missiles. I spent a long time the night before trying to reduce what I wanted to say into a twenty-minute period. The next morning, March 11, 1952, I left Jackie in the hotel, telling her that I would be back at about eleven. When I reported to Gruenther, he told me to be out of Eisenhower's office in twenty minutes. When I went in, General Eisenhower was at his desk. After exchanging greetings, I started to sit down in the chair in front of his desk, but he rose from his chair and waved me over to the sofa on the side of the room and invited me to sit down with him. I attempted to start my well-rehearsed talk, but he interrupted by asking where I had been in Europe, and before I could complete a reply on that, he asked, "When is Al Gruenther coming in?" I told him that General Gruenther wasn't coming in and had told me to be brief and be out in twenty minutes. Eisenhower laughed and went to the door and said, "Come on in, Al, and sit down." Again I started to talk, and after about two minutes Eisenhower interrupted and we all got onto the subject of politics. It was only then that I recalled that this was the day of the New Hampshire primary. It soon became obvious that Eisenhower was more interested in talking politics and what his political philosophy was than in hearing about guided missiles and atomic weapons. To make a long story short, others joined us for lunch and we then continued into the afternoon. I soon became more interested in hearing Eisenhower's ideas than in trying to brief him on weapons, but every now and then he would ask me a specific question about some missile, then return to some aspect of politics or policies. It was a day to remember and a real insight into our future president and his philosophy of government; his views on inflation; and budgeting for the military (in this regard he chided me several times by questions about why development of a missile cost so much). He was concerned about inflation; he wanted to make a nickel worth five cents again. He asked about several of our leading scientists and touched on their liberal views. He talked about Korea and the necessity for ending that war—"it should not be allowed to drift on," he said. He discussed how he was beginning to realize what a dirty game politics can be; he stated, "They are even beginning to attack Mamie."

The next day, March 12, 1952, the *International Herald Tribune,* Paris edition, had the usual article on the front page on Eisenhower's activities. The headlines were, "Taft Leads Eisenhower on Scattered Returns in New Hampshire Vote." The article stated: "General Eisenhower spent the New Hampshire primary day in staff conference, preparing for a trip today to Germany.

"The general's principal visitor of the day was Maj. Gen. Kenneth D. Nichols, chief of research and development for the United States Army and also member of the Military Liaison Committee with the

Atomic Energy Commission. Although there was no statement on what the two generals discussed, it was indicated that Gen. Nichols reported to Gen. Eisenhower on progress on the development of tactical atomic weapons.''

On November 1, 1952, Los Alamos successfully exploded their first thermonuclear device, ''Mike,'' with a yield of 10.4 megatons—another major jump in the power of destructive weapons. The thermonuclear era had begun. The test, occurring so close to the election was not announced, but the White House, the Pentagon, and the State Department were discussing how to take psychological advantage of this tremendous stride in weapons development.

Frank Pace was involved in these discussions, and on November 28, 1952, he asked me to write my personal views on the political and military implications of the hydrogen bomb. He gave me three hours to do it. (See the Introduction, pp. 9–12, for the text of the paper I wrote.) Since the outbreak of the Korean War, I had been disturbed because we had failed to use atomic weapons. We had not even threatened to use them when the war first started, nor later, when the Chinese entered the conflict. I sincerely felt that we should demonstrate to the USSR and China that we were not just stockpiling atomic weapons but also that we had the will to use them when our security was threatened. I knew that many individuals in the United States opposed such thinking for idealistic, moral, or other reasons. I was encouraged, however, when Arthur Compton asked, ''Nick, why aren't we using the atomic bomb to stop the war in Korea?'' I told him of the effort I had made to convince General Matthew B. Ridgway and General Vandenberg to recommend to the Joint Chiefs and Truman that we use the bomb. I asked Compton if I could tell them that he supported the use of the bomb in Korea. However, he preferred that I not use his name. But he stated, ''In your position you should continue to advocate the use of the atomic bomb in Korea to stop the war.'' I was pleased to hear that I had his moral support and thoroughly understood why he and many other individuals prefer not to advocate use of atomic weapons publicly unless they are in a position of responsibility that demands such a personal decision. Even many of our military leaders hesitate to go on record unnecessarily. During the Korean War, after the successful Inchon landing, MacArthur sent a staff officer to Washington to present his plans for moving north of the Thirty-eighth Parallel. At that time I attended such briefings in Army Plans and Operations. I asked, ''What will MacArthur do if the Chinese come in?'' The staff officer replied, ''MacArthur will ask for atomic weapons.'' Whether he did or not, I don't know. When the Chinese entered the war I was busy with my new assignment to guided missiles and no longer attended the G-3 briefings. In reading *American Caesar* by William Manchester, I can find no specific reference to MacArthur's ever recommending the use of atomic weapons in Korea. However, there is in another book a statement, ''. . . the chief executive told a press conference on November 30 that nuclear bombs might be used against the enemy and seemed to indicate that the decision would be MacArthur's. That brought Clement Attlee hurrying over from London.''[6] Later, a clarifying statement was made that only the president could authorize the use of atomic weapons.

Considering that we were acting on behalf of the United Nations in this conflict, greater consideration had to be given to the firm stand the United Kingdom was making against our using atomic weapons. Eisenhower finally resolved the issue correctly. In *Mandate for Change* he states, ''The necessity to use atomic weapons was suggested to me by MacArthur while I was President-elect.'' After he became president he considered various alternatives and Eisenhower's ''feeling was— that it would be impossible to maintain'' our ''military commitments— did we not possess atomic weapons and the will to use them when necessary.'' Spurred by ''the lack of progress in the long-stalemated talks—and the nearly stalemated war,'' Eisenhower decided ''to let the Communist authorities understand that, in the absence of satisfactory progress, we intended to move decisively wtihout inhibition in our use of weapons, and would no longer be responsible for confining the hostilities to the Korean Peninsula. We would not be limited by any world-wide gentleman's agreement. In India and in the Formosa Straits area, and the truce negotiations at Panmunjom, we dropped the word discreetly, of our intentions. We felt quite sure it would reach Soviet and Chinese Communist ears.''[7] Thereafter the negotiations for an armistice moved slowly forward to a truce signed on July 27, 1953. Eisenhower had demonstrated that the threat to use atomic weapons, made by a president of the United States, with credibility and under the right circumstances, could stop a war, even though it might not gain a true peace. Apparently the Chinese, the USSR, and the Korean Com-

6. Mosley, *Marshall* (New York: Hearst Books, 1982).

7. Eisenhower, *Mandate for Change* (Garden City, N.Y.: Doubleday & Company, 1963), pp. 180–81.

munists believed him. Credibility undoubtedly is a major factor in making any such threat effective.

Expanding on my views expressed in my 1952 memo to Frank Pace, we must continue to learn how to deter all wars. The existence of nuclear weapons has deterred major war between the USSR and the United States for forty years. However successful we have been in deterring war between the superpowers to date, we have not deterred minor wars between small nations or domination of unwilling smaller nations by the USSR.

Deterring a major war between the USSR and the United States has created a massive buildup of weapons and requires coolheaded leadership to be sustained. A possible disaster could occur if the Soviet Union gains such superiority that the leaders in the Kremlin think that our president can be bluffed into acceptance of unreasonable terms by the mere threat of nuclear attack or by the threat of a demonstrative attack on a single target in the United States. Such a situation would be much like a no-limit game of stud poker. Our offensive strength is pretty well known. If we had a strategic defense as a hole card, the Kremlin leaders would realize they might not have sufficient superiority to bluff their way through to a low-cost victory. A defense never can be perfect. It does not need to be. If we can develop a defense that is about 60 or 80 percent effective, it will be a most important factor, and a difficult one for an enemy to evaluate. Aggressive wars seldom are started unless the aggressor's leaders believe they can win a relatively low-cost quick victory. Historically, introduction of new offensive weapons enjoys an initial advantage, which decreases with time as defensive weapons and defensive tactics are developed to counter them. The art of war never is static. Hitler demonstrated many of the possibilities of psychological war in his use of the term *Schrecklichkeit* ("frightfulness") to terrorize his enemies. The Soviet Union and the United States now have the weapons that give real meaning to *Schrecklichkeit,* and we must continue to learn how to cope with the psychological and political aspects of these destructive weapons. I have never lost hope that our world leaders finally will master the situation. Unfortunately, this does not depend on us alone but requires all the major leaders in the world to recognize and take the necessary steps to achieve peace. A great obstacle to success is that many Soviet leaders believe they can actually win a thermonuclear war. They cite the tremendous losses of World War II as partial proof that they can lose tens of millions of men and still come back to win. Moreover, at the present time they are taking many of the psychological and political means available to discourage us from defending ourselves. It is incorrect to say that no one can win a thermonuclear war. That may be correct if all-out thermonuclear war actually occurs. However, a Soviet thermonuclear threat or attack properly combined with psychological war, such as "Better Red Than Dead," may destroy the will to fight and thereby achieve a low-cost victory. I hope we never forget this possibility and also that it is much more difficult for us to influence the Soviet people because of their totalitarian government than it is for the Kremlin to reach American minds in our open society. The Kremlin has a tremendous advantage in waging such psychological warfare. At present they are weakening our determination to protect our basic interests by misinformation, terrorist attacks, and other forms of psychological warfare.

As expressed in my 1952 memo to Pace, I am a firm believer that defense should be part of our deterrence plan. As deputy director of guided missiles, I worked hard to get the Nike I ground-to-air missile and also Air Force and Navy air-to-air missiles into production and established as a reasonable defense against airplane atomic bomb attack. We did that. Much of it is now out of date. Today we need a defense against IBM attack. When as chief of R&D I urged Army Ordnance to develop an antimissile missile, I heard repeatedly that you can't hit a bullet with a bullet, so it is impossible to develop such a missile. But we did develop an ABM system. So did the USSR. These systems are not too good, to be sure, but a start.

However, our so-called intellectual elite, rather than continuing to develop better defensive weapons, led us first to an agreement to limit the ABM to one system for each country. We then unilaterally abandoned our system planned to protect a major IBM site, whereas the USSR built an ABM system to protect Moscow; they have taken civilian defense measures to protect their population; and they are building a radar system that has the potential to be developed into a system to protect a large percentage of the nation. The Soviet Union has not ignored defense. We have concentrated on trying to limit the size of the two nuclear offensive arsenals that are already far too large, and even if both sides cut by 50 percent, they still would be too large. Many of our leaders continue to prefer relying solely on mutually assured destruction. With new developments and actions the USSR is now taking, MAD may be impossible for us to attain. Fortunately, we now have a president who believes the fundamental objective is to reduce offensive weapons to zero and to achieve an effective strategic defense. However, the Soviet Union objects to our following in her footsteps to achieve better security for the nation by a combination of offense and defense. They derive too many psychological advantages from the present situation. The media and the opponents of the proposal for a strategic defense initiative, SDI, have labeled it "Star Wars" to try to discourage acceptance. Actually, both sides have been engaged in research and development in this area for about twenty years. I believe, that President Reagan is on the right track for accomplishing better security and also a real reduction in the potential threat of thermonuclear warfare. There is a good probability that our SDI program can achieve a reasonable defense. It does not need to be perfect. It will cost money and effort and time—how much, we do not know. Many experts, politicians, and news media will claim it cannot be done, or should not be done. Every new technical advance, civilian or military, has proponents and opponents among scientists, politicians, financiers, and the media. SDI does not have all the frightening aspects that atomic and thermonuclear warfare present, so I hope that support for SDI can be established. I do not know about all the approaches being made to achieve success in SDI. I do know, however, that the state of the art has advanced at a terrific rate in many of the fields that will be involved. For example, the Army has scored a direct hit on an incoming ballistic missile and destroyed it. Our advances with the laser, computers, and electronics have been even more impressive. No one can say at this point what the best combination of competing systems will be, but most important, the SDI should be supported to the extent necessary to achieve an optimum rate of progress.

I would hope that we could achieve a defense successful enough to cope with a Soviet threat to use a small number of missiles to try to destroy our will to retaliate, or a similar threat from some terrorist. Also, the defense must be successful enough to discourage a Soviet first strike designed to eliminate our means of retaliation.

As success in defensive measures is achieved, chances of continued successful deterrence between the Soviet Union and the United States should increase. A reduction of warheads under such a situation that can be verified would be more attractive to both parties because a successful first strike is improbable. Even a reduction to zero warheads might eventually be possible.

At that point, hopefully, the USSR and the United States might learn to trust each other and rely on negotiations to find ways and means to occupy the same planet without resorting to war. Until that trust is established we would need to build more conventional weapons as a deterrent. We have avoided a major confrontation between the United States and the Soviet Union for forty years. SDI gives us a better chance for continued deterrence for another forty years, than continued agreements on numbers of warheads. SDI is designed to destroy offensive weapons of war, not masses of people, not cities, not whole nations, and not the very structure of civilization. We should give it a good try.

Since writing the above discussion a summit conference between Ronald Reagan and Mikhail Gorbachev was held in Iceland on October 11–12, 1986. At this conference complete elimination of nuclear ballistic missiles (and possibly all nuclear warheads) was discussed, but the conference was terminated by Gorbachev insisting on a revision of the ABM treaty that would kill the SDI.

The most interesting aspect of the media discussion after the conference has been the recognition that our nuclear strength has countered the Soviets' massive strength in conventional weapons and that if the nuclear deterrent is eliminated the Soviets will have the capability to sweep through Europe. Also recognized is the fact that the cost of increasing NATO's strength in conventional weapons and manpower to deter a Soviet attack may be more than the NATO nations, including the United States, are willing to support. This puts us right back to the situation that faced Truman and Forrestal in 1948–49 when we first began to rely mainly on nuclear weapons because they were cheaper than conventional weapons (more bang per buck).

Another aspect that is being given more publicity is the difficulty of establishing any method of verification that the Soviets will accept and that will assure our leaders and the public that any cheating will be detected. Can we ever trust the Soviets? Will the SDI make up for the lack of trust?

During my 1952 visit to Paris, Al Gruenther suggested that I should be thinking about an assignment in Europe, that I shouldn't stay in the Pentagon forever. I certainly agreed with him.

In November I was very happy to see Eisenhower elected president, even though I knew that would terminate sooner or later the Keller regime in guided missiles. The volume of our work gradually was slowing down. I spent more and more time on Army R&D. Also, I was thinking of wider fields to expand my experience.

Shortly after the New Hampshire primary, General Ridgway replaced General Eisenhower as supreme Allied commander in Europe. In the spring of 1953, Eisenhower, as president, appointed Ridgway as chief of staff of the Army, and Gruenther was selected to replace him. I thought that might be a good opportunity to take Gruenther's advice and get transferred to Europe. I wrote to Gruenther and discussed the idea with General Ridgway, who smiled and told me that as chief of staff, he thought I should stay in Washington, although as supreme Allied commander in Europe he might have thought differently.

Lewis Strauss had supported Eisenhower for president, and after the election, Eisenhower asked him to be his atomic energy adviser and to find a suitable man to become chairman of the Atomic Energy Commission. However, President Eisenhower finally decided that Strauss should take the job himself. Strauss was appointed, and the Senate confirmed the appointment on June 27, 1953. On July 17 Lewis invited me to lunch and asked me to become general manager of the AEC. He said that he had talked to President Eisenhower about ways and means to improve relations between the military and the AEC, and both felt that if I would take on the job, that would help. One of the conditions for accepting the position was that I resign or retire from the Army. I told Lewis that I was very happy with my career in the military. I had been more successful than I had initially hoped, I was still the youngest major general in the Army, and the future looked bright. As a career officer, I would do as I was told but would not volunteer for the job. We then discussed other considerations. I wanted to be sure that President Eisenhower was ready to push for development of commercial atomic power and would support the many changes in the Atomic Energy Act and the declassification of information that would make it possible. Further, I wanted a clear understanding that any differences we had in policy would be discussed in private and resolved, and that at commission meetings I expected him to back my recommendations. The major difficulty was the requirement that I retire. I said, "I cannot afford to resign, and retirement requires approval as being in the best interest of the government." We departed with the understanding that Strauss would initiate action on the matter.

Shortly thereafter, I informed General Ridgway about my conversation with Strauss. I made it clear to Ridgway that I preferred to remain in the Army but was willing to leave the decision to higher authority. If higher authority decided that I could be of more value to the government as general manager of the AEC, I would request retirement and accept the position. I asked General Ridgway to determine what I should do. He commended me on my attitude and said that although he would not like to see me leave the Army, he would find out what the decision should be. I learned later that Secretary of Defense Wilson opposed the move. He had questions concerning my industrial experience and felt that a qualified civilian should be found. I believe that President Eisenhower and Lewis Strauss resolved that issue. General Ridgway soon told me to proceed by applying for retirement and that in the request I should state, "Retirement is requested on the basis that higher authority has determined that my services will be of more value to the United States if I accept the position of general manager of the Atomic Energy Commission than if I remain on active duty in the Army," and it would be approved. I would retire as a major general with twenty-four years' service.

Keller had concurred with my leaving the Army. It gave him an additional excuse to resign and also to recommend that his office be eliminated. He resigned effective September 17, 1953. The position was eliminated, but a few years later a similar position was created, and Eger Murphree accepted the position. He asked me who could help him as I had assisted Keller. I recommended Colonel A. W. Betts, Corps of Engineers. He did a fine job and later became chief of R&D. By that time, the office called for a lieutenant general, and I was happy to see him promoted. Likewise, the rank criterion for chief of the AFSWP had been advanced to call for a lieutenant general. I had the jobs at a time when I was too young to be so advanced, but I was very happy to see that the military had recognized that the importance of these two organizations deserved higher rank.

I retired on October 31, 1953, terminating my military career with considerable regret. I always enjoyed my work as an officer in the Corps of Engineers and remain very proud to have been part of such an elite organization. At my somewhat delayed retirement review (September, 1956) I was awarded another DSM for my service from 1948 to 1953, and thirty years later I was most pleased to be selected for the Chiefs of Engineers Award for outstanding public service.

15

WASHINGTON MERRY-GO-ROUND, 1953–55

ON NOVEMBER 2, 1953, I TOOK the oath of office as general manager of the U.S. Atomic Energy Commission. The primary mission of the AEC was the security of the United States. The first afternoon I was briefed by the director of military application on the emergency plan for transfer of additional weapons to the military in the event of war.

I felt right at home with the staff. A large percentage of the AEC employees were former officers or civil service employees in the MED. Walter J. Williams was deputy general manager; Richard W. Cook was director of the Production Division; Brigadier General Kenneth E. Fields was director of the Military Application Division; Larry Hafstad, whom I knew quite well, was director of the Reactor Development Division; Curt Nelson was manager of the Savannah River Project; Sam Sapirie was manager at Oak Ridge; and David Shaw was manager at Hanford. There was a vacancy to be filled as manager for the Albuquerque office. The one key employee I missed the most was Vanden Bulck. Van still was in a key position at Oak Ridge, but the AEC had selected a new man for comptroller, Don S. Burrows, an excellent man for the job; but no one, in my opinion, ever could compare with Vanden Bulck as chief administrative officer. Shortly after I arrived at the AEC I convinced Harry Traynor to work for me as my personal assistant.

Another individual, new to me, whom I learned to rely upon was Jesse Johnson, in charge of uranium ore procurement. Phil Merritt, who had worked under Ruhoff for ore procurement during the war, still was in the New York area but reported to Johnson, who headed the

AEC Division in Washington because of his far wider experience. Ore procurement still was a key project: We needed more to fulfill our military requirements, and we expected that commercial development of atomic energy would further increase our demands for ore. Jesse was a pleasure to work with. He knew his field, and he knew how mining and availability of ore responded to price. He concentrated on his own work and wanted to know from me and the Commission only how much ore we needed and the price he was authorized to pay.

I knew all five of the commissioners and liked and respected them all. They were Lewis L. Strauss, Thomas E. Murray, Henry D. Smyth, Eugene M. Zuckert, and Joseph Campbell. During my last six years in the Pentagon, I always had at least two bosses and at times as many as six to report to. I never had serious trouble getting along with any of them. In the Pentagon all my bosses had responsibilities other than directing mine, which were of relatively minor importance compared to theirs. The result was that multiple jobs and multiple bosses gave me more freedom of action. In the AEC, each commissioner was by law accountable for all of the commission's responsibilities. Strauss had additional responsibilities as adviser to President Eisenhower and also met with the Security Council. Five commissioners might be justified for the AEC regulatory responsibilities, but even for regulation, just one would be better. To prescribe that a major operating organization such as the AEC should have a five-person, full-time commission creates an inefficient and almost unworkable organization. A part-time commission with a full-time chief executive as chairman would have been far more efficient. To compound the difficulties, the AEC had to keep the Joint Congressional Committee on Atomic Energy fully informed and to deal with the MLC on the military application of atomic energy. There were just too many "checks and balances." When I first gave consideration to accepting the position as general manager, I was very apprehensive about how to make it work, and I discussed it at great length with Strauss. Having been successful in the Pentagon with many bosses, I finally decided to give it a good try.

As soon as it appeared that I would take the position, Strauss invited me to meet with the commission. Each member gave me a warm welcome. With a twinkle in his eye, Zuckert asked me, "Nick, are you coming over to the AEC to continue your fight for military custody?" I laughed and replied, "No, we won that one by adding a name. The military now has availability of weapons, the commission retains custody. I am completely happy with the solution." Zuckert then commented, "I just wanted to be sure you are satisfied, because I am." Zuckert had been assistant secretary of the Air Force and understood the issue thoroughly.

Immediately after reporting as general manager, I appeared before the Joint Congressional Committee. There I received a much warmer welcome than I anticipated. I thought that some of them might object to having a retired Army officer as general manager, but instead they seemed pleased to have me back in the atomic energy field. I remember in particular Melvin Price's comment: "Welcome back. I am most pleased to have someone take on the job who is experienced and can proceed from a running start. We have many things to accomplish."

After meeting with each of the division directors I decided to visit the field installations to see the status of operations. Walt Williams was fully qualified to carry on during my absence. Strauss suggested that Joe Campbell accompany me (he had just been appointed a commissioner) and that we take our wives along. I set up a schedule with Pete Young as pilot to visit: Idaho, Hanford, Berkeley, Albuquerque, Los Alamos, Denver, Kansas City, and Burlington, November 8–18, 1953; the Savannah River plant, November 22–24; the Portsmouth, Paducah, and Oak Ridge gaseous diffusion plants, and the Lockland, Fernald, and Mounds plants, December 4–10. It was great to be back with many old friends, to see the status of the widespread AEC activities, and to hear firsthand accounts about the successes and the problems that were occurring. The trips also gave Jackie and me the opportunity to become better acquainted with the Campbells. We enjoyed their friendship for many years. Campbell had been treasurer of Columbia University, and I had many dealings with him during the war and also during our negotiations to organize Brookhaven National Laboratory on Long Island. Initially Strauss had announced that his period as chairman would be short and that Campbell would succeed him as chairman, but things didn't turn out that way.

One of my goals as general manager was to advance the development of commercial atomic power. Lewis Strauss was as eager as I to achieve this goal, and he felt he had the complete backing of the president. We knew that we would need an act of Congress amending the 1946 Atomic Energy Act. Declassification of many aspects of atomic energy would be necessary, and we could expect opposition from the military. Also, plutonium and enriched U-235 had to be made available for commercial use. Authorization for expenditure of government funds for research and development and for subsidy of experimental plants was necessary. A method to permit access to information that remained classified needed to be developed so that private interests could participate in the development. Government policy needed to be established concerning whether electric power atomic plants be government-owned, privately owned, or be a mix of the two.

To develop better relations with the Joint Congressional Committee, which would have to formulate the changes necessary in the Atomic Energy Act, Strauss initiated the practice of having Congressman W. Sterling Cole, the chairman of the JCC, join us for breakfast about once a week to seek his political advice on what Congress would support and what they would oppose, and to advise on ways to obtain our objective.

The use of radioactive isotopes in the medical and industrial fields was a beneficial by-product of the Manhattan Project that was being developed and was being supported by both private and government funds. Many believed that the medical use of radioactive isotopes as a diagnostic tool was a more important potential contribution to man's welfare than atomic power. The development of thermonuclear power for producing electric power was another project ardently supported by Strauss. Everyone recognized that the development was a very long-range project but that success would mean a virtually inexhaustible source of energy.

Our major objective was, of course, to develop a deliverable hydrogen bomb and improve fission weapons to meet the needs of the military. This required testing in Nevada and in the Pacific. Fields was Director of the Military Application Division, and I relied on him to accomplish our mission. I could not have had a more competent man. He had replaced McCormack in 1951. I had participated in Fields's selection and helped Gordon Dean, then chairman of the AEC, get him assigned, although at the time I was no longer a member of the MLC. Gordon Dean had routinely asked the Army Personnel Division if Fields could be made available but was turned down. Dean invited me to lunch and asked my advice. He said he knew I favored Fields, but how could he get him assigned to the job? I told him to forget the Army staff and go directly to General Marshall, secretary of defense, and request his assignment. He did.

Part of the reason for my selection as general manager was the need to improve the relationship between the military and the commission. This relationship again deteriorated after Robert LaBaron had succeeded Webster as chairman of the MLC. LeBaron's devious way of stating and achieving his objectives had confused and antagonized many military officers and several of the commissioners. On one occasion when I had lunch with Gordon Dean, he commented, "I find little difficulty understanding and getting along with the men in uniform, but I have great difficulty with LeBaron." I told Dean I couldn't help him, that when I was on the MLC, I had equal difficulty understanding LeBaron. To me he seemed to argue in circles and argued about everything. While on the MLC, I did not care for LeBaron, and apparently the feeling was mutual. Many of my activities in the AFSWP required policy decisions and approval not only of the three chiefs but also by the MLC and the secretary of defense. As to be expected, I sought the approval of the three chiefs first because I reported directly to them and also I would have discussed the matter with my Navy and Air Force deputies who were also on the MLC. On one policy matter LeBaron sent for me and complained, "Here is another policy paper I am asked to approve. I find that you have already received the approval of the three chiefs of staff and all the military members of the MLC. There is nothing I can do but approve it. I am supposed to originate policy, not you."

After I moved to guided missiles, I had little contact with LeBaron, but now as general manager I needed to contend with him again. As a result, both Strauss and I maintained our personal contacts in the Pentagon as a way to discuss some policy matters prior to discussion with the MLC. Strauss assured me that LeBaron would be leaving soon, but it was not until the following August that he left. Meanwhile, our direct contacts helped clarify some issues.

A secondary goal I had was to halt the constant increase in number of AEC employees. The AEC was rapidly becoming another bureaucracy, and I always have believed that an organization functions best when there is a shortage of personnel. When there is a surplus, many are fighting for more responsibility and more personnel to justify their position and promotion rather than concentrating on their work. Voluminous papers are written and passed from desk to desk as substitutes for just getting the job done. Soon I had an opportunity to rectify this in part.

An early technical objective was to select the type of reactor that would be most suited for demonstrating an electric atomic power plant.

With these few priorities in mind, I was looking forward to accomplishing them, with the anticipation that other good ideas would come along as I became more familiar with the work under way at the various AEC field installations. However, the path to achieve my objectives was not as clearly defined as I had anticipated. In the Manhattan Project, I was happy to be the "doer"—Groves cleared the way. Likewise, in the Pentagon, my main job pertained to operations; I was just on the fringe of policy formation, and some distance from the political level. However, in the AEC, political problems arose that diverted my attention and that of the commission to time-consuming activities that accomplished little or nothing toward reaching our main objectives. In the first of these diversions I became involved in the most controversial event of my entire career.

At 1:15 P.M. on December 3, 1953, Strauss called me to his office; he had been to the White House. On his desk was a letter William L. Borden had written to J. Edgar Hoover concerning Oppenheimer. I had read the letter and realized something had to be done about it. Literally "the fat was in the fire." Borden, an attorney, had performed very capably in the responsible position of executive director of the Joint Congressional Committee on Atomic Energy. However, the Republicans had won control of both the House and the Senate, and he, being a Democrat, resigned from this position. On November 7, he sent a letter to the FBI. Borden was seriously apprehensive about the derogatory information in Oppenheimer's personnel file. In his letter Borden emphasized the wide scope of Oppenheimer's activities and his access to information in highly secret national security matters. He outlined the factors that gave him concern about Oppenheimer. He listed Oppenheimer's association with Communists, which included the following: monetary contributions to the Communist Party; that his wife, his younger brother, his mistress, and many of his friends were Communists; association with Communist organizations; recruiting Communists for the early Berkeley wartime project; and contact with Soviet espionage agents. Borden continued with reference to Oppenheimer's giving false information to General Groves and employing communists at Los Alamos. Borden referred to Oppenheimer's support of the H-bomb project until August 6, 1945. He cited that Oppenheimer was instrumental in slowing down activities on the H-bomb project and other military atomic energy projects after 1945.

Borden then stated:

From such evidence, considered in detail, the following conclusions are justified:

1. Between 1929 and mid-1942, more probably than not, J. Robert Oppenheimer was a sufficiently hardened Communist that he either volunteered espionage information to the Soviets or complied with a request for such information. (This includes the possibility that when he singled out the weapons aspect of atomic development as his personal specialty he was acting under Soviet instructions.)

2. More probably than not, he has since been functioning as an espionage agent; and

3. More probably than not, he has since acted under a Soviet directive in influencing United States military, atomic energy, intelligence, and diplomatic policy.

Borden further stated, "It is to be noted that these conclusions correlate with information furnished by Klaus Fuchs, indicating that the Soviets had acquired an agent in Berkeley who informed them about electromagnetic separation research during 1946 or earlier." His closing was, "The writing of this letter, to me a solemn step, is exclusively on my own personal initiative and responsibility."

There was essentially nothing in the letter that I had not heard before from one source or another except for Borden's conclusions, in which I did not concur. After World War II I had been questioned several times about Oppenheimer's earlier Communist associations. Whenever his record was reviewed by Army or Air Force security, they raised the issue of not clearing him. Eventually the matter would reach the secretary level and on several occasions I was called in and asked if I knew about Oppenheimer's questionable record and did I recommend clearing him. Invariably I answered, "I probably know Oppenheimer's record as well as or better than anyone else. He could be considered a security risk by almost any reasonable security standards, but I think that in spite of his record he is loyal to the United States. If you consider his services essential, as Groves and I did during the Manhattan Project, I recommend that you clear him. He is one of our most outstanding scientists."

At the meeting with Strauss that afternoon the matter was not that simple. Security regulations had been changed, and politics complicated the question. Senator McCarthy was running wild on Communists in government and was beyond control by anyone. Communists in government had been a major issue during the presidential campaign. Eisenhower had promised to weed out subversives and accordingly had issued order 10450 in April 1953 to eliminate security risks from government services, and all government agencies were to review personnel files to determine those who might be security risks. Eisenhower emphasized that just being a "security risk" was the basis for not granting clearance. It was not necessary to prove disloyalty. The Atomic Energy Act was similar; under it, it must be determined that clearance of an individual will not endanger the common defense or security. As a result, well over a thousand individuals had left government service either as a result of a hearing or by resignation. Further compounding the problem was the fact that Attorney General Herbert Brownell, Jr., on the same day Broden signed his letter, had given a political speech reviving the issue of Harry Dexter White, who, at the time of his death, had been under investigation for espionage by the House Un-American Activities Committee. The Republicans had claimed during the election campaign, and Brownell reiterated the claim, that both Roosevelt and Truman had appointed White to high government positions despite warnings from the FBI. The situation in regard to the Oppenheimer case was too similar to be disregarded. The letter having been written by a Democrat, it all looked like a political trick to trap Eisenhower. Many years later, Borden convinced me that he had no political motives in writing the letter; he said he was expressing his sincere convictions that he felt obligated to put in writing.

Strauss reported that he had attended a meeting with the president to discuss the matter and that Eisenhower had told the meeting that the letter and the FBI file on Oppenheimer that Hoover had prepared for the president had disturbed him very much. He ended the meeting by directing that a "blank wall" should be placed between Oppenheimer and any further access to secret or top-secret information until a hearing had been completed. Later in the day, Strauss had received an action copy of a memorandum from the president to the attorney general confirming that verbal directive. We had to decide on our course of action. There were not many options left. The president had already made the basic decision. As a result, we soon agreed that we would comply with Order 10450 and follow our own procedures. The first step was to prepare a letter listing all the derogatory information about Oppenheimer. William Mitchell, who was our general counsel, was responsible for doing this. He was to give consideration to Borden's letter but was to base the list of derogatory information on what was in our files or on what might be furnished by the FBI. I did not intend to sign the letter until Oppenheimer returned from Europe and until Strauss and I had an opportunity to discuss the matter with him. Oppenheimer might decide to resign. That would make the matter moot.

Since both Strauss and I were about to leave Washington on business trips, he gave instructions to my deputy, Walter Williams, to inform the AEC commissioners of the directive. I did nothing more until I returned to my office on December 11. I met with Strauss and learned that the commission, attended by all members, had held a meeting on the tenth and had approved Mitchell's preparation of a letter to Oppenheimer outlining the derogatory information about him.

The next day, Saturday, I went to the office to clear my desk of the paperwork that had accumulated during my trip. During the morning Charles Bates of the FBI called to inform me that Oppenheimer would arrive in New York from London at 8:25 A.M. Sunday. I informed Bates that Strauss planned to invite Oppenheimer to come to Washington. At 11:45 A.M. I called Mitchell and in his absence spoke to Harold P. Green, an attorney in the legal section, to see how they were progressing on the letter to Oppenheimer. They seemed to be making reasonable progress. I emphasized that I wanted a thorough and accurate letter and to be sure to include the Chevalier incident. On Monday I conferred with Mitchell and Green about the letter. On Tuesday I again conferred with Mitchell, and met with the commission to inform them about the contents. They approved the letter. The final letter listed the derogatory information concerning Oppenheimer's support of Communist organizations, his association with Communists, the employment of Communists, the Chevalier incident, his veracity, etc. I was not happy with the inclusion of a reference concerning Oppenheimer's opposition to the hydrogen bomb development. Green had inserted it on his own initiative on Sunday, the day after he talked with me. I would not have done so. His argument for including it pertained to the question of veracity. That appeared to be a sound reason so I finally concurred with the draft but worried about the possibility that we would be accused of trying to control the thinking of our scientists. I was assured that the way it was written would avoid such an accusation. I had my doubts. It was a fine grammatical point concerning paragraphing, but it lost its significance when *The New York Times* subsequently printed the letter with the reference to the hydrogen bomb divided into four separate paragraphs instead of combined into one, as Green had drafted it. I don't know who was responsible for this editing—Oppenheimer, his attorneys, or *The New York*

Times—but the change supported the adverse effect that I had feared, and judging by subsequent events was deliberate.

Before we presented the letter to Oppenheimer, the FBI also checked the letter to be certain it did not compromise any of their confidential sources.

On December 21, 1953, Strauss and I met with Oppenheimer. I had agreed with Strauss that he should do most of the talking and that I would write a memo for the record concerning the meeting.

Strauss explained to Oppenheimer the reason for the meeting: the President's Order 10450; that a former government employee had written a letter to the FBI calling attention to Oppenheimer's record; that the FBI had informed the president, who had ordered in writing that we take the necessary action to resolve the matter. We were complying with Order 10450. A letter containing the derogatory information and explaining our procedures and the options open to Oppenheimer had been prepared for my signature; I had not yet signed it. After Oppenheimer read the letter, he consented to including the sentence, "This classification (Confidential) has been discussed with Dr. J. Robert Oppenheimer and he presently desires that it be maintained."

Mitchell had advised us not to be the first to bring up the alternative of resignation. We should allow Oppenheimer to do it. He obliged us by so doing and asked if the case would be closed if he resigned. Both Strauss and I assured him that it would be. Oppenheimer further discussed the merits of resigning versus a hearing but did not indicate any decision. He said that he would advise us the next day of his decision. I noticed that both Strauss and Oppenheimer were composed and polite. Oppenheimer betrayed no emotion other than regret at the possibility of severing his relations with the government.

That evening Oppenheimer called me at home and asked me for my advice concerning the course he should take. He specifically asked, "Should I resign?" Although I hoped that he would resign, I replied, "Oppie, even though I am your friend I am also now officially your accuser, so I no longer can advise you on this matter. You will have to make the decision." Perhaps if I had been certain that he did not have a recorder or another person listening on the line I would have given him my advice. However, Mitchell had strongly advised me at no time to advise Oppenheimer to resign.

The next day, at Strauss's request, I called Oppenheimer at 12:30 P.M. at Princeton and asked him if he had made his deicision. He said he was not quite clear and asked if he could call Mr. Strauss or me later in the day. I told him that if his decision was to accept the letter, we must get the letter off by courier shortly after 3:00 P.M. that day. He asked if I had any further thoughts on which course of action would make the best sense. I told him only that I would hate to have to make such a decision and that he had to make up his own mind. He promised to call by 3:00 P.M. to give his decision.

At 1:05 P.M. he called and said he would like to come to Washington on the next train and see Strauss and me the following morning with his answer, either to receive the letter or to write one. I told him we would be ready with the letter if that was his decision. We made the appointment for 9:00 A.M. I reported this conversation to the commission at 2:45 P.M.

The next day, in Strauss's office, Oppenheimer gave us his decision, and I gave him the letter suspending his clearance. The next step was to inform the Joint Congressional Committee. I called Sterling Cole, the chairman, and Senator Hickenlooper. We also informed certain individuals who regularly had contact with Oppenheimer, such key scientists as Von Neumann, Rabi, Bradbury, Bethe, and members of the commission staff.

We now had a respite from the Oppenheimer case. We needed to wait for Oppenheimer to reply to my letter within thirty days and request a hearing before a personnel hearing board, if he desired one, or if he did not, our procedures prescribed that I decide concerning clearance. Instead of replying within thirty days, he requested a delay, which was granted.

The respite provided time to get going on other objectives. I met with Donald Quarles, the new assistant secretary of defense for research and development, and Strauss and I met with Admiral Arthur Radford, chairman of the Joint Chiefs of Staff, as part of our program to establish more direct contact with the Pentagon. I met with my key staff members to inform them how I proposed to operate with them and with the commission. I prepared for budget hearings, hearings to modify the law to permit commercial atomic power, worked on finding replacements for certain key positions, tentatively selected the reactor to be developed to demonstrate the prospects for commercial atomic power, and started action to declassify some reactor information.

It would have been pleasant and rewarding to be able to continue to devote ample time to my primary responsibilities, but yet another political diversion came along.

*   *   *

The TVA was requesting additional federal funding to build new generating capacity to meet growing needs for Memphis, Tenn. Eisenhower opposed the use of federal funds to provide electric power to a local metropolitan area on a tax-free, low-interest-rate, low-cost basis. He thought it unfair to use taxes paid by all of the country for such a local use.

Joseph M. Dodge, director of the Bureau of the Budget, and Rowland Hughes proposed that the AEC renegotiate our power contract with TVA for reduction of 600,000 kilowatts of power and negotiate with Edgar Dixon of Middle South Utilities and Eugene Yates of the Southern Company to build a new plant across the Mississippi near Memphis to supply a like amount of power. I had our staff study the proposal, and on January 4 I informed Strauss, "If we purchase power from an outside utility to replace TVA power, the increased costs to us would be $4 million to $6 million per year. The preferable solution would be to have TVA procure the power directly."

At a luncheon with Strauss, Hughes, and Williams, I informed Hughes that I saw no way that I could justify the AEC spending several millions per year more for power to release to TVA power that we had under firm contract with them. I then added that although I personally agreed with the political objectives, I thought that the way they were trying to accomplish these objectives was lousy politics, it would stir up the public versus private power issue and be a direct attack on the sacred cow, the TVA, and further they would be crucified by the Democrats when they tried to justify the action before Congress. I was told very firmly by Hughes, "Politics is not your function."

Shortly thereafter, Sherman Adams, the president's chief of staff, informed me that Eisenhower was personally interested in proceeding as proposed by Hughes, and Adams requested that I cooperate to the extent legally possible. I agreed and did my darndest to see that the contract with Dixon-Yates was beyond reproach. (Actually one contract was with the Mississippi Valley Generating Company, but everyone continued to use the name Dixon-Yates.)

On the commission, Murray, Smyth, and Zuckert were opposed to the concept, and Strauss and Campbell were for finding some way to support the Administration.

By April we had a satisfactory proposal from Dixon-Yates to build a coal plant. I recommended to the commission that we send it to the Bureau of the Budget as a satisfactory proposal but inform them that the AEC would need instructions from higher authority concerning the higher costs as well as specific instructions to proceed with the contract. The commission approved the letter, but before the vote Gene Zuckert said to me, ''Nick, I have been wondering for a long time how you were going to get out of this one, and now I find that I can vote to approve your recommendation.''

On June 17, 1954, the long ordeal of congressional hearings started. A week later, it was obvious that there would be serious political repercussions from the Dixon-Yates contract. This was reported to the president. I have a note in my diary, ''The president is firm that we go ahead, so far as we are concerned we have our orders.''

The Democrats made Dixon-Yates a big political issue in the 1954 congressional elections, and there is no doubt in my mind that the adverse publicity was a factor in the Republican defeat. The Democrats won control of both the House and the Senate. In 1955 the Democrats would have the chairmanship and majority control of the Joint Committee for Atomic Energy. If the Administration was to gain approval of the Dixon-Yates contract it would have to be obtained while the Republicans still controlled the JCAE.

The president decided to proceed. Accordingly, the commission approved the contract, by a majority vote. In the JCAE the final result was a party-line vote of ten to nine in favor of a waiver of a thirty-day provision, and the contract was in effect. Strauss considered it a great victory and was very proud that work on the plant now could be started. However, it turned out to be a costly victory: He had forever alienated Tom Murray, who in many ways had the same philosophy about private enterprise and national security as Strauss had. They should have been friends and allies on many common objectives; instead, they were enemies on anything that had political implications. Further, Strauss had irritated many of the more conservative Democrats and definitely had increased the animosity of Senator Clinton Anderson that eventually became intense hatred and a vendetta.

In 1955, with a Democratic majority in the JCAE, Senator Anderson, now chairman, reopened the Dixon-Yates hearings in an effort to force the AEC to cancel the contract. Murray and Anderson raised the issue that the commission, general manager, and staff were spending too much time on Dixon-Yates to the detriment of our other missions. On February 10 I was testifying, sitting directly in front of Clinton Anderson, when suddenly I saw him turn a fiery red, staring at something behind me. I thought he was having a heart attack or a stroke and then realized it was just rage. I looked behind me and saw our security people stacking cartons on the floor. Anderson shouted, "What is this all about?" I replied, "I don't know." Strauss then announced that he wished to refute the charge that the commission had neglected its responsibilities and had been diverted by Dixon-Yates. He pointed out the small pile of commission papers that involved Dixon-Yates, as contrasted to the large stack of cartons that involved other work. The spectators and many of the joint committee laughed or smiled, but not Anderson or Murray. Strauss again had won a tactical victory but had further irritated Murray, and the public attempt to humiliate Anderson made him more determined to get Strauss.

The Dixon-Yates contract still was in force and most of the issues between the TVA and the AEC were resolved while I still was general manager, but eventually the city of Memphis came up with an alternative solution to its power needs and the Dixon-Yates (Mississippi Valley Generating Co.) contract was canceled. Dixon-Yates put in a claim for damages but lost because of a conflict of interest involving a Bureau of the Budget consultant. The AEC was absolved of any involvement in the conflict of interest. Thus ended a time-consuming political fiasco.

During the interval while Oppenheimer was preparing his reply to my letter of December 23, 1953, many additional actions had to be taken. Normally the general manager was in complete charge of selecting a hearing board and making the final decision concerning clearance after receiving the board's recommendations. However, due to the national importance of this case to the government, to the scientific community, and to the national security system, the commission insisted on approving my actions step by step, and Strauss kept the White House informed. J. Edgar Hoover assigned Charles Bates to assist the AEC. In view of the fact that Oppenheimer was retaining Lloyd Garrison, an outstanding New York attorney, as his chief counsel, we decided to employ a more experienced outside counsel to present the derogatory information, to question the witnesses, and to aid the board. On January 23 I interviewed an attorney from Minneapolis who had worked for me after the war, who was an experienced trial attorney, and who due to his MED experience was familiar with security regulations. He was an excellent prospect but had difficulty freeing himself from his work load. Meanwhile, William P. Rogers, assistant attorney general, called Strauss and recommended Roger Robb, an effective trial lawyer. I assigned C. Arthur Rolander from the security staff to assist Robb.

Bates also helped and had the responsibility to see that no FBI sources were compromised.

On January 19 I met with Garrison and Strauss. Garrison requested an extension of time, and it was granted. He also proposed that the AEC replace the hearing board. The matter was discussed, and many disadvantages were pointed out to Garrison, who requested that he be cleared. He accompanied me to my office to meet with James Kelehan, assistant general manager for administration, to arrange for his clearance and to get the necessary forms. He later called to see if Herbert Marks could be cleared, and I told him to have Marks fill out the necessary forms. Sometime later, Garrison informed me that he was not applying for clearance. I thought that his decision not to apply was a mistake.

On February 19, Garrison called to request another extension, to March 1, for filing reply. I asked him to confirm this in writing. He said he would, so I told him he could consider the extension granted.

Isidor Rabi, chairman of the general advisory committee, was very disturbed about the situation. He was actively seeking a solution that would avoid a hearing. He proposed that we just wait until Oppenheimer's contract expired on June 30, 1954, and not renew it. To do this, Oppenheimer would have to concur, and we would need to find a way to discourage other agencies from requesting that he be granted an AEC clearance. I had some hopes that this might work.

On March 1, Strauss informed me about his conversation with Garrison in regard to Rabi's suggestion that it was possible Oppenheimer might forgo a hearing. Therefore, Strauss delayed the time of presentation of the letter for another day.

Shortly thereafter, Garrison called and asked if Strauss had informed me of their telephone conversation. I replied that he had and that Strauss had approved the letter being a day late. Garrison said he would be seeing Oppenheimer the following day and would want to do a little deliberation in the afternoon so that it possibly would run over into Wednesday, and asked if further delay could be arranged if necessary. I told him it could.

All this encouraged me to hope that a hearing might be avoided, but late Tuesday, March 2, Garrison asked if he and Marks could see me for about five minutes at 5:45 P.M. I agreed.

Mitchell and I met with Garrison and Marks. Garrison said Oppenheimer would reply to the letter, which would mean proceeding with the hearing, and had asked him to tell us how greatly he would like to relieve both himself and the commission from this distressing situation, but he did not see that he could. Marks said Oppenheimer was terribly sorry that he didn't see any way to disengage himself. Garrison asked if delivery of the letter to us could be delayed until Friday morning, since they had to have it typed and signed; and I approved this further delay. Then I called Strauss to tell him that the Rabi plan was dead.

On Friday, March 5, Oppenheimer's reply was delivered to me. The letter was long and carefully composed. It covered in considerable detail his early life, education, prewar experience, his assignment as director at Los Alamos, postwar assignments, and contributions to U.S. security and international relations. He gave his versions and answers to some but not all of the derogatory information.

On March 26, Mitchell and I met with Garrison and Marks, who requested that we consider transferring all or part of the hearing to New York City for the convenience of Oppenheimer, his family, and his witnesses, and to permit John W. Davis to assist him. Garrison also delivered PSQ's and requested clearance for himself. We had given him the PSQ forms on January 19, and if he had returned them, there would have been plenty of time to complete the FBI investigation, but he had informed me later that he had decided not to request clearance. Now he had changed his mind again and wanted an emergency clearance. I contended that the only emergency was of his own making but said I would ask the FBI to expedite their investigation. A few days later Strauss called; he had heard about Garrison's request for emergency clearance. Strauss said he thought we should make it perfectly clear to Garrison that we offered to do this last January and that we wouldn't give any special consideration at this time; he added that we should not give him emergency clearance and should not hold up hearings pending his clearance. I assured Strauss that my actions, already taken, had been in accord with his thinking.

On the 26th, I asked Mitchell to inform Garrison that we would not transfer the hearing to New York City.

By the time we received Oppenheimer's reply, we had selected the hearing board. Of all the individuals suggested, I considered Gordon Gray the most suitable candidate and possible chairman. I had known him when he was secretary of the Army during the Truman administration. Now he was president of the University of North Carolina and was a Democrat of towering integrity. He was somewhat familiar with the Oppenheimer record, being one of the secretaries I had advised after the war concerning Oppenheimer's clearance. The other two board members were Ward V. Evans, a scientist and professor emeritus of chemistry at the University of Chicago, and Thomas Morgan, a retired executive of Sperry Gyroscope.

Prior to accepting the responsibility of chairmanship, Gordon Gray asked me what I considered the key issues. He reminded me that I once had told him I believed Oppenheimer to be loyal. I confirmed that I still considered Oppenheimer loyal and that I thought the Chevalier incident was the only issue that might create any doubt about his loyalty. I hoped that the board would be able to get to the whole truth of the matter, that to date we had conflicting statements and at best only half truths. Then I voiced my reluctance to include Oppenheimer's opposition to the H-bomb in the letter because it might imply that we were trying to control the thinking of scientists. I let Gray know that I hoped he would accept the chairmanship. It was a difficult, onerous assignment; no matter what he decided, it would be controversial, but considering all the circumstances, the task had to be done.

Word of a hearing began to leak out to involved individuals. Groves phoned me to find out what had precipitated it. He had been asked to meet with Garrison. I explained that a letter from Borden had precipitated the hearing and that I saw no way to avoid it. He asked me what he should advise Oppenheimer or Garrison. I told Groves, ''Tell Oppie to be very truthful about all matters—maybe you and I will learn the truth about the Chevalier matter. Is he protecting Frank?'' Groves thought that should he be asked to testify, it would be more appropriate to be called as a government witness rather than as an Oppenheimer character witness. Groves ended the conversation with, ''You are in a no-win situation.''

John Lansdale, who had served as General Groves' security chief, believed that I could have aborted the hearing. But he, like others, did not seem to understand that once President Eisenhower had ordered a hearing to determine the validity of Borden's charges, the AEC had little choice but to proceed unless the president reversed his directive.

Shortly before the hearing began, I met with Robb, Rolander, and Bates, and we discussed evidence supporting the various items of derogatory information. I called to their attention that the information in my letter, Oppenheimer's reply, and the FBI report about the Chevalier incident seemed to miss several aspects I remembered to be in the record in 1943. I recalled reference to three contacts, microfilm and also remembered listening to a recording I thought was made when

Colonel Pash interviewed Oppenheimer. Bates said he didn't recall such a recording. I suggested it might be in one of the several book boxes containing the records of our most critical security risk cases that Groves had sent to the FBI before our turnover to the AEC. The very reason I had placed emphasis on the Chevalier case was my memory of this recording. The next day I learned from Bates that he had found the recording in the FBI files and had turned it and a transcript of it over to Robb. Bates agreed that the recording indicated that the Chevalier incident was much more important than indicated by the FBI interview, after the war, of Eltenton, Chevalier, and Oppenheimer.

The hearing finally got under way on April 12, 1954. The previous week had been a busy one with the Dixon-Yates case, the Pacific tests, and congressional hearings. Work had piled up. Late in the day I had a long call from Groves. According to my diary note of April 12, Groves said he had been interviewed by Garrison and that Garrison had also called to say he would call that day to arrange for Groves' appearance before the board. Groves asked if the commission was going to call him, and Garrison said he didn't think they were planning to. He said Garrison asked what his answer would be if he were asked if he were in the commission's place, would he bar Oppenheimer? Groves said he would bar him. Garrison told him that Groves' recollection of the Chevalier incident did not agree with what Oppenheimer said— Groves told him that Oppenheimer didn't tell him the whole truth. Groves also told Garrison he hoped that whatever Oppenheimer or his wife were asked, they would tell a straightforward, truthful story and not omit anything. Groves asked if I thought it okay to be called by them, and I told him I thought it much better that way.

At 6:10 P.M. Rolander and Robb brought me the transcript of the first day's hearing. I had a few other things to clear up before going home, so I decided to violate one of my cardinal rules and took the transcript home to read.

That evening I retired very early to read the transcript. Jackie, who kept a watchful eye over my well-being came into the bedroom.

Her memory of our conversation is as follows: "What's bothering you lately?" she asked. "You must have a heavy load on your mind." I replied, "As a matter of fact, I have." Helpfully she asked, "How would you like a bit of brandy as a relaxer?" "That would be great," I answered. Shortly she presented me with a tray containing two glasses of brandy—not brandy snifters, but very small, pale blue Mexican glasses that hold barely an ounce. I stared at them in disbelief. "What's the matter, Nick, are you okay?'' Jackie mumbled. "Sure. But do you remember where we got these glasses?'' She said, "Of course I do. Oppie brought them to us one evening when he had dinner with us at Oak Ridge.'' I then decided to tell her, "Isn't it ironic! This transcript I'm reading is classified confidential, but it is not military information. It concerns Oppie. He is in trouble. Eisenhower ordered us to have a hearing to determine if he should still retain his AEC clearance.'' She asked, "May I read it? I would like to.'' I agreed, "Yes, you may. If we do decide not to clear Oppie, I would like you to know why. The hearing started today. This is the first day's transcript.'' I finished reading the transcript and handed it to her along with a copy of Borden's letter. After a while she asked, "Do you agree with Borden? Do you think Oppie was a spy?'' I answered, "No, I don't think he was a spy or gave out any information to Russia. However, if he should turn out to be the best darn spy in history, my face would be red—I wouldn't have a leg to stand on considering that I know all the indiscreet things that are in his file. I don't know how it will turn out, but I expect the board will have a tough time clearing Oppie under present rules.''

The next day *The New York Times* published the full story in an article by James Reston. It included my letter and Oppenheimer's reply. The separation of the H-bomb information into separate paragraphs increased the impression that having the wrong opinion is a factor in making one a security risk. Garrison later stressed this.

At the beginning of the second day's hearing, Gray was critical of Garrison and Oppie for implying the previous day that they were trying to keep "fingers in the dike" when they had already given the information to Reston on Friday night. I read the transcript about the publicity discussion. I thought, What a way to start off the hearing with this board. But it soon became apparent that they had deliberately decided to take their case to the media and to try to gain the support of public opinion.

A considerable number of very distinguished individuals testified as character witnesses in support of Oppenheimer. These included Bush, Conant, Karl Compton, Gordon Dean, Bundy, Fermi, Keith Glennan, Lilienthal, John J. McCloy, Frederick Osborn, Pike, Rabi, Von Neumann, and others. Robb in his cross-examination concentrated on showing how little they knew about his Communist associates. Groves in his testimony supported his decision to clear him during the war on the basis that Oppenheimer was essential to the success of the project. He praised Oppenheimer for his "magnificent performance." On cross-examination by Robb, Groves indicated that he was not completely satisfied with Oppenheimer's explanation of the Chevalier incident. Then Robb asked, "General, in light of your experience with security and in light of your knowledge of the file pertaining to Dr. Oppenheimer, would you clear Dr. Oppenheimer today?" Groves first gave his interpretation of what the Atomic Energy Act requires. He then stated, "In this case, I refer particularly to association and not to the associations as they exist today, but the past record of the associations. I would not clear Dr. Oppenheimer today if I were a member of the commission on the basis of this interpretation. If the interpretation is different, then I would have to stand on my interpretation of it."

Teller also was a damaging witness. He testified that he "would feel safer if the security of the country were in other hands." And in response to a direct question from Gray, "Do you feel that it would endanger the common defense and security to grant clearance to Dr. Oppenheimer?" Teller responded, "I believe that is merely a question of belief and there is no expertness, no real information behind it, that Dr. Oppenheimer's character is such that he would not knowingly and willingly do anything that is designed to endanger the safety of this country. To the extent, therefore, that your question is directed toward intent, I would say I do not see any reason to deny clearance.

"If it is a question of wisdom and judgment, as demonstrated by actions since 1945, then I would say one would be wiser not to grant clearance. I must say that I am a little confused on this issue, particularly as it refers to a person of Oppenheimer's prestige and influence. May I limit myself to these comments?"

Oppenheimer himself was his own worst witness. Regarding the Chevalier incident, he first testified in accord with the interview he had with the FBI in 1946.

I was somewhat surprised when I read his direct testimony, in view of the fact that Garrison had recognized that Groves' recollection of the Chevalier incident did not agree with Oppenheimer's story. In cross-examination, Robb asked him questions concerning the interviews he had with our security officers in 1943: Johnson and Pash on August 26, Lansdale on September 12, and Groves in the autumn; also, Oppenheimer had an interview with the FBI in 1946. Oppenheimer testified that he had invented a "cock and bull story," that he had "lied" to Pash, that his statement that "Eltenton had attempted to approach members of the project—three members of the project—through intermediaries" was untrue. In answer to Robb's question "Why did you do that, Doctor?" he answered, "Because I was an idiot." Then he further explained, "I was reluctant to mention Chevalier" and "no doubt somewhat reluctant to mention myself." Robb then led him through references to "microfilm," how "information would be transmitted through someone at the Russian consulate." By this time Oppenheimer apparently recognized that Robb had a transcript and reluctantly responded or agreed with Robb's questions. Oppenheimer admitted that he had lied to Lansdale and finally admitted that he "told not one lie to Colonel Pash, but a whole fabrication and tissue of lies . . . in great circumstantial detail. . . ." Later, in regard to telling General Groves that Chevalier was the Mr. X in his account to Pash and in answer to Robb's question "In order words, you lied to Groves, too?" Oppenheimer answered, "No, I told him the story I told Pash was a cock and bull story." However, telegrams from Lansdale and also my office dated December 12, 1943, showed that the only thing Oppenheimer revealed to Groves was identification of X as Chevalier. There still were references to three contacts. Robb asked, "Does that indicate to you that you told General Groves that there weren't three contacts?" Oppenheimer answered, "Certainly to the contrary. I am fairly clear." Robb next asked, "You think General Groves did tell Colonel Nichols and Colonel Lansdale your story was cock and bull?" He answered, "I find that hard to believe."

Never in any of my many conversations with Groves did I ever get any indication that Oppenheimer had told anyone that his story concerning Ellenton and Chevalier was a "cock and bull story." We discussed many times that Oppenheimer had not told us the whole truth; one reason for this might have been that he was protecting his brother Frank. The first time I had ever heard the term "cock and bull story" in this matter was when I read the transcript of the hearing. I certainly do not believe that Oppenheimer told Groves that it all was a "cock and bull story" in the autumn of 1943.

The hearing continued until May 6, when it recessed to permit the three members to contemplate the matter. On May 27, the board reported its finding that J. Robert Oppenheimer was a security risk. Gray and Morgan supported the finding, whereas Ward Evans dissented, basing his dissent primarily on double jeopardy. Robb was greatly surprised with Evans' vote. From time to time throughout the hearing, he had commented to me that Evans was most critical of Oppenheimer and feared Evans might disqualify himself by some of his remarks.

The majority, with Robb's assistance, wrote a very well-thought-out report. It discussed the problem of secrecy and security, freedom of the individual, and the need for a nation to protect access to information and to limit access and employment to those who were trustworthy in order to protect national security. They recognized that the hearing not only involved Oppenheimer's clearance but also was putting the security system on trial.

They found that Oppenheimer did delay the initiation of the effort to develop the hydrogen bomb and that his conduct was disturbing. They found that Oppenheimer's character and associations did not meet government standards for clearance. Concerning veracity, they remarked that he "was less than candid." They found no evidence of disloyalty.

Robb, Mitchell, and Rolander helped me prepare my recommendations to the commission. In reviewing the transcript of the proceedings, I had been mentally prepared to accept the findings of the hearing board. I thought that their decision might go either way. I had hoped that Oppenheimer might tell a convincing story about the Chevalier affair that would eliminate the small but persistent shadow of doubt that remained in my mind about the importance of it. I had not thought out what I would do if they had recommended clearance on the basis of his wartime contribution to our success, recognizing there is no specific provision in the Atomic Energy Act that provides for giving consideration to such a factor. However, as a result of discussion with them, I believed that at least two of the commission members might support clearance on the basis of such a consideration.

I was pleased that the board considered Oppenheimer to be "a loyal citizen." Concerning the board majority voting against recommending reinstating his clearance, I generally agreed with their findings, but in my letter recommending that his clearance should not be reinstated, I emphasized my greater concern about the veracity of Oppenheimer. Whereas the board found Oppenheimer "less than candid," I stated, "In my opinion, Dr. Oppenheimer's behavior in connection with the Chevalier incident shows that he is not reliable or trustworthy; his own testimony shows that he was guilty of deliberate misrepresentation and falsification either in his interview with Colonel Pash or in his testimony before the Board; and such misrepresentation and falsification constitutes criminal . . . dishonest . . . conduct."

I also pointed out that the complete record of the Chevalier incident was not considered by the Atomic Energy Commission in 1947 because it was not made available to them by the FBI.

I based my findings on considerations relating to Oppenheimer's character and associations. I emphasized that at no time had there been any intention on my part or the board's to question any honest opinion expressed by Oppenheimer. Further I stated, "In reviewing the record, I find that the evidence established no sinister motives on the part of Dr. Oppenheimer in his attitude on the hydrogen bomb, either before or after the President's decision. I have considered the testimony and the record on this subject only as evidence bearing upon Dr. Oppenheimer's veracity. In this context, I find such evidence is disturbing."

In regard to Oppenheimer's value to atomic energy or related programs, I found "that throughout World War II he was of tremendous value and absolutely essential. Secondly, I believe that since World War II, his value to the Atomic Energy Commission as a scientist or as a consultant has declined because of the rise in competence and skill of other scientists and because of his loss of scientific objectivity probably resulting from the diversion of his efforts to political fields and matters not purely scientific in nature. Further, it should be pointed out that in the past 2 years since he has ceased to be a member of the General Advisory Committee, his services have been utilized by the AEC on the following occasions only: October 16 and 17, 1952. September 1 and 2, 1953. September 21 and 22, 1953."

In my conclusion, I stated that in addition to weighing Oppenheimer's past and future contributions against the security risk, I gave consideration to the nature of the Cold War in which we are engaged with communism and Russia and the horrible prospects of hydrogen bomb warfare if all-out war is forced upon us. In consideration of all these factors, I recommended that "Dr. Oppenheimer's clearance should not be reinstated."

The commission split four to one on the decision. Henry Smyth voted for clearance. Thus, out of the nine individuals considering the hearing testimony, only the two scientists voted to reinstate Oppenheimer's clearance.

Immediately after the hearing board issued its findings, Oppenheimer's counsel released their answering brief, which stressed the inconsistency between the board's finding that Oppenheimer was loyal and its conclusion that he was a security risk. They stressed that if

Oppenheimer was a security risk because he had opposed the H-bomb, that would discourage scientists to advise the government in the future. They stressed that the distinguished character witnesses were the best bases for judging character. They made no mention that Oppenheimer had admitted lying to the security officers.

As Oppenheimer and his counsel had hoped from the beginning, he was winning the battle for support of public opinion. Many individuals in government recommended releasing the transcript of the hearing, but it was not until Zuckert lost a copy of the transcript while returning to Connecticut on June 4 that Strauss finally got the commission support. Although the transcript was found, the commission voted four to one to release it. Smyth voted against releasing it at that time. He wanted to wait until the commission had reached its decision, which it had not done on June 14, when the release was approved.

The release of the transcript brought to the forefront (temporarily, at least) the Chevalier incident and the fact that Oppenheimer had admitted lying to the security officers. The commission released its decision on June 19, 1954, including the position of each commissioner.

There were no winners in the Oppenheimer hearing, except possibly the preservation of the security system for protecting classified data. For Oppenheimer, it was an undeserved tragedy, although he was not blameless. Clearance was denied, but public opinion supported Oppenheimer. Many people, even today, believe that Oppenheimer was punished for his opinions on the H-bomb. It divided the scientists, some supporting Oppenheimer and others supporting the government. Teller became ostracized by many in the scientific fraternity. The hearing should have educated the public to distinguish between "security risk" and "loyalty," but somehow it did not. There should never have been a hearing, but considering the spirit of the time, McCarthyism, and the political campaign concerning "Communists in government," I have never been able to come up with a procedure whereby it could have been avoided, once J. Edgar Hoover had called the president's attention to Borden's letter. Oppenheimer was a victim of the political climate of the times. Except for Borden's action, the probability was good that Oppenheimer gradually would have lost his position of great responsibility in national defense circles. Many thought that his scientific opinions were becoming biased by his moral sense of guilt and his desire to influence political decisions. Strauss had already taken steps to curtail his use as an adviser to the Department of

Defense and the National Security Council. This type of action is normal with the change in administration. Political advisers and those who try to get in that field (as distinguished from those who confine their advice to their own expertise) frequently find they are no longer wanted by a new administration.

There is no question in my mind that Strauss felt that Oppenheimer should be eliminated as a government adviser not only because he didn't like his advice but also because he thought he was a serious security risk verging on disloyalty. Strauss did not initiate action against Oppenheimer on the basis of presidential Order 10450 until Borden's letter precipitated action and the president ordered such action. During the hearing, I had the impression that Strauss wanted a fair hearing. His approval of Gordon Gray as chairman is the best indication of that. Gray was not only a man of great integrity but also a Democrat and could not be expected to yield to any political influence. I knew that the FBI had set up some surveillance of Oppenheimer as a normal precaution as soon as Hoover had forwarded the case to the president. I did not learn until recently that about December 17, 1953, Strauss learned that Hoover planned to discontinue surveillance unless they received a letter from Strauss. As a result, Strauss wrote a letter requesting that the FBI set up full surveillance. The FBI tapped Oppenheimer's telephone, and teams of agents followed him wherever he went. The surveillance produced a day-to-day record of his activities—including, in many instances, plans he made with his attorneys.[1]

Groves and I had arranged to tail Oppenheimer and others during World War II and had tapped telephones at Los Alamos, Oak Ridge, and other places, but that was during wartime. I doubt that I would have concurred with Strauss in the request to the FBI if he had asked my advice. He did not inform me about it or ever show me any of the daily summaries he received from the FBI. A few years ago I heard rumors of Strauss's action, but I did not believe them until more specific information was released by the FBI.

Many times I have been asked by friends why I supported clearance for Oppenheimer in 1943 during World War II but supported not clearing him in 1954.

There is considerable difference between the situation in 1943 and that in 1953. In 1943 Oppenheimer was considered absolutely essential

---

1. Richard Pfau, *No Sacrifice Too Great: The Life of Lewis Strauss* (Charlottesville: The University Press of Virginia, 1984), p. 157.

to the success of the Manhattan Project, whereas in 1953 he was no longer essential to the work of the AEC.

In 1943 Oppenheimer was considered a security risk, and the principal derogatory information was his Communist associations; but the Soviet Union, a Communist country, was an ally then (although the policy was not to inform the USSR about our project), whereas in 1953, the Soviet Union was recognized as our potential enemy.

In 1943 the Chevalier incident had occurred, but we did not know about it at the time Groves decided to clear Oppenheimer regardless of his security record. When the incident was revealed voluntarily but piecemeal by Oppenheimer, we doubted if he was telling us the whole truth, but both Groves and I thought that he probably was trying to cover his brother Frank. Otherwise, veracity was not a serious issue. In addition, I informed Oppenheimer that we were tailing him when away from Los Alamos, and he knew we were censoring mail and tapping telephones at Los Alamos. We thought we had the situation under control. Whereas in 1953, we had more reason to worry about his veracity, Oppenheimer had told a different story to the FBI about Chevalier after the war. In addition, there was some question about his veracity involved in his opposition to the hydrogen bomb.

In 1943 there were no hard-and-fast rules about security clearance. Army security refused to clear Oppenheimer, so Groves took over our security and set his own rules, which he or I waived whenever we thought a man essential to our work. However, in 1953, Eisenhower had established more rigid rules concerning a security risk and made it clear that loyalty was not the only issue. Also, Communists in government had been a campaign issue, and Eisenhower had promised to rid the government of Communist security risks.

Finally, during the hearing, Oppenheimer's own testimony showed that the veracity was a more important issue than we had thought in 1943. This issue coupled with his past and continued association with Communists made it impossible to consider Oppenheimer not a security risk under existing atomic energy law and presidential Order 10450. To decide otherwise required adaptation of the concept that if you are sufficiently distinguished, rules should be waived.

Oppenheimer was sufficiently distinguished. He was a genius. His past contributions to the United States warranted an exception, but the then-current issue of communism coupled with veracity made it not only unwise but also politically impossible.

*   *   *

In spite of the Oppenheimer and Dixon-Yates diversions, we had time for other work. I always have admired the leadership Strauss demonstrated in initiating and accomplishing other objectives, such as the Atoms for Peace program, legislation encouraging and permitting the commercial development of atomic power, the AEC program for cooperative development of commercial power, improving relations with the military, strengthening the AEC organization, achieving the successful development of thermonuclear weapons, increasing production of plutonium and U-235, and increasing research on thermonuclear energy. In addition, he showed great leadership in handling the many problems that arose due to testing fission weapons at the Nevada test site and the thermonuclear testing in the Pacific. This is the type of work I enjoy, and I am thankful for the constant support that Strauss gave me within the commission for my part in carrying out the program to achieve our objectives.

Just before the Oppenheimer problem arose, Strauss was working on what became known as the Atoms for Peace program. Strauss's chief adviser and assistant in initiating and carrying out this program was Professor Isodor Rabi, who at the time was chairman of the General Advisory Committee. Strauss and Rabi were a rare team; they differed in personality, ways of getting things done, education, and background, but both were determined to find some way to utilize the peaceful aspects of the atom to benefit our nation and the world. Their thinking and objectives differed somewhat. Strauss realized the need for disarmament and the hopelessness of any atomic confrontation between the Soviet Union and the United States. He realized the appeal of simple slogans like "Ban the Bomb" but considered far more important the necessity to win national and world public support for the United States in its efforts to get a verifiable and effective disarmament agreement, and the need to deter war by having an adequate stockpile of weapons. He was suspicious of the USSR and did not trust the Soviets. Rabi, having been a scientific leader in the development of the atomic bomb, had supported Oppenheimer in his recommendation to the commission not to develop the hydrogen bomb. Rabi realized even more emphatically the hopelessness of any atomic confrontation. Being more idealistic than Strauss, Rabi felt that getting scientists together to talk and to strive for common objectives might be a step toward better understanding and toward peace. He believed that scientists were more likely than statesmen and politicians to find a common basis for peace as an alternative just to piling up more and better weapons. Many of the ideas supported by Strauss were originated by Rabi. Both had the respect of Eisenhower, who also was seeking a better approach than deterrence to avoid war with the USSR.

The first step was Eisenhower's speech at the United Nations on December 8, 1953. Strauss helped write it. Eisenhower proposed that the United States, the USSR, and Great Britain contribute fissionable material to an International Atomic Energy Agency, such material to be used for peaceful purposes in medicine, agriculture, and to produce electric power. The proposal offered a hopeful alternative to the present course toward a destructive use of atomic weapons.

Early in 1943, Rabi proposed to Strauss that the United States promote the idea of an international conference of scientists to discuss the peaceful uses of atomic energy. The idea gradually evolved into the First International Conference on the Peaceful Uses of Atomic Energy, held by the United Nations at Geneva in July 1955. The conference was a huge success. Considerable information was declassified; journalists, industrialists, government officials, and scientists participated in large numbers.

As a part of the Eisenhower proposal, the International Atomic Energy Agency was formed, with headquarters in Vienna, Austria. From a public-relations point of view, the peaceful atom had its day—for a while. International politics intervened, however, and the agency never achieved the expectations of its original proponents.

In searching for a reactor to demonstrate the production of electricity, it soon became obvious that the pressurized water reactor that Rickover was developing for use on a Navy aircraft carrier was the best choice. The carrier program had been canceled. Initially I had intended to wait until we found a man from industry to be a director of industrial research and development, but with Rickover having a going organization and a reactor project under way, that now had no specific use to justify it. I changed my mind and decided to proceed with it. Dr. Hafstad, director of our reactor program, and Rickover were in accord. They proceeded to develop a proposal so we could present it to the commission. On January 4, 1954, I told Strauss that I had decided on the Rickover-Westinghouse pressurized-water reactor and would select a utility to participate in a demonstration program. The commission was informed and agreed with the general approach to invite individual utilities or groups of utilities to propose how they would cooperate, such as providing the conventional part (the turbines and generators) of an atomic power plant. We would consider all types of proposals. The field was wide open. By February 16, proposals were coming in. I told our information division that I did not want any release on the proposals. I was not going to inform Congress until all proposals had been analyzed and the commission had approved my recommendation. For all queries (and there were many), we should state that we have several proposals and are evaluating them.

After I had given up hope of finding and convincing a suitable industrial man to take on the position of assistant general manager for industrial research and development, I requested and received commission approval to assign Alphonso Tammaro to the position. He had worked for me in the MED and I knew he was imaginative, aggressive, intelligent, and a live wire who could get along with industry. I informed Tammaro he had a new job and he was to evaluate the proposals, aided by director of research and development Hafstad, Rickover and whomever else he needed. I found that Hafstad and others were pleased with the appointment. By Wednesday, March 10, all proposals we expected were in. Tammaro had discussed the best ones with the organizations making the proposals and informed me he was ready to make his recommendation. That day I had defended our budget before the House Appropriations Committee, and they had pressed me for information on the proposals we had received. I told them the staff was evaluating the proposals and that I knew very little about them. When I returned to my office at 5:00 P.M., Tammaro was there with his recommendation to select Duquesne Light Company of Pittsburgh. I informed Strauss that I proposed to recommend the selection of Duquesne at the 8:30 A.M. commission meeting the next day. The commission approved the recommendation. At 10:00 A.M. I was back with the House Appropriations Committee and reported that we had selected Duquesne Light's proposal and were preparing to announce on Sunday that we were negotiating with them. The committee had considerable difficulty understanding how such an important decision could be made so quickly. The next day, Smyth (Strauss was out of town); Fields, Dr. John Bugher, head of our Medical Section; Tamaro; Hafstad; Edward Trapnell in Public Relations; and I met with the Joint Congressional Committee at their request to discuss the current tests in the Pacific and the reactor program. They were informed about the selection of the Duquesne proposal; also, there was discussion of our five-year reactor development program.[2]

Negotiations with Duquesne Light proceeded rapidly. With Westinghouse and Bettis Laboratories and Duquesne Light all being in the Pittsburgh area, there was good communication among the participants of this joint venture between industry and government. Westinghouse was responsible for the reactor and Duquesne was responsible for the conventional part of the plant. Tammaro and Rickover remained in overall charge of the project for the AEC.

In August, President Eisenhower decided to give a talk in Denver on Labor Day on atomic energy. He wanted to tie the talk in with a ground-breaking ceremony at Shippingport, Pa., for the PWR atomic power plant. The president wanted to wave a wand in Denver at the appropriate time to signal an unmanned bulldozer to move the first dirt at Shippingport for the ground-breaking. I told Rickover that he should coordinate the ground-breaking ceremony with presidential press secretary James Haggerty's arrangements for broadcasting Eisenhower's speech. He should arrange for TV coverage at Shippingport, and I expected that whenever Eisenhower waved his wand the unmanned bulldozer would push dirt ahead of it. He asked what role the AEC Information Division would have. I told him to pick one of them to work for him. I added, "You are in absolute control of all our responsibilities for this occasion, and if the bulldozer doesn't push ahead and move dirt when Ike waves his wand, you are fired." He smiled and said, "Don't worry, I'll still be around here after the event."

I was invited to Shippingport but decided to spend the holiday with my family and watch the whole event on TV. It was a good speech and the bulldozer moved forward as Eisenhower waved his wand, and pushed about a six-inch layer of dirt. The whole thing went off perfectly. I told Rickover that I had operated a bulldozer, and as an amateur, I frequently dug too deep and stalled the thing. I asked, "How did you arrange that an unmanned bulldozer would not do that?" He laughed and said, "I anticipated that possibility so I buried two railroad rails six inches beneath the surface and the bulldozer blade just rode on top of the rails; it couldn't dig in." It pays to pick the right man and give him full responsibility.

In addition to initiating action to get a commercial power plant

2. Shortly thereafter, Hafstad left the AEC, and the Reactor Division was headed by Kenneth Davis, with Louis Roddis as his deputy.

under way, on January 12, 1954, I met with Hafstad; Colonel James B. Lampert, head of the Army reactor program, and Don Burrows, our comptroller, to organize a joint AEC-Army program to develop the Army power package program. Lampert would be in charge of both the AEC and the Army activities and arrange for funding. Having participated in initiating the program in the Army as chief of Army R&D, it was convenient to be in position in the AEC to ensure AEC cooperation and get the program under way. Strauss and I arranged directly with Donald Quarles, in the Defense Department, how to fund the project, and Quarles and I coordinated our testimony before the House Appropriations Committee. With Strauss's backing, the commission readily supported the project and the organizational setup. It was now up to Colonel Lampert to get the job done. He was a superior executive and very tactful, so he had no trouble working with the AEC, the Army, and the Defense Department. While still in the Army, I had arranged with the chief of engineers that Lampert should stay with the project until it was completed successfully. He was in charge of the project from 1952 to 1957, for which he received a DSM

To initiate commercial development of atomic power it was necessary to enact changes in the Atomic Energy Act and to declassify large areas of classified material. This effort had to be coordinated with the White House, the Defense Department, and, of course, Congress. Our Legal Division, under Mitchell, had to do most of the work. Strauss, of course, coordinated our activities with the White House, and I established direct contact with Quarles and Frederick Seaton in the Defense Department and with General Herbert Loper, now retired from the Army and chairman of the MLC. Generally we had the cooperation of both Republicans and Democrats on the JCAE, but of course there were some political differences on ownership of fissionable materials. However, on the whole the 1954 act went through with less controversy than expected. On August 3 I accompanied Strauss and Smyth to the White House to witness Eisenhower's signing the 1954 Atomic Energy Act. We now were in a position to cooperate with industrial companies and utilities to develop commercial power either for use by private investor-owned utilities or by publicly owned utilities such as the TVA.

The commission now was faced with developing regulations and licensing procedures for privately owned and privately operated atomic power plants. Harold Price, initially under Mitchell, as largely responsible for developing proper procedures for licensing and safeguarding the public. He later was designated as director of Civilian Application. Insurance turned out to be a big problem and was solved by the Price-Anderson Act, which provided for government insurance. For most new plants, we anticipated the need for some government support, particularly in research and development. This activity came under Tammaro's office of assistant general manager for research and industrial development. We had much to learn about safety, regulation, licensing, insurance, the economics of atomic power, and utility systems. While I was general manager, I believe we gave it a good start on the road to success, and contrary to the opinion of the antinuclear opponents, I believe that in the United States nuclear power today is the safest, most economic, and cleanest method for producing electricity if it is properly managed by industry and reasonably regulated by government

Both Strauss and I felt it necessary to develop a stronger, more efficient organization to accomplish our objectives. We first developed a faster way to get commission decisions. One of my first acts was to control papers submitted to the commission. The staff was responsible for developing all the facts and factors influencing a decision, but in most cases I retained the final responsibility for determining the recommendation to be made. I also urged Strauss to avoid endless commission discussion for the purpose of getting a unanimous decision in a collegiate manner.

In developing a stronger organization, my first major selection turned out to be a poor one. Our AEC manager at the Santa Fe operations office was leaving at the time I became general manager. My first choice to replace him was Brigadier General Kenner Hertford, who had experience in the AFSWP at Sandia, but he was not ready to retire. My interim selection was highly recommended but it turned out he just could not get along with scientists, a common failing with many military officers who otherwise are highly successful. I eventually convinced Hertford to retire and take on the job, and his success is best demonstrated by the fact that he continued in the position for about eight years.

Most of the other changes were made by selecting personnel from within the AEC to move into more responsible positions as vacancies occurred and to give and to expect them to take full responsibility to get the job done. By the time I left, I was quite content with practically all of my key people.

I also was able to reduce the number of AEC employees. Early in 1954 I had to defend the AEC fiscal year 1955 budget for funds and personnel before the House Appropriations Committee. The budget had largely been prepared before I arrived. It called for a 10 percent increase in personnel. When we came to that item, I asked the chairman, Congressman Philips from California, to defer it until after lunch, and I invited him to have lunch with me. At lunch I asked him to ask only a few questions about personnel and let my staff answer them, and when he marked up the bill he should reduce the personnel by 5 percent instead of increasing it by 10 percent. He asked me how I expected to achieve that. I replied that if I told him or anyone else I probably would not be able to get away with it. I indicated that I wouldn't appeal the 5 percent cut. But I said he shouldn't make any big deal of it—the less attention it attracted the better. Immediately thereafter I called Samuel R. Sapirie manager at Oak Ridge and asked him how many more personnel spaces he would require if I abolished most of the New York operations office that was coordinating supply of feed materials for Oak Ridge and Hanford and if he took on the responsibility for doing that function. His comment was that he probably could reduce the total of his own personnel by a few because he felt keeping New York informed of his needs and seeing that they did it right took more men than if he had control. I told him that that is what I thought the situation was. I then added that of course when I transferred the function to Oak Ridge and abolished the personnel slots in New York, I intended to offer the personnel the right to transfer to Oak Ridge. Sapirie said, "God, Nick, I don't want any of them." I replied, "Don't worry; I doubt if any New Yorker will want to transfer to Tennessee." They didn't.

I also told Jesse C. Johnson, director of the division of raw materials, to consolidate his New York and Washington uranium procurement offices in Washington and that I expected him to operate with fewer people. He was delighted and commented that he had previously recommended this action but could not get approval.

I expected the responses I got. During the war the feed materials program was organized by Crenshaw and Ruhoff. Because of the experimental nature of the production plants, there were many individual plants and operations scattered around, from Iowa to the East Coast, that required coordination. When the AEC expanded production to meet military requirements, they wisely consolidated the operations, mainly in two plants; however, no one then determined that much of the need for coordination had essentially been eliminated. As usual in government operations, no one wanted to take the onus of telling employees they no longer are needed. On March 23, 1954, I informed the commission of my plan for reassignment of the major responsibilities of the New York operations office to Oak Ridge to provide for closer integration of the production program. I also pointed out that we were over the hump on construction of new plants and that the decrease in personnel required for the Construction Division would provide for an increase in the Operations Division. Their only criticism was that I had proceeded so far that my actions were practically irreversible. I was told that in the future such policy matters should be discussed with the commission prior to action.

While I was general manager, the membership of the commission changed. Campbell left to become comptroller general of the United States. Vacancies also were created by Zuckert and then Smyth leaving because their terms expired. William Libby, John von Neumann, and Harold S. Vance were the replacements. Murray stayed on and until his term expired remained a thorn in Strauss's side. Eisenhower, in spite of the urging of Murray's supporters, refused to reappoint him.

Our priority mission was to develop and produce better weapons in larger quantities. Our production plants were being expanded; new weapons, including the H-bomb, were being developed by Los Alamos and by Livermore in California and being tested at the Nevada and Pacific test sites. Under the capable management of Fields, as director of military application, Bradbury as director at Los Alamos; and Edward Teller, at Livermore, the H-bomb was making progress. We were in a race with the USSR, more of a race than many of us realized at the time. In 1950, shortly after the British notified the United States that Klaus Fuchs was a spy for the Soviet Union, Loper and I had written an evaluation of Fuchs's spying activities. If Fuchs had transmitted all the information he had access to and if the Soviets had properly evaluated it, they might well be ahead of us. I felt that we should develop and produce H-bombs as expeditiously as we could. We had lost time in getting started. By June 1951, the major technical controversy between Oppenheimer and Teller over whether a thermonuclear bomb was technically feasible was eliminated by the Teller-Ulam approach to the problem. The Greenhouse atomic tests at Eniwetok convinced even Oppenheimer that the H-bomb was technically feasible. Teller, however, was frustrated by what he considered a lack of emphasis and resigned from Los Alamos and returned to the University of Chicago.

This started a movement for a second weapons laboratory, and Livermore with Teller as director was established before I became general manager. Los Alamos successfully developed a thermonuclear device that was a great success. On November 1, 1952, the device yielded over ten megatons but could not be considered a deliverable bomb, since it was far too large. However, a new idea made it possible to develop a smaller device. Production facilities were hastily installed at Oak Ridge in one of the old Y-12 buildings, and on March 1, 1954, in Operation CASTLE, the first thermonuclear device that was small enough to be deliverable was exploded, with a yield of fourteen megatons, much larger than expected. As seemed to be my luck, I again had to miss an historic event. I was too busy with the Oppenheimer case, selecting the first demonstration reactor proposal (at Shippingport), budget hearings, Dixon-Yates, negotiations with Canada on purchase of plutonium, and a multitude of other problems. However, although I could miss the tests, I could not avoid the problems that shifting winds had caused. We had our first real case of massive fallout. The resulting publicity alerted everyone to the danger and extent of fallout of radioactive materials from the hydrogen bomb. Dr. Bugher first alerted me to the fact that more than two hundred inhabitants of the Marshall Islands had to be evacuated from their islands and needed medical treatment. At first we minimized publicity about the fallout, but on March 10, all hell broke loose. I have a diary note, "Called Dr. Bugher to ask if we have anything factual from our own people re Japanese fisherman. He said we had nothing except confirmation that the same sort of white ash was seen at Rongelap [an atoll in the Marshalls]." That morning a news report revealed that the Japanese ship *Fukuryu Maru* had returned to port with its entire crew of twenty-three men suffering from radiation sickness. They reported they saw the flash when the H-bomb exploded, felt the shock, and about two hours later a white ash fell on them. Later they were sick. Our problem was to get facts. Reports started to spread that all parts of the earth might become contaminated, that the H-bomb was out of control. The public for the first time was exposed to vivid accounts of the potential horrors of a thermonuclear war, a combination of massive destruction and danger from radioactive fallout. It seemed that statements issued only raised new questions, and the news media competed in finding new and better scare stories. As photographs and movies became available, comparisons were made in each city, showing maps of the city and the area covered by the H-bomb blast.

We had similar problems, on a lesser scale, with testing of fission bombs at the Nevada test site. There was, of course, much less fall-out—but it was closer to home. Testing was well under way when I became general manager. Many of the tests were sponsored by the Defense Department to determine the weapons effects on test structures and military equipment. At many of the tests, troops were in forward areas to determine the reaction of troops to use of atomic weapons and to develop training procedures and tactics for use of atomic weapons in tactical situations. Correspondents and government officials were permitted to view many of these tests. The commission devoted many hours establishing safety rules, permissible levels of exposure to radiation, security of information rules, public relations, rights of correspondents, and determining the division of responsibilities among the AEC, the Defense Department, and the test manager. It was a dangerous business, and safety considerations had to be balanced against military requirements for weapons deemed necessary to provide for national security during the Cold War with the USSR.

Each test series had to be approved by the president. Although the area around the test site was sparsely settled, there were small communities that had to be protected by limiting shots to times when the wind directions and atmospheric conditions were optimum for safety. Plans had to be made for evacuation of people if wind conditions suddenly changed and radiation monitoring detected dangerous conditions. As a result of all these factors, many shots had to be delayed in the interest of safety. Another determination that had to be made was the division of testing between Nevada and the Pacific. Pacific tests required much more logistical support and time of scientists to conduct and acquire the test results. As a result, only large-yield devices were tested in the Pacific. Everyone realized that with the existing frame of mind of the public, it would take only a single and unforeseeable incident to preclude testing in Nevada or even in the Pacific. Public relations was a major problem. Unusual weather was blamed on the tests. At one point, General Groves called me and asked, "What are you doing to avoid weather changes caused by the shots?" I told him, "I don't think we are causing major changes in the weather. We are studying the weather effects and believe that there are some local effects but no major general effects." He laughed and said, "It doesn't matter what you think or determine; the public will blame you every time they complain about the weather." There also were problems of animals such as cows, horses, and sheep dying. In many cases it was determined that death was due to other causes. Whenever it appeared that radiation had been the cause, the AEC paid for damages.

At the time I became general manager, the whole issue was under review. Could we continue to use the Nevada test site? Fields recommended that we continue to test at Nevada. On February 17, 1954, the commission approved planning and preparations for Nevada tests in 1954 and 1955. However, they deferred approval pending reference of the problem to the General Advisory Committee and the Advisory Committee for Biology and Medicine. Murray dissented because he thought final approval should have been given. Throughout my period as general manager, Murray was my most ardent supporter for stressing the urgency of improving our capability in atomic weapons vis-à-vis the USSR. On June 30, 1954, having received the reports of the two committees, the commission approved Fields' recommendation to continue testing at Nevada. In October, the commission again reviewed the safety precautions that would be taken for the Teapot test series. Again, on February 11, 1955, they reviewed and approved new radiological safety criteria for protecting the public during tests. On February 23, 1955, Senator Anderson, chairman of the JCAE, sent a letter raising the question of whether the new safety considerations so delayed testing that we should consider moving more of the tests to the Pacific. The loss of Senator Anderson as an ardent supporter of testing in Nevada was a real threat to our test plans. Strauss also indicated that he had serious questions about the Nevada site. My reaction was that we had improved safety standards enough to continue testing and that delay waiting for perfect weather for the big shots would not mean an overall delay in the series; we had contemplated delay. Commissioner Murray supported me. He commented, ''We must not let anything interfere with this series of tests—nothing.'' He also argued that moving to the Pacific would not mean a delay of thirty to sixty days, as Strauss thought, but a delay of a year. Murray was right. The next move was up to me, to prepare a letter to answer Senator Anderson. On March 1, the chairman of the AEC sent him the letter indicating that we were continuing with tests at Nevada because of urgency but would study other solutions to the problem. Senator Anderson's reversal was a clear indication that we were losing public support for testing. The March 1, 1954, test of the H-bomb in the Pacific had made radioactive fallout a worldwide issue.

On March 27, 1954, I received a call from C. D. Jackson, of White House public relations. My notes read, ''Said this is just thinking out loud and has no White House overtones whatsoever. Said he has no idea whether H-bomb tests are scientifically or militarily necessary and if they are not he would strongly urge calling off further tests and maybe having the president make some statement. Told him we are aware of the seriousness of this, also told him there are terrific needs for continuing tests because if we stop testing we stop development. He agreed.'' I never knew for sure if Eisenhower had asked Jackson to call me with the suggestion to stop testing or if it was Jackson's idea. Anyway, I placed a call to Admiral Radford, chairman of the Joint Chiefs of Staff, told him about Jackson's call, and also advised him that a science committee was meeting with the president and that there was a bare possibility that someone might get to the president with strong arguments to cancel the tests. Later requests I received from the White House from Colonel Goodpaster indicated that President Eisenhower was becoming more seriously concerned.

I concluded that sooner or later there would be a ban on open-air testing. Actually, when it finally came, underground testing, although more expensive, had certain advantages and was almost as efficient in interpreting results as open-air testing. However, by that time the public was thoroughly alarmed about radiation; many wished to ban not only nuclear weapons but nuclear power as well. The wartime birth of nuclear energy combined with the alarm caused by radiation resulting from weapons testing has been a millstone around the neck of commercial nuclear energy, making it extremely difficult for the public to appraise properly the risks and safety of nuclear power as compared to alternate methods for producing electricity.

At the time I considered retiring from the Army and accepting the general manager's position, I made a thorough study of what the change meant to the financial future of my family. I was confident of future success in the Army, and the military retirement provisions were good. If I transferred to the AEC, the law then required that I forfeit all my Army retirement pay while employed by the AEC. To accumulate retirement pay from the AEC, I needed to stay with the AEC for at least five years, and if I stayed at least eight years, the possibility of combining the two retirements had some very favorable provisions. So it appeared that unless I stayed eight years or more I was doing little toward providing for retirement. In September 1954, I reappraised the situation. Dixon-Yates, the Oppenheimer hearing, and political aspects of the job did not appeal to me. The feud between Strauss and

Murray was unpleasant, particularly because I admired both men and thought the feud unnecessary. Also, the commission from time to time thought I was exceeding my authority, and required advance approval for many actions that clearly were the function of the general manager. In spite of all this, I greatly enjoyed my activities and responsibilities with the AEC, but I had serious doubts that I wanted eight or more years of it. On September 13, 1954, I told Strauss that I had decided not to try to make my present job a career and that I would therefore be talking to some individuals concerning future employment but that I would not aggressively seek new employment and would give him adequate warning. On January 3, 1955, I informed Strauss that I planned to leave the AEC on May 1, and although he did not like it, he agreed that January 15, 1955, was the proper time for me to forward my letter of resignation.

On January 28, a story appeared in the *St. Louis Post-Dispatch* concerning my resigning from AEC and that I had an offer for $100,000 a year. That was real money in those days. I was besieged with reporters trying to reach me. On the next day, at a JCAE hearing, Senator Albert Gore raised the issue of my accepting a $100,000 a year job and stated that I had no right to use my AEC experience to get a high-paying job. I rather sharply interrupted him and replied that I had not received any $100,000-a-year offer but that if I could not use the experience I had acquired in the thirty years I had been on the government payroll, I would not even be able to dig a ditch—I had learned to dig a foxhole in the Army. At that point Congressman Melvin Price leaned over toward me and said, "Calm down and keep quiet. You have a lot of friends on this committee. They will defend you on this issue." Among other favorable comments made, Senator Anderson advised me to hole up in some hotel room and answer no inquiries about the $100,000-a-year news story—if I didn't have such an offer, with all the publicity I soon would receive one equal or better.

On April 27, 1955, I made my last general manager's monthly report to the commission. On April 29 I received the AEC's Distinguished Service Award. I was very happy that I could turn my responsibilities over to the capable hands of General Fields, whom the commission had selected to succeed me.

16

MONITORING THE FATE OF NUCLEAR POWER, 1955–86

ON MAY 1, 1955, I EMBARKED ON my new career. I was not interested in any of the informal offers of positions with corporations involved or hoping to become involved in the commercial development of atomic power. Several of these offers were extremely attractive, even better than the false newspaper report. Practically all of them required moving to another city, and Jackie and I had decided that we wanted to remain in our roomy, comfortable house in Washington and eventually live on the farm we had purchased earlier near Sugarloaf Mountain in Maryland. I was seeking neither maximum compensation nor prestigious position, and certainly no glory of any kind. Above all, I wanted independence and freedom of action. So I opened an office as a consultant, operated with no staff except for my secretary, Miss Olsson, and accepted as clients only those primarily in the atomic power or related fields, or in other areas of particular interest to me. In the discussion of my consulting career, I will cover only a limited number of events and activities that directly relate to the development of nuclear energy. I was, however, engaged in many other activities that involved various military and industrial problems. I served on the Army Scientific Advisory Panel and Department of Defense committees. I was elected director of the Atomic Industrial Forum and headed several study committees for U-235 production problems and private ownership of U-235 production plants. I also served as director on the boards of several corporations.

My thirty years in the consulting field were challenging and rewarding. I thoroughly enjoyed the freedom to engage only in activities

that interested me and the freedom to travel both at home an~~oad~~
whenever I thought it necessary, desirable, or just fun. As a co~~nt,~~
my personal responsibilities were limited, but I was able to ~~de~~
valuable advice and assistance to my clients who were making ~~r~~
advances in the field of weapons and, perhaps more important, in
field of nuclear energy for producing electricity. To cover all my var
activities during these three decades would require another volume.
will spare you that. However, I hasten to point out that the relative
brevity of this chapter does not mean that the commercial development
of nuclear power is not every bit as interesting and important as the
development of weapons. I plan only to discuss the degree of success
we achieved in the United States, why nuclear power is more successful elsewhere, why we got into trouble in the United States, and what
the fate of nuclear power will be. Needless to say, being a consultant
in this interesting field kept me very busy. I enjoyed it.

I was already licensed as a professional engineer in the District of
Columbia and the state of Maryland. The timing was right. Many
corporations were desiring to enter the atomic power field. The type of
talent and experience I possessed was in demand. I did not need to
prepare a brochure explaining what I had to offer. They had the questions. I had many of the answers. I could be selective. I turned down
many firms requesting my services. In fact, one very good friend who
was a director of a large firm advised me not to be so hasty in turning
down requests for consultation just because I thought the firm might
have no basis for being involved in atomic energy. He had recommended that his firm engage me to determine whether they should enter
the field. As a result of his request, I did explore with them what
opportunities there might be to utilize their know-how and experiences
in the atomic field. My answer remained the same, except that I developed it with their top management, I earned a consulting fee, and I
learned about the activities of that corporation. The management and
Board of Directors were satisfied that they were not missing an opportunity. However, I did not like this type of work. I preferred contracts that had a long-time future potential and were in the development
field.

Initially I established an office in downtown Washington, but later
I bought a new home that had a very convenient library for my office.
Most of my work involved advising top management on major policy
decisions. I used their staff to develop the basis for any recommendation and the course of action to pursue. Most staffs were willing to

cooperate with me because they knew that if they gained my support, top management generally would approve.

Most of my work was in the atomic energy field. I had determined to my satisfaction that the pressurized water reactor was the most promising for the immediate application of atomic power, so I accepted a long-term contract with Westinghouse as a major client, and I enjoyed working with them on their domestic and international projects for almost twenty years. They were experienced in the field, having designed and fabricated Rickover's pressurized water submarine plants and also the Shippingport demonstration plant.

Bill Webster, chairman of the board of New England Power, asked me to be a consultant for the Yankee atomic power plant, the demonstration plant he was proposing to build in New England. This project had an ideal setup for atomic power development. Twelve New England power companies formed the Yankee Atomic Power Company to design, build, and operate an atomic power plant at Rowe, Massachusetts. The owner companies contracted to purchase at cost all power produced. Westinghouse was selected to design, develop, and fabricate the atomic reactor components and furnish the turbine generator and other electrical components of the plant. Stone and Webster was selected to design the overall plant and to manage construction. The AEC was willing to support research and development for the plant. The Yankee organization had people trained in the atomic power field at government laboratories and was competent to supervise Westinghouse and Stone and Webster activities and to operate the plant. Bill Webster of Yankee, Charles Weaver of Westinghouse, and I recognized that I was faced with a potential conflict of interest in consulting for both the designer and manufacturer, Westinghouse, and with the owner, Yankee Atomic Power Company. However, it was agreed by all that I would refrain from getting involved in any dispute between the two unless I was specifically asked to do so by both parties. To complicate this issue further, Webster asked me to meet with a group of financial people who were to provide the funds for financing the project. I met with them in New York City. It was my first contact with a group of conservative financiers. After the usual formalities, the first question was, "How accurate is the estimate? Is it within fifteen or twenty percent of what the costs might be?" I responded, "As an engineer I wouldn't even call it an estimate. It is a 'guesstimate.' In this new field it is impossible to anticipate all contingencies; the plant is not yet designed, and many components are not yet developed. Development problems may increase costs. More important, we must meet safety requirements that are being established by the AEC. New research results may reveal new problems that could double or triple costs of certain parts of the plant. I have been responsible for or involved in the construction of practically all of the reactors built by the government to date and have a good idea of costs, contingency factors, and overruns in estimates. For Yankee, Webster and I have agreed that one of our goals is to build the plant within our estimate, but we might be off by a factor of two or more. The present total figure is only a guesstimate, but I hope we can make it. We have individually checked Westinghouse's and Stone and Webster's estimates for the uncertain areas, and we have the benefit of fixed prices for many conventional items. We finally agreed that fifty-seven million dollars appears adequate to build the one-hundred-seventy-five-thousand-kilowatt electric plant.''

This was followed by a series of questions delving into my experience and problems we had encountered. They asked about the experience of Westinghouse and Yankee personnel. Finally, they thanked me for meeting with them, and the chairman also thanked me for speaking so frankly about potential problems.

Early the next morning I called Bill Webster and said, ''Bill, I think I have ruined your prospects for financing. I doubt if that conservative group will risk a nickel on Yankee.''

''Like hell you have,'' he said. ''You talked yourself into another job. They okayed the financing provided that you will act as the independent engineer for the project and prior to each drawdown on the loan you will certify that the project is still within the estimate.''

''Bill, how can I say that I am the independent engineer? I am a consultant to Yankee, the owner. I am a consultant to Westinghouse, the designer, manufacturer, and supplier of all the major components. I have only a secretary, no engineering staff. I would have to rely on Stone and Webster, the designer and construction contractor, for any technical assistance I need.''

''Nick, I told them all that and they still insist that they want your signature on the certifying letter. My attorney tells me that you can avoid any legal problems by stating this relationship in your letter each time you certify to the adequacy of funds. Consult your own attorney and take on the job; it is a good field to be in; you are being paid for your integrity, reputation, and experience.''

''Okay, Bill, I'll do it provided Stone and Webster will agree to do such staff work as I request and work with me.''

Bill replied, "I have already talked to them about it, and F. W. Argue, their executive vice president, will be cosigning each certificate with you."

I should not have been surprised by Webster's lack of concern about the potential conflict of interest involved. On September 30, 1949, as he was leaving the position of MLC chairman, he wrote a letter commending me to the chief of staff of the U.S. Army. It is one of the finest letters I have in my file. He included the statement, "I consider him one of the most capable men I have ever worked with. His most outstanding qualities are probably his near perfect intellectual honesty and his ability always 'to keep his eye on the ball.' "

Shortly thereafter, when it was publicly announced that the Yankee plant was under way, Admiral Rickover called me and said, "General, I just read the announcement about Yankee and your participation in it. What makes you think that you can build a plant twice the size of Shippingport at one half the cost? It is none of my business, but that low cost figure is impossible to achieve. I called because you have always been very frank with me and I hate to see you ruin your reputation."

I replied, "Thanks for calling, Rick, I really appreciate it. But we are going to make that estimate. We are taking advantage of all the good development work that you have done for Shippingport and for submarines. Bill Webster and I think that on many of the more conventional components we can use normal utility standards that will save us money without any sacrifice of safety and still achieve the reliability required for a commercial plant. But I do want to thank you for calling." He replied, "Well, don't say I didn't warn you."

Working with Bill Webster again was, as in the Pentagon, a pleasant and educational experience. We were friends, and we respected each other. We saw eye to eye on most things, and when we differed, we each learned more by arguing it out point by point until the issue was resolved. Moreover, there is nothing that compares with the pleasure and satisfaction of working on a successful project. Construction was done during a slight recession, prices were somewhat better than we anticipated, we each had contingency money in the parts of the program that we were more personally familiar with, and we had no major unexpected contingencies. We achieved our goal to construct the first privately owned pressurized-water plant below estimate. The final capital cost was approximately $45 million, as compared to our estimate of $57 million.

Work with the Yankee organization led to participation in the Connecticut Yankee plant. The same organizations participated—Yankee, Westinghouse, and Stone and Webster—with the owner in this case another group of utilities, in the Connecticut area. The project was headed by Sherman Knapp, chairman of Connecticut Electric Power Company. This was a slightly larger plant: 582,000 kilowatts and estimated to cost $100 million. The Yankee plant started operation in 1961 and the Connecticut Yankee plant in 1968, and they still are operating. Although both were considered experimental plants and were not expected to be competitive with coal and oil, later they became competitive because of inflation and, in particular, because of the large increase in price of coal and oil. For an atomic plant, the major cost factor over the life of the plant is the capital cost invested in the plant. The cost of fuel is a minor factor, whereas in a coal plant the cost of coal over the life of the plant is the major cost factor. As a result, once completed, the atomic plant is less subject to inflation. However, if inflation and interest rates increase rapidly during the construction period, and if construction is excessively delayed, the excess capital costs of the atomic plant become too great to be compensated for by the more economic fuel. The "antinukes" appraised this factor correctly and concentrated their opposition efforts on various measures to increase costs, such as promoting excessive and constantly changing regulations, intervention at every stage to cause lengthy and a continuous series of hearings to force a longer construction period and to cause plants to stand idle, increasing interest costs, while awaiting operating permits.

As a result of the antinukes' effort, overregulation, and in some cases poor management, we now find it takes almost twice as long to build an atomic power plant in the United States as in France, Japan, Taiwan, or South Korea. I am referring to essentially identical boiling-water or pressurized-water plants of basic U.S. design. In 1981 I worked with Walker Cisler, chairman of the board of Overseas Advisory Associates, Inc., advising Taiwan on their energy needs. General Electric personnel were assisting Taiwan Power place in operation Kuoching 1, a 948-MWE boiling-water plant. The first of its kind of this particular plant had been started in Mississippi two years before the Taiwan plant, but it still was at least two years from completion. Certainly our system must be faulty when a United States-designed plant can be constructed successfully by Taiwan Power, using their own construction force, much faster than we can do the same job at home. The major difference is overregulation, constant changes in regulations, and delays caused by interveners in the United States.

I also visited Maanshan 1 and 2, 890-MWE pressurized-water plants of Westinghouse design under construction. The production schedule there should have amazed me, except that I knew we, too, were able to work that fast back in the 1960s. I was greatly impressed with the Taiwan Power superintendent of construction and the appearance and orderliness of the job; and they were on schedule. I asked him where he acquired his construction experience. He replied, "I worked for Bechtel building nuclear plants in the United States." We can teach people how to do it, but we no longer can do it ourselves. Why? With the completion of Maanshan 6 in May 1985, Taiwan now can supply more than 50 percent of their electrical demand with nuclear power, and costs for producing power are lower there than for oil or coal.

In May 1985, I spent a short time in South Korea, and while there I confirmed that South Korea, like Taiwan, has continued an aggressive, ambitious nuclear power program, which calls for twelve units. Three units of 1,790 MWE were already operating and in 1984 generated 21.9 percent of South Korea's electricity. Six more unclear units of 5,476 MWE capacity were in various stages of construction. Two of these units went on line in 1985 and 1986, and by 1991 four more units under construction should be completed. With the leveling off of the increase in demand for electricity and with the decrease in price of oil and coal, there were no specific plans for going beyond nine units. At that time, South Korea had purchased six pressurized water plants from Westinghouse, two from Framatome, and one pressurized-heavy-water plant from Atomic Energy of Canada. South Korea hopes to standardize on one or at most two designs. Also, South Korean engineers and industry are participating more and more in design, construction, and supply of equipment.

In establishing their energy program, South Korea, like Taiwan, decided to rely on a mix of nuclear, coal, and hydro. They have continued with this mixed program with considerable success.

France provides an even better example of how other countries can use our assistance to develop nuclear power plants, improve on the design, cut costs and construction time, and retain public confidence and government support to such an extent that in France's case, it now produces more than 60 percent of its electric power in pressurized-water nuclear plants that have proven to be much cheaper than oil plants or coal plants. These nuclear plants also are more reliable, safer, and cleaner than oil plants or coal plants. France has exploited to her advantage the possibilities of nuclear power, whereas we have handicapped and essentially ruined our nuclear industry by a series of what by some are labeled "institutional failures."

I had the pleasure of participating in the growth of the nuclear industry in France. Starting with the 1955 and 1958 Geneva conferences on Atomic Energy, the United States demonstrated leadership in making available the benefits of peacetime applications of nuclear energy. Congress cooperated in authorizing supply of enriched uranium for nuclear power plants. We negotiated bilateral treaties for cooperation with various countries and with Euratom. Our industrial participants in nuclear energy had the support of our government in seeking contracts abroad for supplying nuclear power plants.

As a consultant to Westinghouse, and as chairman of the Board of Westinghouse International Atomic Power Company in Geneva, Switzerland, I participated in this effort. In France, an American entrepreneur, an agent for Westinghouse, Robert Schasseur, organized Framatom to facilitate a cooperative effort between Westinghouse and Belgian and French industries to build a nuclear plant for France and Belgium. As part of the Euratom program it was required that the purchase of any United States-designed nuclear plant would involve the transfer of U.S. industrial know-how to European industry, and in turn a royalty would be paid to Westinghouse under a licensing agreement. France had its own nuclear power program for utilizing gas-cooled nuclear reactors. The French government invited me to visit all their nuclear energy facilities. In fact, they also revealed to me from time to time their progress with military atomic weapons. I never knew the reason for this but assumed they wanted some informal channel to keep the United States informed.

Framatom over the years grew in size, competence, and importance, and the French industry and government gradually acquired ownership. Électricité de France, after acquiring experience with the first French-Belgian pressurized-water plant, utilizing Framatom and intially Westinghouse for design and manufacture, concentrated on building and standardizing pressurized-water plants. Their effort has been highly successful. Today France is placing major reliance on nuclear plants to produce electricity. They build coal plants only to the extent necessary to keep their small number of coal miners content. Coal-produced electricity is much more expensive than nuclear power but is cheaper than oil. Although France has effective control over safety requirements, it is not conducted as an adversarial contest among Électricité de France, Framatom, and the government. They are all striving for adequate safety. There are some antinukes and they may demonstrate, but they do not have the right to hold up construction or operation of plants by endless intervention tactics requiring endless hearings for resolution of issues, piece by piece, over and over again. Today France leads the world in percentage use of nuclear power, and Framatom now competes successfully with Westinghouse worldwide for design and construction of pressurized-water plants.

France also leads the world in development, design, construction, and operation of the plutonium, fast-breeder type of plant. The first electric power produced in the United States by a nuclear reactor was by a fast-breeder experimental plant designed by Walter Zinn. I participated in the Enrico Fermi fast-breeder experimental plant built by Detroit Edison and a group of utilities as part of the demonstration program. Walker Cisler, chairman of the board of Detroit Edison, organized the project and created the Power Reactor Development Company to own and operate the plant. I participated as a consultant to Detroit Edison, and later I became a member of the Detroit Edison Board of Directors. Unfortunately, a mean, personal type of political controversy handicapped government support of the project. After an accident at the plant, with no injury to anyone and in spite of the fact that the plant was repaired and put back in operation, our government, because of the political controversy, did not support continued operation. The Japanese offered full support, but Walker Cisler did not consider it appropriate to accept it.

The French also participated in the operation of the Enrico Fermi plant and, taking full advantage of our experience, including our finding the cause of the accident, built an improved version of the plant. They have continued an aggressive development program. I visited Cardarock, where they have constructed their experimental plants. I also visited the Phoenix 233-MWE fast-breeder demonstration plant at Gard when it was almost completed. Again I was impressed with the French approach to the problem of properly merging adequate consideration of safety measures with the practical problem of attaining a reliable economic plant. I have not visited the Creys-Malville 1,200-MWE commercial-size fast breeder at Isers. It is the largest fast breeder to date and began operation in 1986. France is well in the lead. Scientists and engineers in France continue to have more influence on government and public reaction than the news media and movie stars like Jane Fonda.

The United Kingdom at one time was well ahead of France in the development of the fast breeder. I went to Dounreay, Scotland, several times to observe British experimental work and the 250-MWE demonstration plant. They have been working on the design of a commercial-size plant but have not yet initiated construction. However, they are working with France and West Germany and will participate in a joint venture to build the next breeder in West Germany.

During the World Energy Conference in Moscow in 1968, I had asked to visit the Soviet 600 fast-breeder plant in the Sverdlovsk region. I had been assured that my request would be approved after I reached Moscow, and each day I was further assured that approval was imminent, but the reason for the delay was revealed to be an unanticipated event. Jackie and I attended the impressive reception held at the Hall of the Congresses in the Kremlin. About five thousand delegates from all over the Communist and the free world attended. It was quite a public-relations achievement for the Soviet Union to have succeeded finally in holding such a conference in Moscow. The reception was the highlight for demonstrating Soviet goodwill and hospitality. Following the reception, we attended a beautiful presentation of *Swan Lake*. Leaving the Kremlin, Jackie and I elected to walk back to our hotel and enjoy the moonlight. In a euphoric mood, Jackie commented, "Anyone who can put on such a hospitable reception and such an impressive presentation of *Swan Lake* can't be all bad." I almost was inclined to agree.

Early the next morning I was awakened by a call to join Walker Cisler at breakfast; there I learned the chilling news that the Soviet Union had invaded Czechoslovakia. I also learned that our request to visit the fast-breeder plant probably would not be approved, due to the crisis. I considered that the size of the invading force was far larger than necessary. This disturbed many of us. Consultation with our acting ambassador led to the decision to complete the conference, but to leave it up to the individual U.S. delegates to decide on whether to go on one of the postconference excursions that many had signed up for. My trip to the fast breeder canceled and my next appointment elsewhere in Europe being over a week away, I asked Jackie where we should go during the interval. She answered, "Why, Paris, of course." So I anticipated the rush and made reservations for a flight to Paris. The next day the Soviet Intourist office was swamped. The day following the end of the conference the airport was a mob scene of delegates swarming to get on any available flight or trying to find the flight on which they had reservations. Judging by the comments over-heard, the USSR had in one stroke lost most of the goodwill they had acquired by the conference, the reception, and the *Swan Lake* perform-ance at the Kremlin.[1]

Japan also has a great interest in the fast breeder. They, like France, have no large resources of fossil fuel or uranium. The fast breeder is the obvious answer for Japan's longer-range need for electric energy. I have visited Japan three times, the first in 1957. Initially the State Department disapproved of my going to Japan, as they had previously disapproved a proposed visit by General Groves; I was more persistent than he had been and the White House finally gave the word to the State Department to approve my trip. One of my objectives at that time was to open up the competition for an atomic power plant that Japan was planning to build. Someone (I suspect the British) had convinced many of the Japanese scientists and engineers who had visited both the United States and the United Kingdom that the American water reac-tors had a positive temperature coefficient. When I learned that that was the main reason for Japan's reluctance to deal with General Elec-tric and Westinghouse, it was relatively easy to convince the chairman of the Japanese Atomic Energy Commission that we would never build any reactor having a positive temperature coefficient. A positive tem-perature coefficient means that the power increases with temperature and thereby creates the possibility of a runaway reaction. After my meeting with him, he announced at a press conference that he was giving further consideration to U.S. plants. However, my success came too late; the chairman died shortly thereafter and negotiations had proceeded too far for a British gas-cooled plant, so they got the order. Succeeding plants have been water-cooled American-type plants.

During that 1957 trip, at Westinghouse's request, I also met with Mitsubishi, which later became a Westinghouse licensee for pressur-ized-water plants, and with other associated firms. I also addressed the Japanese Atomic Energy Forum. On each occasion I made certain that everyone understood that I had been district engineer of the Manhattan District and had played a key role in producing the Hiroshima and Nagasaki bombs. I did not want any misunderstanding˘about my pre-

1. The Soviet Union finally succeeded in operating its fast breeder in 1980.

vious role. No reference ever was made by anyone in Japan about the use of A-bombs against Japan, except for my interpreter, who at one of the luncheons called my attention to one individual, stating, "He received the burn scars on his face and hands at Hiroshima." On my return flight, I sat next to U.S. Ambassador to Japan, John Moore Allison and told him about my visit and that the State Department had opposed my visit on the basis that the Japanese might resent my presence in Japan. He commented, "They should have asked me for advice. The Japanese would consider that you had only done your duty as an Army officer and respect you for it."

I went to Japan again in 1960 and 1981 to become familiar with their program. In 1981 they emphasized their interest in developing a fast breeder and reprocessing of spent fuel and pointed out to me the obstacles that the United States was creating to slow down their progress. Japan now has twenty-nine units, mostly pressurized water, of 19,000 MWE capacity on line. Their goal is to produce more than 50 percent of their electricity by nuclear power.

En route to Japan, in 1957, I went to India as the U.S. representative to attend the inauguration of the Indian Atomic Energy Establishment and dedication of the Indian research reactor. I knew Homi Bhabha, the top nuclear scientist in India, and was pleased to participate in the event. In addition to the main event, India invited the forty international guests to participate in a grand tour of India to see the thorium plant at Cochin in South India and other major attractions in India. Due partly to Homi Bhabha's influence, I was considered to be the top-ranking delegate. Usually about seven or eight of the delegates would be lodged in the governor's compound as a special honor. I always was included, but for each occasion, different delegates from the group were selected. Initially, Drs. D. V. Skobeltzyn and V. S. Emilianov, the Soviet delegates, refused to be a guest of the governor unless both were invited. But after being chided by other delegates who also had not been invited, they finally decided that one, Skobeltzyn, should accept. François de Rose from the French foreign office also was invited for the second time on that occasion and remarked to me that this would be the first time that he had ever had the opportunity to discuss issues with a lone Russian, since they always went everywhere in pairs. The three of us, who had all been together working on international control of atomic energy at the United Nations ten years earlier, had a very interesting evening together. It was a joy to discover that Skobeltzyn had a very affable and friendly disposition when not under the scrutiny of another Russian. We all enjoyed the opportunity to discuss divisive nuclear issues in a very frank manner, but the evening did not lead to any novel way to resolve our differences.

During the visit, I also met Prime Minister Jawaharlal Nehru of India at the home of Dr. Bhabha's mother. That gave me the opportunity to suggest that when India decided to build an atomic power plant, they should consider the United States as well as the United Kingdom as a supplier.

Upon my return to the United States, I learned that General Electric had protested my appointment as the U.S. delegate for the Indian inauguration. Undoubtedly it gave Westinghouse an advantage, but as it eventually turned out, General Electric was the main beneficiary. In 1960, when India decided to build an atomic power plant, Homi Bhabha invited me to come to India to discuss the matter. I first went to Bombay to meet with Bhabha and a few of his staff and some political dignitaries. I had little trouble convincing Bhabha that water reactors had many technical and economic advantages compared to the gas-cooled British reactor. Bhabha asked me to meet with Nehru and explain the advantages of opening competition to the U.S. suppliers. My meeting with Nehru was one of the most extraordinary hours of my life, even though I did most of the talking. Bhabha told me that Nehru was very busy and that he had had difficulty getting a fifteen-minute appointment. Nehru remained at his desk throughout our meeting. As I talked to him, he looked away from me and over his right shoulder, I thought absorbed with some major government problem. It was rather distracting. At the end of fifteen minutes, I stopped talking. No reaction from Nehru; he continued to look over his right shoulder. Bhabha motioned to continue, which I did, and at the end of another fifteen minutes, the same thing was repeated. At the end of the third period, I thought my effort was becoming rather futile, but Nehru slowly turned his head, facing me, and started to ask a series of questions that surprised me. He not only had been listening but also was well informed about atomic possibilities. He concluded by stating that Homi and I seemed to agree on most matters, and he told Bhabha to open up the competition to include U.S. reactors. He then surprised me by asking what I thought about the capabilities of the Indian research group under Bhabha. I told him that in my short acquaintance with them on my two visits, I had been most favorably impressed and also added that Bhabha was considered to be among the world's most capable scientific leaders. Nehru then shifted to Bhabha and asked, "Can you develop an atomic bomb?" Bhabha assured him he could and in reply to Nehru's next question about time, he estimated that he would need about a year to do it. I was really astounded to be hearing these questions from the one I thought to be one of the world's most peace-loving leaders. He then asked me if I agreed with Bhabha, and I replied that I knew of no reason why Bhabha could not do it. He had men who were as qualified or more qualified than our young scientists were fifteen years earlier. He concluded by saying to Bhabha, "Well, don't do it until I tell you to."

On the first round of bidding, G.E. had the lowest price. I met with Bhabha in Paris, and he accepted my proposal to continue negotiations with both suppliers. Again G.E. was low. I had an appointment to meet with Bhabha again in Paris, but he never got there—his plane crashed into a mountain in Switzerland and he was killed. G.E. was awarded the contract.

Returning to the subject of the fast breeder, I must say that the U.S. program has faltered badly and at present very little is being done. It is unfortunate because the fast breeder offers the best prospects for a long-term source of energy. Already there is aboveground depleted uranium from the gaseous diffusion plants that in terms of fuel for generating electricity in a fast breeder is greater than all of our coal reserves. I believe it still is being carried on the government books as of zero value. The comparative cost of mining and transporting our coal reserves would be in the trillions of dollars.

One of the principal arguments against the fast breeder is that it utilizes plutonium as the core material to sustain the chain reaction necessary to generate power, while within a blanket of depleted uranium, the U-238 is converted to plutonium in quantity greater than the plutonium burned in the core (that is why it is called a breeder). This overall reaction assures a continued supply of plutonium for continued operations. The only new supply of material necessary is the depleted uranium that we have stored in vast quantities at the Oak Ridge, Paducah, and Portsmouth plants. It has no other important use except in a fast breeder. Should we waste this valuable resource of fuel that is already mined, in a usable form, government-owned, and completely paid for?

The fear of plutonium, created by the antinuclear opposition and aided by the media, has politically stymied our fast-breeder program.

The opponents worry about more plutonium being generated, and possibly being stolen by terrorists or diverted by other nations for use in weapons. They refuse to recognize that once plutonium is in a reactor, it is not easy to steal by anyone; it is radioactive and requires time and facilities to recover it. Although the breeder generates more plutonium than it burns, the plutonium produced becomes a source of fuel for the same plant as well as for additional plants to meet continued growth in requirements for electricity. Our present water reactors also produce plutonium, but the plutonium is not too good for weapons. Except for USSR, most countries produce plutonium for weapons in specially designed reactors. The USSR uses dual-purpose reactors, producing both plutonium and power, such as the Chernobyl MKBR. We have just one dual-purpose reactor, at Hanford. Plutonium is reasonably safe to handle, and methods and procedures already are developed for safeguarding it. The fact that plutonium is good bomb material stymies rational thinking in the state of emotion that exists in the United States. Not only the fast breeder can use plutonium; France now is planning to recycle plutonium in its pressurized-water plants pending the buildup of a fast-breeder demand for plutonium. I can think of no better way to dispose of plutonium than to burn it in a reactor and generate electricity.

As previously discussed, France and the USSR are proceeding with the development of the fast breeder for production of electricity. Japan is moving at a slow rate. West Germany and the United Kingdom are cooperating with France, and the next fast breeder will be built in West Germany. The United States tries to discourage others from developing reprocessing of spent fuel and the breeder. Instead of this we should resume our effort to develop fast-breeder plants that eventually will be economically viable. As a leader in the field, we would have more influence on establishing methods to control it and safeguard the plutonium. Overall, fast breeders are far less likely to incite war than a continued struggle among nations to control essential supplies of oil. The problem is how we should make rational judgments on issues of this kind that are charged with emotion to the point where all facts are not properly considered. I often tease my wife about her claim that the French are more rational than anyone else. Maybe she is right; France seems to have resolved this issue better than we have and is pursuing a very rational course to develop the fast breeder. Yet we also think of the French as being very emotional; maybe they have learned to separate, or possibly combine, their two outstanding traits to good advantage.

The political aspect of our antiproliferation policy was illustrated at the start of the Argentine nuclear program. In February 1968 I visited Buenos Aires at the request of Westinghouse International. The Argentine government was planning to build a nuclear plant, and it appeared that an award would be made to a West German firm for a heavy-water nuclear plant. Westinghouse felt that my military background would assist in gaining an audience with the admirals and generals then running the Argentine government. It did. I had no difficulty meeting with all the key cabinet officers. They were all most friendly and courteous to a military associate. Most of them informally agreed with my claim that a pressurized-water plant was most economic for generating electric power. But they all asked the same question: "If we order a U.S. plant and we are supplied enriched uranium for fuel by the United States, will the United States insist on control of what we do with the plutonium generated?" My answer, in accord with our antiproliferation law, was, of course, "Yes." Then they indicated that they were going to accept the West German proposal for a heavy-water plant fueled with natural uranium from their own sources, and Argentina would retain control over the plutonium generated. The West German proposal also was supported by a favorable financial agreement. Proliferation is a political and diplomatic problem and cannot be solved by ineffective trade restrictions. In particular it cannot be solved by our refraining from going ahead with reprocessing of spent fuel and the fast breeder when other countries are pushing ahead to meet their own power needs and to gain an advantage in the future international market for fast breeders.

In the United States, early success created great optimism—perhaps too much optimism—concerning the prospects of nuclear power during the 1960s and the first half of the 1970s. Also, like the interservice rivalry to attain dominance in the guided-missiles field, there was a race in specifications by the designers. To counter the disadvantages of the increased capital costs of nuclear energy, it soon became apparent that if the size of the plant is increased, the unit cost of electricity is lowered. So if Westinghouse offered a larger plant at a lower kilowatt-hour cost, General Electric, Combustion Engineering, and Babcock and Wilcox offered one of the same size or larger, with greater savings. No one correctly interpreted the added complexity of these larger plants, which would add engineering, construction, and licensing delays. It meant bigger turbines and generators, more crowded and larger piping concentrated in the small volume of the containment structures, and more expensive and harder-to-build containment structures. Adequate quality control was more difficult to attain in the larger plants. With the benefit of hindsight, it is easy to see that excessive optimism caused the industry to expand the size too fast. We should have built more Connecticut Yankees. It was a successful plant but was too small to compete with the theoretical advantages of bigness. Now it is competitive, whereas some of the larger plants are not. France followed a wiser course of building more duplicates of smaller plants before embarking on the larger ones.

To make the matter worse, more and more of our smaller utilities wanted to get into the act. The larger plants were too large for their systems, and some of the utilities lacked the talent to supervise the design and construction of the plants and to organize and train personnel for operating them. There was a shortage of adequate personnel. Accordingly the good architect-engineers and construction firms were stretched too thin to perform well on all their contracts. The industry just grew too fast for its own good.

The Nuclear Regulatory Commission also suffered from the tremendous expansion of the industry. There was a shortage of competent personnel, overregulation of details, and continual changes in regulations that required extensive and costly backfitting. The NRC assumed complete responsibility for details of safety, and the utilities relied too much on the NRC. The conventional wisdom was that if the NRC didn't object, it must be safe enough. Safety should be more of a cooperative effort; basically the primary responsibility should lie with the utilities to see that the reactor manufacturers develop safe systems with the NRC checking to see that they are adequate.

Overregulation and frequent changes in regulations made it virtually impossible to standardize plant design and build duplicate plants. Also regarding standardization of plants, which everyone recognized as desirable, each utility had its own ideas about plant design, and particularly they wanted a custom design for the control room. Everyone acknowledged that standardization was the answer to many of the industry's problems, but it seemed that everyone also created obstacles. Even at Detroit Edison, we found it to be impossible for General Electric to furnish two identical plants—there were too many changes in regulations. We finally canceled the second plant. France has set a good example of the advantages of more standardization.

The energy crisis of 1973 compounded the problem. For several decades, the electric requirements doubled every ten years. Constant expansion of generating capacity was necessary to supply growing requirements and to replace obsolete equipment. The oil crisis in 1973 inflated the price of oil, followed by coal, to such an extent that it paid to conserve electricity. I remember very well when a member of our Detroit Edison Board who also was a General Motors senior executive announced that General Motors had initiated a one-year conservation program that would reduce the kilowatt-hour requirement per car produced by 15 percent. He added that greater conservation might follow in succeeding years. It shocked all of us; but even worse, we soon found out that many other consumers had the same idea. The result was that our plans to double our capacity within the next ten years, which was required by previous load-growth predictions, were in error. With considerable remorse, I headed a review committee and recommended canceling all except one of our nuclear power plants under construction or in the design stage. Practically all utilities followed a somewhat similar pattern—to cancel plants and stretch out the construction program for others. In addition to a drop in rate of load growth, inflation, high interest rates, and financing problems forced utilities to stretch out their programs due to lack of funds.

When I first went on the Board of Detroit Edison, Walker Cisler announced at each annual stockholders meeting that as a result of increased efficiency of new plants, we could increase dividends, cut electricity rates, and increase the wages of employees. That era is past. During the inflation era, longer construction periods combined with higher interest rates, changes in regulations, and backfitting caused major increases in the capital cost of the plant, so that even though the plant is more efficient from a technical point of view, a rate increase is necessary. Because of drop in load or low grow rate and the delay or failure to get state regulatory commission approval of increased rates, less profit is available for dividends. Many utilities were having engineering success but were approaching financial disaster. It has been a tough era for utilities and an equally tough era for nuclear equipment suppliers with no new orders for plants.

The oil crisis, increased inflation, energy conservation, and high interest rates all have caused this severe crisis for utilities and the nuclear energy industry, but the worst was yet to come.

On March 28, 1979, at 4:00 A.M., at Three Mile Island, Middletown, Pa., a series of events began at Unit 2 that captivated the attention of the world's TV audience for weeks. A feed-water valve that had inadvertently been left closed by a maintenance crew caused a turbine trip; this in turn tripped an automatic steam relief valve, and the reactor automatically shut down. All safety equipment operated as it should have. However, after the shutdown, an automatic relief valve failed to close, permitting reactor coolant water to be discharged into a drain tank and finally onto the floor of the reactor containment building. The operators did not correctly diagnose that the relief valve still was open. An indicating light showed that power had been supplied to close it, but there was no signal to indicate whether it had actually closed. This was an error in design and a partial excuse for the operators' errors that followed. There were other instrument readings available whereby they could and should have made the right diagnosis of what was happening, and if they had done so anytime within the next hundred minutes, they could have turned another valve and all the subsequent trouble and damage to the plant would have been avoided. All the safety system provisions automatically activated successively. The automatic injection system started injecting water into the coolant system to keep the reactor from overheating. For more than an hour, the pumps continued to cool the system, even though cooling water continued to flow out of the open safety valve. The operating crew still failed to diagnose what was happening and, due to the low pressure and high vibration of the pumps, decided to turn off the pumps. This was a serious operator error. As a result, the reactor was not sufficiently cooled and started to heat up. Up to this time, no damage had been done to any part of the system, and it could have continued to run safely for a considerable time thereafter. However, for the next forty-five minutes the reactor overheated, and even though the relief valve was corrected by closing another valve (which the operators should have done earlier), the reactor heat recorder remained off scale. This is the period when the damage was done to the reactor core.

Initially the coolant water draining into the reactor building was low in radioactivity, but after the reactor had overheated, the radioactivity of the water started to climb; as a result, at 6:55 A.M. a "site emergency" was declared and communicated to civil authorities. Some of the radioactive water leaked into an auxiliary building. A "general emergency" was declared at 7:24 A.M. From 9:15 A.M. to 10:15 A.M., radioactivity levels of 3 to 9 millirads per hour were measured on the site outside the building, an indication that there was some leakage from the equipment building but still well within tolerance levels for employees for a considerable period of time. This low level of radiation was not a danger to the general public.

Sixteen hours after the first incident, forced circulation through the reactor had been established and reasonable pressures and temperature of the cooling were being maintained. The immediate emergency was over. There was serious damage to the reactor core that would take months to evaluate, but there had been relatively little radioactivity outside the containment, and certainly nothing approaching the disaster to the public feared by the antinukes had occurred. Most experts have concluded that the radioactivity released will have negligible overall health effects, but the public is skeptical, and their skepticism is fanned by some scientists and the media.

The accident destroyed public confidence in nuclear power, the nuclear industry, safety of nuclear reactors, utility management, and the competence of the Nuclear Regulatory Commission.

The TV media and the press covered every aspect of the event. Although they may have attempted to give a balanced view, many visual scenes used in the TV reports bordered on the irresponsible and the sensational.

I was particularly annoyed with the repeated reference to the danger of the hydrogen bubble in the reactor vessel exploding and rupturing the vessel. Some important individuals in the NRC added to the confusion about this, so the media are not alone to blame. I thought we had eliminated the possibility of such an accident long ago. I called several of my more expert friends and verified my understanding that hydrogen is maintained in the reactor vessel to ensure that there always is some excess hydrogen to suppress the formation of any free oxygen in the reactor vessel caused by radiolysis. This avoids any possibility of explosion. With no oxygen there can be no hydrogen explosion, even though there is a hydrogen bubble. I believe that many still believe there was great danger even though the impossibility of it has been explained frequently. The media made a tremendous case for building fear about the hydrogen bubble, but when the impossibility of an explosion within the reactor vessel was accepted, the issue was dropped—but the correct story got little circulation. Apparently good news is not news good for TV.

There was, however, the possibility of hydrogen combining with oxygen in the air outside the reactor vessel but within the containment structure either by combustion or by a low-grade explosion. The containment setup at Three Mile Island's Unit 2, however, was designed to withstand such an incident, and there is good evidence that a low-grade explosion or combustion did actually occur within the containment structure without damaging effects. Apparently this distinction about where hydrogen might combine with oxygen is too complex for the media.

Shortly after the incident, I received a call from Chauncy Starr, chairman of the Electric Power Research Institute (EPRI) and agreed to serve on a committee to investigate the accident at Three Mile Island and make recommendations to the Electric Power Research Institute. The president also appointed a commission chaired by John G. Kemeny to study and investigate all aspects of the accident and to make appropriate recommendations based upon its findings. The NRC instituted a special inquiry to review and report on the accident. And Congress also held hearings on the accident.

The accident was thoroughly investigated and results reported to the public. The accident was the most serious of any that ever occurred in the United States and revealed many weaknesses that needed to be corrected in the overall utility and nuclear industry and in governmental organizations responsible for nuclear development and nuclear safety.

All the findings and recommendations after these various investigations are in the public record, and my remarks will be confined to the items that impressed me as most important and the resultant actions taken by the utility industry to strengthen safety.

The accident never should have occurred. The sequence is important enough to repeat. The mechanical failure that contributed most to the accident was a stuck pressure-operated relief value. It opened properly as a result of overpressure in the cooling system. It should have closed a few seconds after the pressure was relieved, but it did not. The operators assumed it had closed because a light had gone off indicating the electric power that had opened the valve had gone off. If they had known the valve was stuck in the open position, they could have closed a backup valve; the release of coolant water would have been stopped immediately, and there would have been no damage to the core. The incident would have been over. The fact that they failed to diagnose that the valve was open is partly the blame of the operators because there were some inconsistent effects of actions they then took that should have alerted them to the fact that the valve was open instead of closed. More important, however, is the fact that a similar incident had occurred about a year before in the same type of plant. In this incident, the problem was diagnosed and resolved before it became serious. If the manufacturer, the utility, and particularly the NRC had all reacted properly, this first incident would have been promptly reported to all operators of similar plants, along with a correction of incorrect operating and training procedures that the operators at Three Mile Island were following. Recognizing that this was primarily a failure to disseminate information promptly about incidents that could have serious consequences, the EPRI recommended to the owner utilities that they not rely on the NRC, but that the EPRI set up a computerized information and analysis system that would record all incidents; that for all incidents having the potential for serious consequences, an analysis be made quickly to determine what corrective actions both to equipment and to operator procedure would be necessary; and that this information be disseminated to all utilities having similar plants. The system also can be used whenever something out of the ordinary occurs; a call to the computer center can get a prompt answer about any similar occurrence and what can be done about it. Also, a group of experts can immediately be made available to assist in analyzing any new problem. In establishing this information system, the utilities were properly taking on more responsibility for maintaining safety. This is as it should be; too much reliance had been placed on the NRC for this function, and their failure had contributed to the accident.

In addition to the fact that just closing a backup valve would have prevented serious damage to the plant, analysis of the incident shows that if the operators had simply left the plant's high-pressure injection pumps run, there would have been no serious problem for several hours, but concerned with what they thought was too much water, they turned them off. The plant automatic safety system was doing a good job and was thwarted by this operator error. As a result, the EPRI recommended that in case of an emergency, the automatic safety provisions be allowed to operate without interference and that no action be taken until the problem is analyzed properly. To facilitate such an analysis, the NRC now has prescribed that a more highly qualified individual be available at all times the plant is operating, and in addition EPRI has organized to provide more expert assistance whenever a serious incident occurs.

As a result of the operator errors, the EPRI recommended that the utilities set up an industry organization to train operators and to inspect and examine periodically the various plant operations, and report deficiencies in training and operation, and recommend necessary corrective action. This program, tied in with insurance as a means of enforcing recommendations, has been and should continue to be very effective.

Concerning the small amount of radioactivity released from the plant, the Kemeny Report states, "The radiation doses were so low that we conclude that the overall health effects will be minimal. There will either be no case of cancer or the number of cases will be so small that it never will be possible to detect them." This, of course, we were all happy to hear, but even of more importance to the nuclear industry was the finding that the intensity of radioactivity of what was released was lower than previous calculations indicated it should be. I remember very vividly the excitement around the table at one of our meetings of the EPRI Safety Committee when Dr. Norman C. Rasmussen, a leading safety expert, first brought up the question of why the small amount of the release at Three Mile Island was so low in radioactive iodine. That was in 1979. He stated then that if research supported his hunch, there would be a drastic revision downward in our emergency planning requirements for protecting the public against severe accidents. That was the single best news we had about the sequences and the consequences of the accident, but we realized it could be extremely important.

In the past, the possible release of radionuclide of cesium and iodine following any severe accident was considered to be the greatest threat to public health and safety. The 1975 nuclear safety report *WASH-1400* (the Rasmussen Report) postulates that approximately 70 to 90 percent of the radionuclides in the core might escape the site and endanger the public. With the finding that the accident at Three Mile Island released essentially no cesium or iodine, source-term research on *Wash-1400* was initiated.[2]

As a result of independent source-term research by several independent organizations, these organizations agree that the reduction factor for light-water reactors is on the order of one hundred or even one thousand, but they recommend that a conservative reduction factor of ten be used in source-term calculations. If this recommendation is accepted by the NRC, it would call for a reduction in the current requirements for an emergency evacuation zone with a ten-mile radius.

---

2. "Source term" means the amount of radioactivity that would actually escape following any accident.

To date, the NRC has not yet fully determined what to do about these research findings, but in the case of the Indian Point station north of New York City, where interveners attempted to shut down the station because of its proximity to heavily populated areas, the NRC voted not to shut down the plant.[3]

The NRC is considering other specific cases where these research findings may appropriately reduce the consequences of a credible accident. Much more study and research may be necessary before the NRC will be able to formulate new rules and regulations for general application to light-water reactors. In time, the Three Mile Island accident may be given credit by the NRC for proving that the greatest threat to the public of any credible accident has been greatly overestimated. However, more time and a clear, understandable presentation of facts will be required for the frightened public to accept it.

Statistical analyses by expert statisticians in various countries of the comparative risks of using light-water reactors (LWR) versus coal, oil, gas, or hydraulic power for producing electricity consistently favor LWRs if all risks or hazards such as mining, transportation, construction, operation of the plants, possible plant accidents, normal and accidental discharges from the plant, and disposal of waste products are considered. Statistically, coal is by far the most dangerous method of producing electricity. Mining and transporting the vast quantities of coal required are well known as being hazardous, and many individuals are injured or killed; fumes from burning coal combined with certain atmospheric conditions have killed hundreds on several occasions; acid rain and the greenhouse effect of fossil fuels are just beginning to be recognized as potential hazards to life and our environment. Even considering the most improbable nuclear accidents combined with the worst atmospheric conditions for nuclear energy, the vast majority of scientific and engineering societies and other unprejudiced experts have found the hazards of coal as compared to nuclear power more dangerous by factors of ten to a hundred times. They can't all be wrong, yet our public has an emotional fear of nuclear plants and radiation that makes them suspect all such calculations and statistics. We have learned to live with black lung, automobile and airplane accidents, fires, and natural disasters, but we want assurance that our nuclear plants are absolutely safe and have zero risk, and we demand that additional redundant safety factors be added

3. *Nuclear News* (La Grange Park, Ill.: American Nuclear Society), May 1985.

whenever a new safety idea is suggested. I hesitate to guess what the selling price of gasoline would be today if our governments, federal and local, suddenly decided to apply to petroleum comparative safety rules to refining, storage, transportation, distribution, and use that now are prescribed for all phases of the nuclear industry. Practically everything we do in life involves some risk. In some cases the public has demonstrated that it can assess risks versus benefits and decide that the benefits outweigh the risks. Driving an automobile or flying in an airplane are good examples. However in the case of environmental protection, where the government is making the assessment of risk versus benefits and where cost also is a factor, we seem to have great political difficulties establishing national regulations satisfactory to those who want more protection as well as to those who have to pay the bill. It is a matter of degree, and not whether or not we want protection of the environment. In the case of nuclear energy, the problem is even more complicated because of the great fear and the lack of understanding of the public of radiation hazards. Statistics concerning the risks of nuclear accidents and the effects of radiation mean nothing to too many people. They just don't want any nuclear power or any increase of radiation. As a result, we have failed to convince the public on "How safe is safe enough?" However, we soon had an example of "not safe enough."

On April 28, 1986, radioactivity three to fourteen times the normal background level was detected in various locations in Sweden. This was recognized as a warning that a nuclear accident had occurred, probably somewhere in the USSR. Information from the Soviet Union was very sketchy. In the absence of authentic reports, our news media reported any information they could get from any source. The antinukes had "sources": two thousand killed, mass burials, radiation five hundred times normal, etc. On April 29 the official Soviet news agency Tass reported that two persons were killed in an accident at Unit 4 at the Chernobyl nuclear reactor in the Ukraine. A special government commission investigating the causes of the accident is now at the site. The accident resulted in destruction of parts of the structural elements of the building housing the reactor, resulting in damage and leakage of radioactive substances. The radiation situation at the station and the adjacent area now is "stabilized and necessary medical aid is being rendered to the injured." Local residents at the settlement serving the reactor and from nearby populated localities have been evacuated.

Radiation levels are being monitored. This official Soviet statement contradicts news reports of more than two thousand killed in the accident. A description of the reactor was included in the Soviet statement.

The USSR was reluctant to issue sufficient information, so the media continued to rely on interviewees who interpreted the facts or rumors that became available: satellite photos, an intercepted ham operator radio message, etc. Most of the news concerned the radioactive plume that circled the earth and the effects of fallout on farm products, milk, and water supplies. Some of these reports were accurate and some greatly exaggerated, and some incorrectly explained. The satellite photos raised the question of whether one or two reactors were involved and whether the graphite fire was under control.

In early May the USSR released enough information to indicate that eighteen of the more than 197 hospitalized, mainly operators of the plant and fire fighters, had died and that some more still were seriously ill. They also gave more details on when and how many had been evacuated from the vicinity of the plant. From press conferences it could be inferred that at the time of the accident the plant was partially shut down; there was a blockage of coolant water, which resulted in the melting of pressure tubes and cladding, which led to generation of hydrogen, which led to an explosion, which led to a fire in the graphite. Satellite photos showed that the roof over the reactor was gone. Pre-accident photos showed that the roof was not designed to take any significant overpressure. There was no Western-style containment.

The antinukes called for revoking all construction permits for commercial nuclear plants that did not yet have a full operating license. The pronukes pointed out the differences between our plants and the Chernobyl plant.

For almost four months there were too few facts to appraise the cause and consequences of the accident, or to determine what lessons could be learned from it. Meanwhile, we had learned a great deal from the Three Mile Island accident. More than six thousand changes in reactor hardware, operating procedures, and plant management were made after the accident to make our plants even safer. Certainly there was much to be learned from Chernobyl, which apparently was a much more serious accident than Three Mile Island. The USSR's failure to release more information as soon as it was confirmed was inexcusable, but not unusual behavior for the Russians. While they accused our

media of being irresponsible for releasing too much false and uncon-
firmed information, they continued to err the other way.

The issue was resolved as a result of the USSR acceptance of the
International Atomic Energy Agency's offer to review the accident. In
preparation for a meeting scheduled to be held in Vienna August
25–28, the USSR prepared a working paper containing a description of
the plant, the cause of the accident, the consequences of the accident,
and measures they were planning to take to make their plants safer.
The working paper provided a frank statement concerning these subjects and also provided information necessary to compare the safety of
the Soviet RBMK-type plant with the LWR-type plant. A good comparison is of major importance to the United States and much of the
Free World, which is relying so heavily on the LWR.

The following comparison of design difference is based on information issued by the U.S. Atomic Industrial Forum: A Presentation by
John J. Taylor, V.P. Nuclear Power Division, EPRI, to the U.S.
Nuclear Regulatory Commission, on the State of the Nuclear Industry,
on May 15, 1986; The Working Document for Chernobyl Post Accident Review Meeting submitted to the International Atomic Energy
Agency, entitled "The Accident at the Chernobyl Atomic Eneregy
Station and its Consequences"; and my own personal knowledge and
experience.

My summary of fundamental design differences will concentrate on
three areas: reactor differences, safety system differences, and containment differences. My comparison will be the U.S. light water
reactor.

The Chernobyl reactors are light-water cooled, graphite moderated,
pressure tube reactors. These 1000 MWe plants are called "RBMK-
1000's" by the Soviets.

The first obvious difference between Soviet RBMKs and U.S.
LWRs is the use of graphite as a neutron moderator. The RBMK
design evolved in the Soviet program from earlier weapons production
reactors that used graphite and low enrichment uranium. The RBMK
is a dual-purpose reactor, producing plutonium for weapons as well as
electricity. Our one dual-purpose reactor at Hanford is also graphite
moderated but of a different design than the RBMK. For power application U.S. designers selected light-water moderation over graphite
moderation for a number of reasons. First, the obvious safety problems
associated with a potentially flammable material (graphite) inside the

reactor core are eliminated. Second, the unique physical and chemical problems of graphite deformation under irradiation, including the Wigner effects, are avoided. The Soviet design avoided the Wigner effect by a design that provided for a maximum temperature in the graphite of 700 degrees centigrade, a high temperature that is questionable. Third, graphite reactors are huge and very complex when compared to U.S. LWRs. One thousand six hundred and sixty-one individual pressure tubes are embedded in the 1,500 tons of graphite, each with its individual inlet and outlet piping. The large size of the core requires a very complex control system consisting of 211 control rods. Fourth, the few advantages of graphite moderation, very low enrichment uranium, and on-line refueling were not attractive to U.S. utilities, who had little interest in providing plutonium for the U.S. weapons program.

The next major difference between U.S. and Soviet reactor designs is in the defense-in-depth barriers provided to ensure that nuclear fuel and fission products cannot escape the core. Both the LWRs and the RBMK reactors use uranium oxide fuel pellets surrounded by zirconium or zirconium-niobium alloy cladding. These fuel elements are similar, but the next barrier of defense against release is radically different. The Soviets use 1,661 individual pressure tubes, 7 meters (23 feet) long, and 88 mm (3½ inches) in diameter to contain the fuel elements and light-water coolant flowing past the fuel elements. The pressure tube walls are 4 mm thick (.16 inches); whereas the pressure vessel walls on U.S. LWRs are about 6½ to 8½ inches thick, or about fifty times thicker. This means that damaged fuel elements on a Soviet RBMK reactor have a much greater chance of penetrating this second barrier of defense (the pressure tubes) than on a U.S. LWR (a single large-pressure vessel).

These first two Soviet differences, use of graphite as the moderator and thin-walled pressure tubes, combine to form a third distinction, the possibility of creating dangerous hot-graphite/hot-steam reactions caused by a breach in the pressure tube wall.

Although both the RBMK and the LWR designs use light water as coolant, the different reactor configurations create some unique cooling problems for the Soviets. First, with a relatively small water inventory in the reactors, little time is available to respond to changes in the coolant inventory. Moreover, at low power levels, below 20 percent of full power, the RBMK reactor's positive void reactivity coefficient results in a net positive power coefficient; a dangerous condition that makes the reactor very difficult to control. In recognition of this risk the Soviet prescribed operating procedures restricting operation below 700 MW(th).

Although the Soviet ECCS (emergency core cooling system) on modern RBMK design appears to be similar to Western approaches, it appears to lack the degree of redundancy and diversity we have provided in our modern LWRs.

The final defense in depth for the LWR is a full rugged containment vessel about the entire reactor, cooling system, pumps, and control rods designed to contain the maximum credible accident. The USSR term for containment is accident localization systems. The USSR containment consists of several leak-proof compartments, with different design overpressures to contain the reactor space, the downcomer pipes and main circulation pump pressure headers, and the distribution group header and lower communication lines. There was also a pressure suppression pool below the reactor. Compact walls gave radiation protection and containment horizontally about the reactor. Below the reactor there was a steel plate penetrated by 1,661 cooling channels, circulation pipes, and some control rods. Above the reactor there was a 1,000-ton steel reactor cover penetrated by almost 2,000 cooling channels, control rods, circulation pipes, and so forth. This cover plate and the 2,000 penetrations turned out to be a weak point in the reactor containment.

The containment for the RBMK is not the equivalent of the full containment required by the U.S. NRC for our LWRs. In fact the Associated Press stated that: "Ivan Yemelyanow, Deputy Director of the RBMK design organization, 'confirmed that the reactor did not have a conventional containment vessel used in the West to prevent radiation leaks in case of a breakdown.' " Moreover, due to the physical size of the RBMK reactor it is very doubtful that full containment is practicable.

Considering the deficiencies in the RBMK design, I am certain that it would not be licensed by our NRC for construction and operation within the U.S. There are just too many risks of accidents and dangerous release of radioactivity. As one of our safety experts who attended the meeting in Vienna commented to me: "Their design has risk in depth rather than defense in depth."

In this regard it is interesting to note that the British reviewed the RBMK design in 1975 and in a March 1976 report noted similar deficiencies. These included: "The lack of full containment"; "Insufficient protection against the failure of a pressure tube''; ''the reactor has a positive void coefficient,'' ''the high temperature of the graphite core (700 degrees centigrade)—were considered excessively high.'' Accompanying this report, Lord Marshall of Goring, Chairman of the U.K. Central Electricity Board, included in his letter to our Atomic Industrial Forum the comment: ''Broadly speaking, the designers of that day decided that we could learn little from the Russians about pressure tube reactors because their safety thinking was so different from our own.''

IAEA Review of the Chernobyl Accident

The International Atomic Energy Agency (IAEA) was a logical choice to review the Chernobyl accident. On May 6 the USSR expressed its willingness to provide the agency with information. The Working Document for Chernobyl Post Accident Review Meeting held in Vienna August 25–29, 1986, under the chairmanship of the Swiss nuclear scientist R. Rometsch, a former director of the IAEA, and the additional information supplied at the meeting by the USSR and the discussion of the many experts from the Agency Member States and from other international organizations is the basis for a summary report prepared by the International Nuclear Safety Advisory Group (INSAG). The information in the INSAG report is adequate to draw many conclusions about the cause of the accident, corrective action that needs to be taken, and the consequences of an ''accident that is almost a 'worst' in terms of risk of nuclear energy.'' (In the following discussion all quotes are from the INSAG summary report unless otherwise noted. I have used quotes extensively to avoid misinterpretation of the meaning of this international group of experts.)

Description of the Accident

''The accident took place during a test being carried out on a turbogenerator at the time of a normal scheduled shutdown of the reactor. This was meant to test the ability of a turbogenerator, during station blackout, to supply electrical energy for a short period until the standby diesel generators could supply emergency power. . . . Improper written test procedures from the safety point of view and strong violations of basic operating rules put the reactor at low power (200 MW[th]) in coolant flow and operating conditions which could not be stabilized by manual control. Taking into account the particular design characteristics [the positive power coefficient at the lower power levels] the reactor was being operated in an unsafe regime. At the same time, the operators deliberately and in violation of rules withdrew most control and safety rods from the core and switched off some important safety systems. . . .

"The subsequent events led to . . . increasing amount of steam voids . . . introducing positive reactivity . . . increasingly rapid growth in power . . . a manual attempt to stop the chain reaction . . . the automatic trip . . . having been blocked. But the possibility of a rapid shutdown of the reactor was limited as almost all the control rods had been withdrawn completely from the core. . . .

"The continuous reactivity addition by void formation led to a super-prompt critical excursion. The Soviet experts calculated that the first power peak reached 100 times the nominal power within four seconds. . . .

"Energy released in the fuel . . . ruptured part of the fuel into minute pieces. Small hot fuel particles (possibly also evaporated fuel) caused a steam explosion. . . .

"The Energy release shifted the 1000-ton reactor cover plate and resulted in all cooling channels on both sides of the reactor cover being cut. After two or three seconds a second explosion was heard and hot pieces of the reactor core ejected from the destroyed reactor building. It is still uncertain what role hydrogen may have played in this explosion. The destruction of the reactor permitted the ingress of air which led subsequently to burning of graphite. . . .

"Fires began, especially in the hall of unit 4, on the roof of unit 3 and on the roof of the machine room housing the turbogenerators of two reactors. Fire units . . . responded promptly and heroically fought the fire. . . . The fire was finally extinguished at 05:00 on 26 April, about 3.5 hours after it started."

The Cause of the Accident

INSAG states: "The accident was caused by a remarkable range of human errors and violation of operating rules in combination with specific reactor features which compounded and amplified the effects of the errors and led to the reactivity excursion.''

The operator errors were more serious and deliberate than was the case at Three Mile Island. The deliberate violation of operating rules indicates that the Soviets, in addition to also taking the many actions we have to improve training and to check on the competence of operators, need to create and maintain a ''nuclear safety culture.''

Of even more importance are the deficiencies in the RBMK design. The Soviets are taking short-term steps to prevent the complete removal of control rods from the core, to keep more rods within the core; to avoid operating below 700 MW(th); and to provide additional shutdown protection. In the longer term additional steps will be taken to mitigate the problem of the void coefficient. Fixed absorbers will be installed. Higher enrichment of fuel will be used, and fast-acting shutdown systems will be studied.

Concerning these proposed modifications the INSAG report states: ''The brief study . . . cannot provide full confirmation that the intent . . . to make it much more difficult to reach operating conditions that could result in a fast reactivity excursion from any cause, including severe violation of operating procedures . . . has been achieved by these modifications.''

In my opinion continued operation of the RBMK reactors pending completion of these modifications, and even after the modifications have been completed, presents a safety hazard that I believe our NRC would not accept.

Release of Radioactivity

The accident at Chernobyl was the first time that the quantity of radioactivity escaping from a nuclear power plant could be termed a catastrophe. The reactor was large, it had been in operation since 1984, and it contained a radioactive core inventory of one billion curies. When the 1,000-ton plate was lifted by the explosions to a vertical position above the reactor the vertical cooling channels were all open at the top, and like 1,661 Roman candles, spewed radioactive debris and gases out of the top of the plant. The heavier debris blown out of the reactor like large pieces of graphite fell in and close to the plant, and fine particles and gases rose high in the atmosphere to form a radioactive cloud that was carried by the wind around the world.

The release of radioactivity was prolonged. There was an initial release the first day of about 12 million curies, on days four to five it dropped to about 2 million curies, and then due probably to the confinement of heat by the 8,000 tons of boron, carbide, dolomite, clay, and lead dropped on the reactor to seal if off, it rose to 8 million curies on day nine, and then suddenly dropped to about 100,000 curies a day. This drop may have been due to the cooling effect of the nitrogen introduced below the core. The spreading of this radioactivity over parts of the Soviet Union and neighboring countries was the major cause of world alarm.

Effects of Radiation

It is difficult for the public to comprehend radiation and the effects of radiation on health. I have been in the atomic business for more than forty-two years, and although I was exposed to many fine teachers, I still have difficulties with this. A few simple definitions, principles, and accepted effects may help the reader as they have helped me. The term *rem* is a radiation unit. Rem is used to measure the amount of radiation an individual has absorbed. A millirem is one thousandth of a rem. In order to avoid confusion I generally will use millirems for all measurements. We are constantly being exposed to radiation; we call this background radiation. Radiation comes from outer space, uranium in the ground, thorium, and so forth. The average individual receives about 100 millirems each year from background radiation. He receives additional radiation from television, medical X rays, and so on. Radiation is usually expressed as a rate, so many millirems an hour, a day, or a year. The total received in a given time is referred to as a *dose*. For example, if you are exposed to 1,000 millirems per hour for six minutes, you have received a dose of 100 millirems. Correspondingly, if you live in an area where the background is 100 millirems a year, the background is .011 millirems per hour.

Radiation is used to treat cancer. Radiation also causes cancer. There are many other causes of cancer, such as smoking, asbestos, and so forth. Under normal conditions we can expect about 16 percent of our deaths to be due to cancer. This percentage varies from country to country. From data derived from Hiroshima, Nagasaki, and medical X ray experience, we have learned that if an individual receives a dose of radiation of about 1 million millirems, he is almost certain to die in a short time. If he receives a dose of 500,000 millirems he has a fifty-fifty chance of recovering. If he receives a dose of 100,000 millirems he will have symptoms of radiation sickness, such as nausea, but he will soon recover. If one hundred individuals receive 100,000 millirems of radiation each (a collective dose of 10 million millirems), one will die of cancer caused by radiation, in addition to the sixteen who would normally die of cancer. There is a mathematical relationship between collective doses and risk of death by cancer that seems to work for radiation levels of 100,000 millirems or higher. For low levels of radiation there is no proven relationship. There is evidence that the effect drops dramatically below 50,000 millirems. But there is a difference of opinion among the experts on whether or not there is some level of radiation, a threshold, below which radiation has no effect on health. One of the bits of data that indicates there is a threshold concerns varying background levels. The background radiation level in the vicinity of Denver is about twice the average level in the United States, yet the cancer rate is lower than the average rate. There are several other such locations. If there is no threshold the cancer rate would be higher rather than lower than the average rate. Another bit of evidence that there is a threshold is that experimental biologists do most of their work at 50,000 millirems because they do not detect any carcinogeous effect below 10,000 millirems. However, for the purpose of establishing safety standards for workmen, among others, it is assumed that there is no threshold. Very conservative limits have been set for permissible radiation for workmen in nuclear plants, and even more conservative limits have been set for the general public. In estimating the effects of fallout from the Chernobyl accident, the formulas used for the effects of the collective doses assumes no threshold and that the formulas will apply to low-level irradiation. Likewise in other parts of the dose calculations; particularly for internal exposure there are assumptions that are much too conservative. We don't know by how much, but we can be certain that the long-term-effects figures are too high. Hopefully, the Chernobyl accident will be the last time we have an opportunity to collect more data to determine the long-term effects of low-level radiation.

The INSAG Report Continues: Early Acute Health Effects

"Several groups had been exposed to radiation at the Chernobyl power station to such an extent that resulting whole body doses produced various forms of acute radiation syndrome. In such cases the absorbed doses . . . were in the range from about two hundred thousand to about 1.6 million millirems. The groups included operating personnel . . ., emergency squads and to the largest extent the fire brigades fighting the extensive early fires on the site.'' There were a total of 300 persons hospitalized. In most of the 203 ''cases of acute radiation syndrome with lethal outcome the burns had a strong effect on the ultimate fates of the victims.'' (All radiation levels within the quotes of the INSAG report have been changed to millirems.)

Off Site Emergency Response

On April 26 a group of specialists were sent from Moscow to assist local authorities and plant management. While radiological and meteorological data were collected, people in Pripyat were instructed to remain indoors with windows and doors shut. Schools were closed. Potassium iodide tablets were distributed on a door-to-door basis. By April 27 radiation levels reached a value of 1,000 millirems per hour (approximately 80–125 thousand times backgrounds of .008–.012 millirems/hr). "With this rate the lower intervention level for evacuation of 25 thousand millirems whole body dose and even the upper level of 75 thousand millirems would be exceeded. Evacuation was ordered started in the morning of the 27th and it was well organized. No individual from off-site areas had to be hospitalized for radiation injuries, although many attended hospitals for other reasons.

"In the following days, the same protective actions had to be gradually applied to the other population centers in an area of radius about 30 km around the plant. . . . In addition to people, some tens of thousands of cattle had to be evacuated from the area in hundreds of trucks. . . . The consumption of milk and other foodstuffs had to be banned over a considerable area.

"Because of the contamination, the area within a 30 km radius was divided into three zones: a special zone (some 4–5 km around the plant), where no re-entry of the general population is foreseeable in the near future and where no activity besides that required at the installation will be permitted; a 4–10 km zone, where partial re-entry and special activities may be allowed after some time; and a 10–30 km zone, where the population may eventually be allowed in and agricultural activities may be resumed, but which will be subject to a strict programme of radiological surveillance.''

Decontamination

"Heavy contamination, requiring extreme efforts, for its control and removal, has been deposited at the site, both on inner and outer surfaces of units 1, 2 and 3, and in a zone with a radius of 30 km surrounding the site. Approximately half of the released material was deposited within this 30 km zone.

"In this zone the ground and its vegetation, buildings and water were all affected. In addition, some areas outside this zone up to a distance of about 60 km showed significant contamination levels, requiring efforts to reduce them.''

Additional large areas within the Soviet Union and outside countries experienced some contamination. "Although outside the Soviet Union the contamination levels were such that direct decontamination measures were not necessary, measures such as controlling milk, vegetables and other foodstuffs were taken in some countries in order to avoid unnecessary doses to the public.

"Because the ventilation systems [in units 1, 2 and 3] had continued operation for several hours after the accident, the contamination . . . was much higher than would otherwise have been the case. Immediate decontamination was necessary because the radiation levels due to contamination inside the units was measured as 100–600 millirems/hr, and the units had to be kept in safe shut down condition. Various decontamination methods were applied, including spraying with water and various decontamination solvents, steam ejection, dry methods using polymer covers, and the manual washing of surfaces. These methods were very successful, the contamination level being reduced by a factor of 10 to 15, and the dose rate inside the buildings dropping to a measured range of 2–10 millirems/hr.

"Before the decontamination of land and villages in the 30 km zone can begin, the destroyed unit 4, which is still releasing several curies of aerosols per day, has to be fully entombed.'' The ultimate aim of various decontamination methods being proposed "is to allow the reuse of the land for agricultural purposes as soon as possible.

"In these fields there should be strong international co-operation, and all countries which have experience or will gain experience and which can contribute to solutions should be invited by the IAEA to participate."

Short-Term Irradiation and Health Effects Off Site

During the period the radioactive plume or cloud was passing over the USSR and other countries the public below was subject to direct external irradiation from the fallout, and internal irradiation from inhalation of radionuclides. "Deposited radionuclides, particularly I-131 and caesium isotopes, entered the terrestrial food chains, and in the most affected areas some doses will have been received through consumption of contaminated milk, leaf vegetables and fruit before measures were introduced to restrict consumption of these products. . . .

"Estimates of doses were needed . . . these were obtained from environmental monitoring . . . predictive modelling . . . measurements made of I-131 in people's thyroids, particularly for children, and whole body measurements . . . to determine levels of Cs-137. . . .

"The collective dose from external radiation to the 135,000 people who were evacuated was estimated to be 1.6 billion millirems. Most doses to individuals were less than 25 thousand millirems, although some people in the most contaminated areas may have received doses as high as 30 to 40 thousand millirem or more. Doses to the thyroids of individuals from inhalation (and possibly ingestion of contaminated foods) were estimated to be mostly below 30 thousand millirems, although some children may have received thyroid doses as high as 250 thousand millirads. In other regions of the Soviet Union doses to individuals from external irradiation and from intake of iodine were very much lower."

Late Effects

The INSAG report discusses the difficulties of obtaining sufficient information and knowledge to make a prediction of the long-term health effects. The report states: "However, an appropriate and tentative estimate of the number of fatal cancer cases can be made on the basis of collective dose estimates derived so far for parts of the Soviet Union and presented at the Review Meeting. Under the assumption of a no-threshold linear dose-response relationship, the estimate would be equal to the product of the risk coefficient" and the collective dose.

"The . . . uncertainties, and the fact that the coefficient was derived from epidemiological observations at the higher doses for the specific purpose of planning protection, makes it obvious that the use of the same coefficient for assessing consequences in a given exposure situation is to some extent speculative."

Using this questionable method estimates are as follows: "The number of all radiation induced cancer in this group (the 135,000 evacuees) should equal about 160. . . . The collective dose to the thyroid . . . [adds] some ten cases of fatal thyroid cancer [plus a substantially larger number of non-fatal cases].

"To put these rough estimates into perspective it has to be remembered that over 70 years about 20% of those who were evacuated would normally die of cancer, i.e., about 27,000. The estimated 170 additional cancers would constitute about 0.6% of the so-called spontaneous cases. For that part of the population of the European Soviet Union to which the collective dose estimate applies (75 million people), a similar calculation yields a relative increase in the cancer mortality due to external irradiation and internal exposure for Cs-137 that would be within 0.03–0.15% of the natural value. Similarly, the possible increases in the spontaneous mortality from thyroid cancer was estimated at about 1%.

"As these values constitute a very small fraction of the spontaneous incidence (and mortality), the chances of epidemiological detection of these effects are negligible. Only in cohort with mean doses substantially above 10 thousand millirems would some effects possibly be discovered (e.g. leukaemias or benign and malignant thyroid neoplasms)."

Concerning the health effects of Chernobyl, the report states in its summary conclusions: "An opportunity now exists for the world's safety experts to learn from this tragic event in order greatly to improve our understanding of nuclear safety. This accident is almost "worst case' in terms of the risks of nuclear energy. It is to be emphasized that, even under these circumstances, no member of the public had to be hospitalized as a result of radiation injuries. The victims were 300 power plant and fire-fighting personnel admitted to hospitals, of whom 31 have so far died."

The Effect of Fallout on Other Countries

The INSAG report made little reference to the effects of Chernobyl outside the Soviet Union, and data concerning the extent of radioactive fallout is not readily available. Generally speaking, the farther away from Chernobyl the smaller the effect would be. In addition the error in any estimate of cancer deaths based on no threshold would be greater than near Chernobyl because it would be based on even lower levels of radiation. The effect on health caused by direct irradiation probably is negligible except possibly for hot spots caused by heavy rain just as the plume of radiation was passing over. The town of Gävle is the hottest spot that I have heard about. The World Health Organization released figures that indicate Gävle had the highest concentration of the long-life radioactive substance Cesium 137 of anyplace in Western Europe. Some readings in Gävle were as high as one hundred times background, or about 1 millirem per hour. The permissible rate for workmen in Sweden's nuclear plants is 2.5 millirem per hour for forty hours per week. The greatest effect of radiation in Sweden concerned the food chain. Cesium concentrated in farm products such as leafy vegetables, milk, meat, and in fish and reindeer. Sweden, of course, placed a ban on consumption of such contaminated food. Although the real damage probably is more comparable to the contamination of the Rhine river by toxic chemicals released in Switzerland as a result of a recent fire in an industrial plant, the effect of the fallout is that 60 percent of the Swedish population and 80 percent of the Danish population are now against nuclear power. The net result of the fallout in Western Europe, including to a lesser extent France, is to increase the opposition to nuclear energy and renew the debate about the acceptability of nuclear power.[1]

Effect of the Chernobyl Accident on the United States

The radioactive cloud passed over the United States and there was very little fallout that could directly affect health, but continued media

1. Chris Mosey, "We are not as radioactive as you've been led to believe," *Sweden Now*, Stockholm, Sweden: No 4/1986 (Vol. 20).

Major General K. D. Nichols, U.S.A. (Ret.)

accounts of the accident, both the accurate and the inaccurate, have had considerable adverse effect on public opinion. The antinukes have new ammunition to claim that nuclear energy is too dangerous to tolerate. The advocates of nuclear power have sufficient facts concerning the cause of the accident at Chernobyl and the fundamental differences in the design and safety aspects of the RBKM-1000 as compared to our LWRs to be assured that the LWRs are safe. In this regard, the question is frequently asked: Would the LWR containment withstand the explosions that occurred at Chernobyl? The answer is: First, there is no possible mechanism in an LWR for a prompt critical reaction that could produce fuel-coolant interaction similar to that which appears to have occurred at Chernobyl. Second, if an explosion were to occur for other reasons, such as a complete meltdown, the intensity would be much less severe. Third, assuming an equally severe explosion were to occur, the containment would not be completely open as it was at Chernobyl—there might be some cracking and openings, but the structure would still have some effectiveness in reducing radioactive release. But no matter how confident we are, we must continue to check and to recheck our safety concepts and, above all, we must not become complacent.

What Can We Learn from Chernobyl?

We can learn a great deal about the consequences and how to cope with a breach of containment, if we should ever be so unfortunate as to have one. Such an event is extremely unlikely with the LWR but no one can say that such risk is nil. I would hope that the IAEA will be able to participate with the Soviets in the future so that we can learn more about evacuation, monitoring radiation levels, medical care, decontamination, monitoring radiation in crops, monitoring the long-term effects of radiation. A thorough study of all information that could be made available will permit us to define more accurately the potential hazards of such an accident and how to minimize the consequences. Perhaps our general public also may learn if they give consideration to one of INSAG's conclusions: "INSAG remains convinced that if available safety principles and knowledge are effectively deployed, nuclear power at its present status is an acceptable and beneficial source of energy. Although the accident that took place was dramatic and had extensive consequences, it did not exceed in scale accidents of other types that continue to occur, natural and man made. However, there is potential for improvements in the design and operation of nuclear power plants.'' If this conclusion is accepted it might be possible for the general public to rationally weigh the advantages of nuclear power against the risks of the worst potential accident's disadvantages, and compare these risks with the risks of hydraulic power, coal, and oil. Certainly the statement is applicable to our LWRs. There was nothing in the Chernobyl accident that gives cause to believe that LWRs are less safe than we believed before the accident. I believe that the advantages of using the LWRs to produce electricity instead of coal or oil far outweigh any and all disadvantages for most locations in the United States. However, I am not optimistic that those who have an emotional fear of any radiation no matter how low a level will draw the correct conclusion.

The Future of Nuclear Energy in the U.S.

The accident at Chernobyl is bound to affect public opinion about nuclear power adversely. It will be difficult to restore public confidence, but it must be done. We need nuclear power to meet part of our electricity requirements. At present we have about one hundred plants operating, producing 16 percent of our total power. Twenty-seven more plants are under construction, and when they are operating our total production will be 20 percent nuclear. More important, we know that it is possible to design, construct, and operate LWRs that can produce electricity in most areas in the United States cheaper, more safely, and with less overall effect on the environment than coal or oil plants.

We need to continue to demonstrate that our LWRs are safe and reliable. We should plan on correcting the errors we have made. Institutional failures in the United States need to be corrected. We are making progress in utility management, but the financial problems facing the utilities still are serious. The growth of the economy eventually will restore load growth. Additional improvement must be made in the attitude and procedures of the NRC. Basic changes in the law are necessary to prevent the unnecessary delays caused by almost unrestricted and irresponsible intervention by opponents of nuclear energy. We must work to improve public understanding if confidence in nuclear safety is to be restored. That will be the most difficult problem of all. We need more convincing methods of explaining risk assessment. The public must be properly informed about how vital nuclear energy is as a future contributor to fulfill our energy needs. The large area of public uncertainty concerning nuclear waste must be eliminated by successfully demonstrating that nuclear waste can be safely stored. There are several adequate solutions to this, and it is unfortunate that the AEC, the Department of Energy, and Congress have been derelict in not proceeding earlier with construction of underground storage sites. Congress finally has authorized the necessary action. Technical solutions have been known for years. Now it is necessary to select an adequate method from one of several available and proceed with the work.

Conclusion

I remain an optimist. Eventually the United States will join France and others and use more nuclear energy to generate electricity to improve the general welfare of our nation. We absolutely need it, and I expect to live long enough to see general acceptance of my firm belief that for most areas nuclear energy is the least expensive, least dangerous, and least damaging to our environment of all the methods of generating electricity available.

Sometimes it takes considerable time for truth to prevail. Load growth for electricity will increase; most of our utilities are attaining a better financial condition; inflation is lower; interest rates are lower; we have good reasons to believe that the dangers to the U.S. public of a major nuclear reactor accident are considerably less than formerly calculated; operators are being better trained; the Institute of Nuclear Power Operations conducts training and inspects plants for proficiency; the flow of critical reactor information has been improved by the industry; and the public is becoming more aware of the dangers of competing methods of producing electricity, such as "acid rain." Sometime in the future we shall hear that we are threatened by a lack of capacity to meet peak demands for electricity. Initially that need will be met by improved gas turbines, with low capital but high operating costs. When it is obvious to most of the public, the Administration, utilities, state regulatory commissions, Congress, and the news media that we need to expand our base load electric capacity, I am confident that our more rational thinkers will point out that nuclear energy must play a major role and that truth will prevail. Congress will need to overhaul licensing procedures radically; state regulatory commissions will need to improve procedures for improving financial stability of utilities by means such as recognizing interest during construction as a current expense; utilities and manufacturers will need to recognize that standardized plants are to be constructed; as much work as possible should be done in factories rather in the field to achieve the necessary quality control; and plant approval by the regulatory commission must be made prior to construction being approved by the NRC. And once one-step licensing, construction, and operation is approved, the public and the government should recognize that neither construction nor operation should be halted by frivolous court intervention. Intervention should be limited to the period prior to one-step licensing of construction and operation. Safety must remain a primary consideration, but the public must recognize that zero risk cannot be achieved for any major human endeavor. We must recognize relative risks, and we must be satisfied so long as nuclear energy maintains its existing record of being less dangerous than hydraulic power, gas, oil, or coal for meeting our needs for electricity. If we can achieve these goals the inherent advantages of nuclear energy in regard to environmental and economic considerations for a major portion of our electric needs will be obvious. In the United States we can establish proper procedures for fulfilling our public needs, and I expect that sufficient objectivity on the part of all will solve this issue. My only regret is that I may be beyond the age to play an important role in this effort, but I expect to remain a most interested spectator, and I will participate where and whenever I can.

ACKNOWLEDGMENTS

FIRST, I MUST ACKNOWLEDGE that only the outstanding efforts of over 125,000 individuals who contributed to the historic success of the Manhattan Project make the story of my life worth telling. Of this splendid team of diverse organizations encompassing individuals of varied interests, talents, and accomplishments, I had close contact primarily with the top echelon of the leaders.

In preparing this manuscript I repeatedly found that in singling out specific persons I had failed to mention others who made equal or greater contributions to our efforts. I apologize to the hundreds who should have been included. I wish to emphasize, however, that the successful outcome was the result of maximum dedication and cooperation of everyone involved.

I am deeply indebted to A. V. Peterson, his wife, Marie Louise, and to Harry Traynor for their additions and corrections to the manuscript and for their valuable comments; also to my long-time administrative assistant and friend Miss Virginia J. Olsson for her research, proofreading, and careful attention to many necessary details. Dr. Edward Reese was most helpful in locating documents in the National Archives. Dr. John Greenwood, Chief, Office of History, Office Chief of Engineers, assisted me in many ways, provided encouragement and valuable advice, and supplied many photographs from army files. I also appreciate the constructive encouragement, advice, and assistance provided by General Albert C. Wedemeyer in regard to obtaining a publisher. I am grateful to Mr. Bruce Lee, senior editor, William

Morrow and Company, Inc., for accepting the manuscript and agreeing to publish my story.

I also wish to thank Admiral Frederick Ashworth for permitting me to use his account of the bombing of Nagasaki; Jane Larson for providing a copy of Colonel Stafford Warren's (her father) account of his visit to Nagasaki and Hiroshima; Dr. Clarence E. Larson for technical assistance on many aspects of the manuscript; Edward Wiggin, executive vice president of the Atomic Industrial Forum, for making available to me information the forum had on file concerning the Chernobyl accident; John J. Taylor, vice-president, Nuclear Power Division of EPRI, for permitting me access to, and to use excerpts from, his timely report on Chernobyl to the Nuclear Regulatory Commission; and to Dr. Gerald Tape for providing me a copy of the IAEA summary report on the Chernobyl accident.

Throughout the four years I have devoted to writing this book, my wife, Jackie, following the pattern of the previous fifty years, has been a devoted partner, supporting my efforts, giving encouragement when needed, offering constructive criticism when warranted, actually writing parts of the manuscript, and assisting in editing all of it.

BIBLIOGRAPHY

THE MAIN SOURCE of information for the contents of this book is my memory. Particularly my memory of the many events that for one reason or another seemed important at the time. However, memory is fallible, particularly in regard to sequence or timing of events. In this regard I found my personal notes, letters, diaries, telephone and conference records, and travel vouchers very helpful. When there were gaps in my personal records or memory the necessary information was found in historical files, unpublished accounts, biographies, and recorded oral interviews. My autobiography is not intended to be a complete history of any phase or period of my life. However, additional background is added where necessary to better understand why certain decisions were made, why certain actions followed, or to emphasize the thinking, emotions, fears, or aspirations and hopes of the time. I feel fortunate to have participated in an active way in so many significant technical, political, security, and economic developments that have affected in a major way the security and welfare of the world. We are all seeking solutions to the problems generated, and I trust that my account of the past may help others to cope with the future.

After my manuscript is published, I have arranged that my files and personal papers will be turned over to the Office of History, Office Chief of Engineers, to be indexed and made available to historians. The Office of History already has a fine collection of oral interviews and the personal files of many officers of the Corps. I found several of the oral interviews extremely helpful.

Other sources of information utilized were the Manhattan District records at the National Archives, including the Commanding General (Groves) Files, the Harrison-Bundy Files, and the General Administration Files. The Archives also has files pertaining to the Office of Scientific Research and Development (OSRD), and the Armed Forces Special Weapons Project (AFSWP) Manhattan District History.

Official publications utilized as sources of information included the following:

Condit, Doris, *The History of the Office of the Secretary of Defense, Years of Trial 1950–1953,* Volume II (Washington, D.C., to be published in 1988).

Fine, Lenore, and John Remington, *U.S. Army in World War II, The Technical Services, The Corps of Engineers: Construction in the United States* (Washington, D.C.: Office of the Chief of Military History United States Army, 1972).

Hewlett, Richard G., and Oscar E. Anderson, Jr., *The New World, 1939–1946, A History of the United States Atomic Energy Commission,* Volume I (University Park, Pa.: The Pennsylvania State University Press, 1962).

Hewlett, Richard G., and Francis Duncan, *Atomic Shield, 1947–1952, A History of the United States Atomic Energy Commission,* Volume II (University Park and London: The Pennsylvania State University Press, 1969).

Jones, Vincent C., *U.S. Army in World War II, Special Studies, Manhattan: The Army and the Atomic Bomb* (Washington, D.C.: Center of Military History United States Army, 1985).

Reardon, Steven L., *History of the Office of the Secretary of Defense, The Formative Years 1945–1950,* Volume I (Washington, D.C.: Historical Office, Office of the Secretary of Defense, 1984).

In the Matter of J. Robert Oppenheimer: Transcript of Hearing Before Personnel Security Board, April 12, 1954, through May 6, 1954 (Washington, D.C.: United States Atomic Energy Commission, 1954).

In the Matter of J. Robert Oppenheimer: Texts of Principal Documents and Letters of Personnel Security Board, General Manager, Commissioners, May 27, 1954, through June 29, 1954 (Washington, D.C.: United States Atomic Energy Commission, 1954).

Operations Crossroads, The Official Pictorial Record, The Office of the Historian Joint Task Force One (New York: Wm. H. Wise & Co., Inc., 1946).

Other publications utilized as sources of information included the following:

Bradley, Omar N., and Clay Blair, *A General's Life: An Autobiography* (New York: Simon and Schuster, 1983).

Bush, Vannevar, *Modern Arms and Free Men: A Discussion of the Role of Science in Preserving Democracy* (New York: Simon and Schuster, 1949).

Compton, Arthur Holly, *Atomic Quest* (New York: Oxford University Press, 1956).

Dupuy, R. Ernest, *Men of West Point: The First 150 Years of the United States Military Academy* (New York: William Sloane Associates, 1951).

Eisenhower, Dwight D., *The White House Years: Mandate for Change 1953–56* (Garden City, N.Y.: Doubleday & Company, Inc., 1963).

Groves, Leslie R., *Now It Can Be Told: The Story of the Manhattan Project* (New York: Harper & Brothers, 1962).

Manchester, William, *American Caesar: Douglas MacArthur 1880–1964* (Boston, Toronto: Little Brown and Company, 1978).

Mosley, Leonard, *Marshall: Hero for Our Times* (New York: Hearst Books, 1982).

Pfau, Richard, *No Sacrifice Too Great: The Life of Lewis L. Strauss* (Charlottesville: University Press of Virginia, 1984).

Strauss, Lewis L., *Men and Decisions* (Garden City, N.Y.: Doubleday and Company, Inc, 1962).

ABRIDGED INDEX